The Tragic Drama
of Corneille and Racine

The Tragic Drama
of Corneille
and Racine

AN OLD PARALLEL REVISITED

BY

H.T. BARNWELL

CLARENDON PRESS · OXFORD

1982

Oxford University Press, Walton Street, Oxford OX2 6DP
London Glasgow New York Toronto
Delhi Bombay Calcutta Madras Karachi
Kuala Lumpur Singapore Hong Kong Tokyo
Nairobi Dar es Salaam Cape Town
Melbourne Wellington
and associate companies in
Beirut Berlin Ibadan Mexico City

Published in the United States by
Oxford University Press, New York

British Library Cataloguing in Publication Data

Barnwell, H.T.
The tragic drama of Corneille and Racine.
1. Corneille, Pierre — Criticism and interpretation
2. Racine, Jean — Criticism and interpretation
I. Title
842'.4   PQ1779

ISBN 0-19-815779-7

Library of Congress Cataloging in Publication Data

Barnwell, Henry Thomas.
The tragic drama of Corneille and Racine.

Bibliography: p.
Includes index.
1. French drama (Tragedy) — History and
criticism. 2. Corneille, Pierre, 1606–1684 —
Criticism and interpretation. 3. Racine, Jean,
1639–1699 — Criticism and interpretation. I. Title.
PQ561.B37   842'.4'09   81-18691
ISBN 0-19-815779-7   AACR2

Typeset by Syarikat Seng Teik Sdn. Bhd., Kuala Lumpur
Printed in Hong Kong

To Joan
and
our family

Acknowledgements

Some of the material in this book has formed part of a number of articles in learned journals and other publications. I acknowledge with gratitude permission from the editors and publishers to use it here in a different form: *French Studies, Studi Francesi, Australian Journal of French Studies, Studies in French Literature presented to H. W. Lawton, The Classical Tradition in French Literature: Essays presented to R. C. Knight, Mélanges à la mémoire de Franco Simone.* Full particulars are to be found in the Bibliography under my name.

My debts to other scholars are far too numerous to be acknowledged individually here, but they will be evident from a perusal of the Bibliography. I am grateful to all who, by their teaching, their writing, their example, and their encouragement, have lighted my path through what for me is one of the richest and most rewarding of literary territories; but I owe a special debt of gratitude to Professor Roy Knight who first guided my hesitant footsteps and has remained the most trusted of counsellors and friends.

The British Academy generously provided financial assistance and the University of Glasgow granted me two terms' study-leave in order that my work might be brought to fruition. To both I express my sincere thanks.

Finally, but by no means least, I wish to thank my wife and son for their invaluable practical assistance in the preparation of this book for the press and for their affectionate patience and understanding during the months when it was being written.

Un esprit médiocre peut imaginer un dessein vaste et grand, mais il faut un génie extraordinaire pour renfermer ce dessein dans la justesse et dans la proportion. Car il faut qu'un même esprit y règne partout, que tout y aille au même but, et que les parties aient un rapport secret les unes aux autres: tout dépend de ce rapport et de cette liaison, et ce dessein général n'est autre chose que la forme que donne le poète à son ouvrage. C'est aussi la partie la plus difficile de l'art, parce que c'est l'effet d'un jugement consommé.

René Rapin, *Réflexions sur la poétique* (1675)

Contents

	Note	xi
	Introduction	xii
I.	The Dramatic Subject: its nature and its disposition	1
II.	Sources: Tragedy and history	33
III.	'La Vraisemblance': its dramatic function and significance	71
IV.	Plot: Drama in tragedy	93
V.	Simplicity: Situation and dénouement	133
VI.	Simplicity: Peripety and discovery	159
VII.	Rhetoric: Trial and judgement	179
VIII.	Tragic Quality	213
	Notes	251
	Bibliography	261
	Index of proper names	271

Note

1. The spelling of seventeenth-century texts has been modernized; that of earlier texts has been retained in its original form.

2. In the notes, longer titles referred to have been shortened, the full form being reproduced in the Bibliography.

3. The following abbreviations have been adopted for works frequently cited:

Corneille, first, second, and third *Discours*: the three *Discours* in the order in which they are always printed.

Writings: Corneille's critical works in my edition under the title of *Writings on the Theatre*.

Racine, *Principes*: the partial translation of the *Poetics* of Aristotle, published by Eugène Vinaver under the title of *Principes de la tragédie*.

Intégrale: the *Œuvres complètes* of Corneille, as published by André Stegmann in the edition *L'Intégrale* (Seuil).

Picard: the two-volume edition of the works of Racine, edited by Raymond Picard in the Pléiade collection (Gallimard).

D'Aubignac, *Pratique: La Pratique du théâtre*, edited by Pierre Martino.

Other abbreviations are self-explanatory.

Introduction

Ce grand homme a traité la poétique avec tant d'adresse et de jugement, que les préceptes qu'il nous en a laissés sont de tous les temps et de tous les peuples; et bien loin de s'amuser au détail des bienséances et des agréments, qui peuvent être divers selon que ces deux circonstances sont diverses, il a été droit aux mouvements de l'âme, dont la nature ne change point. Il a montré quelles passions la tragédie doit exciter dans celle de ses auditeurs; il a cherché quelles conditions sont nécessaires, et aux personnes qu'on introduit, et aux événements qu'on représente, pour les y faire naître; et il en a laissé des moyens qui auraient produit leur effet partout dès la création du monde, et qui seront capables de le produire encore partout, tant qu'il y aura des théâtres et des auteurs...

Corneille, *Le Cid, Avertissement* (1648)

Students of French drama and literature may be sceptical of the value of yet another comparison between the two masters of tragedy in the seventeenth century. From their contemporaries (Saint-Évremond, Longepierre, La Bruyère) to ours (G. May, G. Pocock) parallels have been drawn. Does anything remain to be said?

It is not my intention here to produce a survey of past and present trends in criticism of the two playwrights. Readers of this book are likely to be familiar with them. Nor shall I, here or in the subsequent chapters, make a detailed or systematic analysis or criticism of the ideas of other commentators. Again, it will be evident that criticism is often implicit in the approach, method, and conclusions to be found here. In order to define that approach and that method, it is however necessary briefly to summarize the suppositions on which earlier parallels have been based: many have been somewhat schematic and have failed to allow the work of one dramatist to throw light on that of the other; what might be called the chalk-and-cheese approach has frequently led to the drawing of stark distinctions, sometimes in terms of approval or condemnation of either author depending on the aesthetic or moral prejudices of the critic.

Until fairly recently, comparisons between Corneille and Racine were based in part on psychological and moral considerations, in part on some aspects of seventeenth-century critical concepts. Some of the earliest parallels drew distinctions between a Corneille interested above all in the 'grand sujet' conceived of in terms of an extraordinary action performed by men and women endowed with heroic and invincible will-power and guided by the light of reason and duty, and a Racine presenting profound psychological studies of characters who were the victims of their own blind passions, vacillating, weak-willed, irrational. Corneille's plays were dramatically exciting, their episodes unpredictable, their outcome optimistic; Racine's developed inexorably towards a foreknown and pessimistic end. One was heroic, the other tragic; one was active and dramatic, the other passive and lyrical or poetic. With variations, these broad distinctions have persisted even if only to be turned upside-down and inside-out, their presuppositions being sometimes questioned to the point of perversity.

The older approaches had the merit of being attempts to face the plays themselves and for themselves, even if the emphasis was on the plays not as plays but as purely literary works. The same could be said for the historical studies in the Brunetière–Lanson–Mornet line, in which some at least of the distinctions and similarities between the playwrights were seen to be accounted for by the circumstances, the society, the moral and philosophical climate in which they lived. On the whole, academic criticism has, in diverse ways, continued to explore scientifically the relationship between the plays and their authors and their background. Paul Bénichou's *Morales du Grand Siècle* was one of the most influential general studies and, as far as our subject is concerned, related the creation and celebration of the hero and his eventual demolition to the change in political, moral, and social outlook which took place with the effective accession of Louis XIV on the attainment of his majority. The plays of Corneille and Racine, among other works of literature, were seen as expressions or symptoms of that change. While on the level of their own moral outlook the characters could without difficulty be fitted into this scheme, it failed to take account of the significance of the part they played in the dramas as such and so begged the question as to how their conduct was presented to the spectator. As far as Corneille and Racine were concerned, M. Bénichou's study did not so much change what had become traditional views as reinforce them, while suggesting that the interest of a great writer lay chiefly in his work's being the expression or reflection of the spirit of his age. The appearance of M. Bénichou's book coincided almost exactly with that of Octave Nadal on Corneille which moved partly in the same direction by relating his plays to the ethic of 'gloire', said to characterize aristocratic society in the age of Louis XIII, and partly in the direction of questioning established views of the primacy of 'amour-estime' for his characters.

Also to the same generation of criticism belongs the work of Georges Couton on Corneille and of René Jasinski on Racine. For these scholars the plays represent contemporary events and preoccupations in fictitious form: they are studied virtually as historical documents. Parallel to this approach is that of Lucien Goldmann, whose Marxist analysis relates Racine to the position of his class in society, while that of Charles Mauron makes of his works the starting-point for a psycho-analytical interpretation — Corneille has only recently begun to be subjected the same treatment (by M. R. Margitić and J. Barcillon), but Jacques Maurens had earlier attempted to see in him the spokesman for a form of neo-Stoic optimism which may have been fairly widespread in his day. Although these critics do not make explicit comparisons between the two playwrights, such comparisons could certainly be based on implications in and interpretations of their work. Another major factor has been the abandonment of the somewhat dangerous category of 'pre-classical' and its replacement by that of 'baroque': in consequence, Corneille has ceased to be considered in retrospect as a precursor of the 'classical' Racine; and his works are seen to belong to a distinct and as it were autonomous aesthetic, traces of which are, so Philip Butler believes, still to be found in the works of the younger dramatist. To some of us, however, the very use of labels such as 'baroque' and 'classical' is suspect in that it is almost an invitation to classify or categorize and to read into the plays distinctive characteristics which an unprejudiced study would not find in them. It is commonly believed that once one has put a work of art into a category one has explained, interpreted, or illuminated it.

Categories are dangerous because they are selected or devised beforehand, the works of art then being made to fit them. The critic's humble task, as I understand it, is that of serving his author by attempting without prejudice to explain and clarify for the easier comprehension and greater enjoyment of the reader or spectator. Systems and categories inevitably distort by imposing on the works of the past anachronistic interpretations based upon fashionable methodologies.

The application of anachronistic criteria is also to be found in approaches which are not bound to any particular ideology or theory. Serge Doubrovsky's large-scale study of Corneille is an attempt to transpose the conflict between 'la générosité' and 'la bassesse' into the key of Hegel's master–slave dialectic; but the two moral frames of reference do not fit or coincide. The latest of the Corneille–Racine parallels, Gordon Pocock's, also applies anachronistic criteria, aesthetic this time: the seventeenth-century concept of *vraisemblance*, directed essentially towards the coherence and immediate acceptability to the audience of the work of art, cannot be equated with naturalism as that term is understood in literary history, and it had more to do with plot construction than with character. Difficulties of a similar order arise when modern theories of the tragic or of myth[1] are introduced (Michael Edwards) or schematic divisions of characters into clear-cut categories of good and bad, sympathetic and unsympathetic (Bernard Weinberg). All these are attempts, in whatever guise, to fit the playwrights into systems of thought or aesthetics which were alien to them. For all its scholarship, a book like André Stegmann's *L'Héroïsme cornélien* leaves one with the uncomfortable feeling that its author had a preconceived notion, not of what Cornelian heroism might be — he is much too honest for that — but that such a thing existed at all and that Corneille subscribed to it and gave it expression in his plays: his heroes are surely as various as the plays in which they appear. Here lies the merit of Robert J. Nelson's study, but it has the disadvantage of not taking the word 'tragedy' seriously. Generalizing abstractions in the titles of critical works deserve to be treated with caution.

It follows from these remarks that, although I am grateful to critics whose premises I cannot share (for, even by provoking disagreement, they have enlightened me), the path I have chosen is essentially a traditional one, but one which has surprisingly been little followed in the particular and rather obvious direction I have chosen.[2] It is evident that Corneille and Racine reflected deeply on their art and were very conscious artists. Like most of their contemporaries — Boileau with his 'démon jaloux',[3] Saint-Évremond writing of poetry that 'tantôt c'est le langage des dieux, tantôt c'est le langage des fous',[4] — they sought to understand not their inspiration or the springs of their creative powers but the ways in which to give them expression. 'Il est constant', writes Corneille in the opening paragraph of his first *Discours*, 'qu'il y a des préceptes, puisqu'il y a un art.' The only ideas of which he and Racine have left any traces concern that art: we know nothing — nor could they be of much interest to the present-day audience or reader — of their political, social or moral opinions, although an infinite amount of speculation has been devoted to their pursuit. Unlike Baudelaire ('l'imagination est la reine du vrai'), they and their contemporaries treated imagination less as a creative faculty, the gateway to poetic truth, than as 'cette maîtresse d'erreur et de fausseté', as Pascal put it. Imagination was distinct from genius, Boileau's 'démon', whose operations were regarded, as they were indeed by Baudelaire, as mysterious, and as such beyond the scope of enquiry and analysis. So they wrote about their art, not about Art. In our post- and sub-Romantic era we have put aside their wisdom and caution and, without its strictures, interested ourselves in versions of the Platonic concept of Art.

But the patron philosopher of seventeenth-century French writers was Aristotle, not Plato, however much we may now think they misunderstood him. It was his thought, interpreted or misinterpreted, that provided their aesthetic framework. I am here thinking of the *Poetics* less as the source of detailed prescriptions than as the distillation of an experience of the theatre and an endeavour to understand the processes by which the dramatist achieved certain emotional effects. From his observation of particular successful examples of tragic art Aristotle derived, in his philosophical way, the principles to which the scholars of the Renaissance mistakenly gave prescriptive authority. Instead of seeking to determine what had inspired the playwright — which the playwright himself might not know — he saw that all that was available for analysis was the product of that inspiration on the Athenian stage. The question he asked himself, therefore, was: how does it work? He did not speculate: he conducted a rational enquiry into the actual evidence.

Corneille and Racine, like many of their contemporaries, made use of the results of that enquiry, not as a set of prescriptions which, if closely followed, would ensure success: that is evidently what D'Aubignac believed when he composed, in 1647, the quickly-forgotten *Zénobie, tragédie où la vérité de l'histoire est conservée dans l'observation des plus rigoureuses règles du poème dramatique*. While it is true that they adopted the conventions of the three unities, of *vraisemblance* and *bienséance* as rules because, as René Bray has shown, after about 1635 they could scarcely have done otherwise if they were to succeed, it is difficult to imagine them setting out to write a play for the sake of showing that they knew and could observe the rules in all their minutiae. Certainly they felt constrained to defend their productions against the attacks of jealous and small-minded critics (or even against one another), but beyond the polemics of *Discours, Examens*, and prefaces we discern an understanding and an adaptation of the fundamental principles of the *Poetics*. These critical writings were doubtless in part the occasion for the dramatists to meditate on the nature of their art — but after it had been produced. In relation to the actual experience, their position is therefore analogous to that of Aristotle's treatise which they often had to use, all the same, as a court of appeal. It is none of my purpose, however, to discover how orthodox or otherwise Corneille and Racine may have been in their practice or in their critical writings: it is my intention to concentrate, as they did, on their art as they understood it, that is, as Aristotle also would have understood it. Since they were very conscious artists, it is reasonable to expect to find them commenting on their art: when due allowance has been made for the polemical element, their comments can serve as useful guides to an interpretation of their plays as they understood them and the craftsmanship which made them what they were and are. This book is not a history, but makes some attempt to take the facts of literary history into account when analysing both plays and critical writings. I shall not retrace the rivalry between Corneille and Racine, but shall be interested in some of its consequences. It goes without saying that Corneille was a pioneer rather than a precursor (which would be to belittle his own undoubted greatness): *Horace* was the first tragedy to which the epithet 'classical' can be applied,[5] or, to avoid the epithet, the first authentic modern tragedy in French. What more natural than that Racine should learn from his elder? Yet this is all too often forgotten in our anxiety to differentiate between the two. It is also

of course natural that Racine, in seeking to outdo Corneille, should
at the same time turn the lessons he learnt from him to his own
account, and adapt them to the exigencies of his own genius. But
the framework within which he wrote was inevitably the one created in *Horace* and developed and diversified in Corneille's succeeding tragedies. Again, it is not my purpose to explore this problem
as such: it has been thoroughly investigated by others. It is evident
that both our playwrights learnt from their many lesser contemporaries as well, if only from their mistakes. What all this amounts to
is that Corneille and Racine shared common ground and that their
works reveal parallels and convergences as well as divergences.

Their critical attitudes and their concept of their art were, in their
broad essentials, inevitably Aristotelian. It is also in the spirit of
the *Poetics* that my enquiry is undertaken. My starting-point is the
apparently banal statement that a tragedy is a work complete in
itself in the sense of having a beginning, middle, and end. For my
present purpose, this is interpreted in the sense that the play develops from an initial situation through the working out of a plot to
a dénouement. Is this too obvious to need stating? It is perhaps so
obvious that, as Eric Bentley has said,[6] critics, and academic critics
in particular, consider it beneath them to state it or to investigate
its implications. That investigation is at once a great deal more
difficult than is generally believed, as I for one have certainly discovered, and the most fundamental aspect of dramatic criticism.
Aristotle saw it clearly when he described plot as 'the soul of tragedy'.
What did he mean? The following comment seems to put it very
clearly:

Without a plot, we would not have an 'imitation of an action' . . . A soul is
not a separate entity . . . It is the living body's form, that without which we
would not have a 'living' body, but inert material only, without purposive
action. By 'soul' we indicate that a body is alive. By 'plot', analogously, we
indicate that the events presented hang together and are functionally interrelated. Plot is the structure and composition of dramatic events.[7]

It follows that that structure and composition give both life and
meaning to the events which in themselves are amorphous and
meaningless as they would be in real life as it is lived. It is through
the plot, therefore, that the dramatist conveys to his audience a
significant action which can be discerned through the coherent
arrangement of the dramatic events. That is why there is a world
of difference between the 'story' of the play and its 'subject' which
that arrangement expresses. The 'story' of the Electra plays of
Sophocles and Euripides may be the same: their subjects are very
different. In seventeenth-century French drama, one can make the

same remark of the Sophonisbe plays of Mairet and Corneille, of the Mithridate plays of La Calprenède and Racine, of the Bérénice plays of Corneille and Racine... That is why the traditional form of source study is often disappointing and sometimes misleading: the sources of the story can be identified, but when we know them we have only the raw materials for an analysis of the particular use to which they are put in the construction of the plot; and those which are rejected can be as important as those which are used, whether modified or not.[8]

The plot, as the animating, moving, developing force of the play, has to begin from somewhere, from an initial situation in which the characters, its agents, are set in certain relationships within circumstances which have already determined them. I shall therefore begin at that beginning and consider to what extent the situation is that of the 'story', and shall in this respect compare the practice of Corneille with that of Racine. Then will follow a detailed consideration of the use made of the sources, the 'story', in a small number of plays:[9] this has an important bearing on the production of an authentic tragic effect by the play as a whole as does the form of the initial situation. I have just suggested that the dramatic events, which develop the situation, must be arranged in a coherent order: it is in relation to the creation of that order that I shall proceed to an examination of some aspects of the central tenet of the so-called classical doctrine, *la vraisemblance*. It is on this dogma that important distinctions between Corneille and Racine have often been drawn: from certain points of view these can be questioned. My discussion of the problem will be directed less towards the usual differentiation of *le vrai* and *le nécessaire* on the one hand from *le vraisemblable* and *la biénseance* on the other than towards the establishment of the internal coherence of the plays. This has obvious implications for the treatment of the sources.

If plot is the 'soul' of tragedy, the source of its life, it follows that it is what makes the play dramatic. Here it has seemed to me necessary to put in question the idea, stimulatingly explored by Professor Georges May, that while Corneille is interested in dramatic effect, Racine is not, and the contention of Eugène Vinaver that the greater the emphasis placed on dramatic effect the less likely the play is to be tragic. I then pursue my analysis of plot in two further directions, each an aspect of another commonly-assumed distinction, between a complex Corneille and a simple Racine. First, the form of their plots is examined in its development to the dénouement and in relation to the point at which the dénouement occurs. Second, on the basis of what seems to me to be a fundamental and far-reaching distinction in this respect, I

consider the function of the endings of the plays. This is then related to the processes and place of judgement and self-judgement
by the characters: although marked similarities exist between the
two playwrights, important divergences are discernible here also.
Finally, these and all the other evidence are brought to bear on the
vexed question of the tragic quality of the plays.

It will be seen that my investigation concentrates on the first of
the six components of tragedy as defined in the sixth chapter of the
Poetics, namely plot. Things happen in plays: the characters exist in
order to do them or to have them done to them. Although in
Aristotle's thinking plot was the most essential component and the
one most characteristic of dramatic art, and although it is evident
not only from their own statements about it but from the consummate skill of their practice as playwrights that Corneille and
Racine also regarded it as of primary importance, it has received
comparatively little serious critical attention. Professor May's
study is an honourable exception. I believe that if we investigate
the various aspects of plot, from initial situation to dénouement, we
shall see more clearly both the parallels and the divergences between the two dramatists, not only in their technique itself (what
they called their art) but also in its implications in the presentation
of their tragic vision. It may of course be that neither of them could
ever present a description or definition of that vision, which is presumably why it is expressed in imaginative terms; but if it defies
rational definition on the level of conscious thought, and if therefore one could only at one's peril speak of a deliberate endeavour to
apprehend it and express it knowing exactly what it is, there is no
reason to suppose that the actual means of conveying that intuition
which lurks inside them is not deliberately chosen. It is in that
sense, in the practice of their art as they understood the term, in
the choice which is represented by the creation of situation, plot,
character, language, that one may speak of conscious intention.
The legitimate object of analysis — the only legitimate object,
perhaps — is the means adopted by the playwrights to give expression to their vision, the means deliberately chosen in order to
work on the imagination of the audience through the arousal of
specific emotions.

It will be evident, therefore, that the aim and scope of this study
are strictly limited. Although it is concerned with some of the
effects produced by the plays as theatrical artefacts — effects which
are still felt today, since many of Corneille's plays and almost all of
Racine's are alive now in the theatre —, it is an endeavour to reach
an understanding, in terms familiar to these writers, of those
aspects of their tragedies which fall under the heading of dramatic
construction. In pursuing my investigation I have endeavoured,

therefore, to avoid anachronistic interpretations and above all a preconceived system and even a clear definition of tragedy itself. I believe that this kind of scholarship and criticism, far from suggesting what would now be a merely archaic account of nothing more than historical interest, is the only sure guide to an authentic, live, present-day reading or performance of the plays. That matters far more, in the end, than scholarship and criticism in themselves. Of course, many other aspects of the tragedies conspire to produce their total effect: all that is offered here is an examination of their most fundamental features without which characterization, staging, and poetry (in the narrow sense) would achieve nothing. In the same way as I have refrained from any attempt to retrace the history of the Corneille–Racine rivalry, I have sought neither to describe the evolution of the playwrights' work in the course of their careers nor to analyse in detail — or even to mention — every one of their tragedies. I have selected for analysis those plays with which I am most familiar for the particular purpose in hand or which seem to me most suitable for the various aspects of the enquiry. In this I follow the precedent of Aristotle. Those who look for grand comprehensive schemes — critical, ideological, sociological, metaphysical — need read no further. All I can offer is a modest approach, an approach which I believe to be only a preliminary, though an essential one, to any consideration of other aspects — especially the literary and psychological — of the plays of Corneille and Racine.

It is time to begin at the beginning.

NOTES
Introduction

1. See the important distinction drawn between myth and mythology by R. C. Knight in 'Myth and Mythology in Seventeenth-Century French Literature'.
2. As long ago as 1939, W. G. Moore was pleading for study of plays from the aesthetic, rather than the historical and psychological, point of view. See 'Corneille's *Horace* and the Interpretation of French Classical Drama'.
3. *Satire II*, l. 71.
4. Saint-Évremond, *A M. le Maréchal de Créqui*, in *Œuvres en prose*, vol. iv, p. 112.
5. See, in particular, R. C. Knight, '*Horace*, première tragédie classique'.
6. *The Life of the Drama*, pp. 3–4. See also S. W. Dawson, *Drama and the Dramatic*, particularly chapter I.
7. Eva Schaper, *Prelude to Aesthetics*, p. 72. Cf. J. Jones, *On Aristotle and Greek Tragedy*, chapter I.
8. See R. C. Knight, 'The Rejected Source in Racine'.
9. I shall not, however, be studying verbal sources as such: they would call for another full-scale investigation.

CHAPTER I

The Dramatic Subject:
its nature and its disposition

Les personnages sont des groupements de gestes tout trouvés
... Il faut se garder de confondre l'étude d'une pièce avec
l'étude des personnages. D'une part en effet, les tensions dramatiques ne se présentent pas forcément comme des conflits
entre personnages ou à l'intérieur de tel ou tel personnage.
D'autre part, dans la mesure où ces tensions et ces conflits
coïncident, il faut encore se garder de confondre le point de
vue esthétique et le point de vue psychologique. Les personnages sont faits pour la pièce et non pas l'inverse. Considérer
la pièce comme simple prétexte à une 'peinture de caractères',
ce serait sacrifier l'esthétique à la psychologie. Et la perspective psychologique elle-même s'en trouverait faussée: car on
serait conduit à négliger ou mésinterpréter ce qu'il y a de
proprement dramatique dans la psychologie du personnage.

Robert Champigny, *Le Genre dramatique*

Now a whole is that which has a beginning, middle and end. A beginning is that which is not itself necessarily after anything else, and which has naturally something after it; an end is that which is naturally after something itself, either as its necessary or usual consequent, and with nothing else after it; and a middle, that which is by nature after one thing and has also another after it.

Thus Aristotle, in his usual careful way, defines in the seventh chapter of the *Poetics* the organic structure of tragedy. Corneille comments on the passage in his first *Discours*, emphasizing the implication of a logical sequence and criticizing the murder of Camille in his own play, *Horace*, but counterbalancing this with an outline of the development of *Cinna*.[1] Development in this context suggests movement from one state of affairs to another: the first state of affairs constitutes the situation from which the play begins. That does not of course mean that the dramatic situation has no antecedents, but that in terms of the wholeness and completeness of the action of the play the dramatist has chosen a particular point from which to begin that action.[2] Thus, in *Horace*, Corneille opens his play at the moment when the forces of Rome and Alba are about to join battle; in *Cinna*, at the moment when Émilie's pressure on Cinna to avenge her father's death by assassinating Auguste can no longer be resisted. In *Andromaque*, Racine chooses the moment when Oreste arrives at the court of Pyrrhus as ambassador of the Greek allies to demand the life of Astyanax; in *Bérénice*, that at which Titus is released from the obligations imposed by his mourning for Vespasian and ostensibly freed to marry the Palestinian queen.

At the same time, however, these beginnings are in themselves definitions of the situations which set in train the dramatic development and, in part at least, of the subjects of the plays, though that word will call for some discussion and definition. If we consider the initial situations, we see that they all consist in a set of relationships between the characters rather than in a historical or legendary event, though the characters are of course caught up in such an event. We also observe, if we compare with it the situations thus understood, that Corneille and Racine have not taken them in every particular from history or legend. In *Horace*, Corneille concentrates our attention on one brother out of the three chosen to be the champions of each side in the war, and to the Roman sister, Camille, engaged to the Alban, Curiace, he has added the Alban sister, Sabine, already married to Horace. The effect of this is of course to tighten the web of family relationships which the war threatens to break and to introduce conflict within them, conflict suggested by the historians, but greatly intensified by the playwright. (An interesting parallel is to be found in the almost contemporary *Marc-Antoine*, of Mairet, who introduces Octavie — unhistorically — into his play, placing the hero between his mistress and herself. Antoine is, however, not greatly torn: as Horace is completely committed to Rome, so is he to Cléopâtre.) In *Andromaque*, the quadripartite relationships between the principal characters are, as such, invented, as is the tangle of unreciprocated passions which are brought to a crisis by the arrival of Oreste. It goes without saying that these radical alterations to the historical or legendary situations are introduced, partly perhaps for the sake of novelty, partly in order to provoke a particular kind of emotional response from the audience, and partly to create a dramatic problem whose solution arouses curiosity.

We are immediately put on our guard against taking at face value statements such as that found in the prefaces to *Andromaque* where, after quoting some lines from the third book of Virgil's *Aeneid*, Racine declares: 'Voilà, en peu de vers, tout le sujet de cette tragédie....'. The subject of the play, defined in terms of its situation, bears little resemblance to that passage, i.e. the alleged source. How does this square with the innumerable protestations, made by both Corneille and Racine (and by some of the lesser dramatists, too) of fidelity to the sources which they quote, sometimes at considerable length? It is possible to find an answer to this question only by making a distinction, which they did not make, between subject, as delineated in situation, and sources, which sometimes provide some elements of the situation, some episodes in its development, and some of the characters and of their relationships. In this chapter, I shall try to consider subject and situation before moving on to study some aspects of the treatment of sources in a few plays of Corneille and Racine. This separation has already been adopted and has yielded interesting results in one particular instance:[3] part of those findings, adapted to my present purpose, is implicit in what I write here.

Of course, the subject is something larger and more significant than the situation which is its essential foundation and starting-point, and it is the subject that matters to the dramatists. Corneille's defence of his art as a tragedian is in reality a defence of his conception of the subject — what he calls 'le beau sujet'.[4] When he was criticized, it was that conception which usually came under attack, either explicitly or by implication. Whatever sacrifices or concessions he felt bound to make to the views of his critics, he consistently refused to give any ground in the matter of subject. The modifications made to *Le Cid* after the Quarrel of

1637–8, extensive though some of them are, appear insignificant when one considers the points on which the dramatist did not yield. The passionate behaviour of Chimène remained unchanged, for any change would have entailed abandoning the subject of the play as Corneille understood it; likewise, the possibility — to put it no higher than that — of the eventual marriage of hero and heroine is safeguarded by the retention of the original dénouement:[5] if the ending had been altered to bring it into line with the conventional moral (and moralizing) preconceptions of Chapelain and Scudéry, the subject of the play would have evaporated and its entire scale of moral values would have been inverted. It is moreover important to note that Chapelain appears to have understood this. After considering ways of constructing the play which would have disposed of his objections, he admits that they would have destroyed the subject as Corneille had conceived it, and so, he says, '... le plus expédient eût été de n'en point faire de poème dramatique, puisqu'il était trop connu pour l'altérer en un point si essentiel....'. In this connection, one notes the recurrence, in Chapelain's discussion, of the verb 'disposer'; it is clearly related to the ways in which the playwright may depart from the story he is using: '... il y aurait eu moins d'inconvénient sans comparaison, en disposant le sujet du *Cid* pour le théâtre, de feindre contre la vérité qu'on eût reconnu à la fin le comte pour père putatif de Chimène, ou que contre l'opinion de tout le monde il ne fût pas mort de sa blessure...'. Furthermore, what Chapelain calls 'la fable' is in fact 'l'invention et la disposition du sujet', or 'la conduite'. It is precisely that 'disposition de la pièce' which makes the subject,[6] and which Corneille refuses to sacrifice.

In the same way, in spite of D'Aubignac's strictures,[7] he never accepted that the last part of Act IV and the whole of Act V of *Horace* were anything more than a technical blemish, although he admitted in the *Examen*, written twenty years after the play, that the murder of Camille is inadequately prepared and that its consequence is that her brother, having escaped one peril, falls unnecessarily into another (this being an offence against the unity of action as defined by the playwright himself[8]). These admissions are tactical concessions, perhaps, but the fact remains that the opportunity offered by the collected edition of 1660 for making alterations along the lines suggested by D'Aubignac was simply not taken, presumably because those alterations would have produced a piece of heroic drama, like many others of its period, but not an authentic tragedy.

Again, the only serious charge brought against *Théodore* concerned the essence of the subject, which involved enforced prostitution, and was therefore morally unacceptable. In all other respects, D'Aubignac writes of it in laudatory terms.[9] But again Corneille neither abandoned nor modified his play, but fell back on the argument that what is moral enough for St Ambrose must be good enough for a French theatre-goer of the seventeenth century.[10] Much later, in 1663, his defence of *Sophonisbe* was a defence of the subject of the play as he had created it, against the asseverations of the same critic to the effect that it was unacceptable in terms of contemporary morality: Mairet had handled the subject more satisfactorily because he had made concessions to that morality by inflicting a kind of punishment through suicide on Massinisse, punishment for bigamy as the Christian world understands it. Corneille simply argues that Mairet had in this way altered the supposed historical subject almost beyond recognition — but so had he, in a different way, by inventing the character of Éryxe.[11]

Not that fidelity to historical sources, at least in the matter of the subject and situation, was in itself considered an indispensable or even valid criterion. Indeed, it is clear that the orthodox critics, to whose precepts the dramatists had at least to pretend that they could accommodate their practice, preferred *le vraisemblable* to *le vrai*.[12] So, like Racine after him, Corneille was often quite prepared to show the points at which he departed from his historical sources, but such admissions constituted on the one hand an opportune demonstration of technical orthodoxy, and on the other a defence of the subject. Racine's diatribe against Corneille in the first preface to *Britannicus* is specious: neither dramatist was above making radical changes to his supposed sources in order to create the dramatic subject he required. Generally indeed, Racine adopted the same kind of tactics as his elder. With regard to Pyrrhus, for example, he played his critics off, in the first preface to *Andromaque*, one against another. If, for some, Pyrrhus was too remote from his Greek counterpart because of his resemblance to the seventeenth-century courtly lover, for others he was altogether too near to the character of antiquity by virtue of his ferocity and violence. The suggestion usually made is that Racine was here primarily concerned, as in other prefaces, with arguments relating to *bienséance* and *vraisemblance*; but, accepting the impossibility of reconciling the claims of critics who emphasized either historicity or propriety, he was, at least by implication, defending the subject of *Andromaque* as he conceived it and the part and character of Pyrrhus as they were essential to it. It is true that the dramatists were faced with a difficult and complex task: since the whole theory and tradition of tragedy required the use of well-known stories and characters, usually drawn from some remote area of history or legend (no distinction was made between the two, and I shall henceforth use the word 'history' indiscriminately for both, as the seventeenth century did), and since at the same time the public had little historical imagination and deep-rooted moral and aesthetic prejudices, it was necessary to find some means of compromise between historical accuracy and contemporary credibility. Straightforward adaptation of Greek or Latin tragedy, paraphrase or translation, such as had sometimes been attempted in the sixteenth century, were out of the question, partly because of the prejudices and partly because the whole basis of the form of tragedy, as by the 1630s it had developed out of tragi-comedy and *pastorale*, required dramatic situations of a kind unknown to antiquity. It is this which accounts for the cavalier and ambivalent attitude towards their sources adopted by our two playwrights, as well as for the total disrespect for such sources shown by most of their popular contemporaries of whom, at least by implication, Corneille and Racine were critical.

The question arises as to whether these two writers found their dramatic situations in the sources they acknowledge. If we are asked, as ordinary theatre-goers or readers, what is the subject of a play, we usually begin by defining the initial situation, the relationships between the characters, and the working-out of that situation and those relationships through the plot to the dénouement. It is only when we describe the development that we are able to suggest character or language. In such an account, we use proper names — we are talking about individuals in particular circumstances. We are thinking about the subject as a story in which those individuals are placed in a situation which involves them in certain relationships one with another, a situation which is usually imposed upon them by forces beyond their control (the war in *Horace*, the death of Vespasian in *Bérénice*) or which they have manipulated in the past (Auguste's proscriptions in *Cinna*, Pyrrhus's betrothal to Hermione in *Andromaque*). But the story, as presented by the dramatist, is by no means a repetition of the one he received from history and, to begin with, the initial situation is rarely that of the sources, particularly because the actual historical event is only the incident which occasions the story of the play: the story concerns those individuals who are affected by and sometimes affect the historical event.

This becomes obvious when we consider some of the facts of literary history. It is a commonplace that, as Corneille[13] observed, two or more dramatists (Sophocles and Euripides) may make use of the same story (that of Electra), treating it very differently, yet preserving its most essential features. The same was true, of course, in seventeenth-century France were theatrical rivalries were rife: Corneille himself in 1663 took up again the story of Mairet's *Sophonisbe* (1639); he and Racine coincided in 1671 with the Bérénice plays; Racine's *Mithridate* (1673) followed that of La Calprenède (1635)... Although in such cases the outcome of the story may be the same in several treatments of it, differences of treatment are such as to involve different initial situations and different plots. On the other hand, the situation concerning one set of historical personages may more closely resemble that of a play built around quite different personages. What, then, is the subject? Is it the story, in whatever version, of a particular group of individuals? Or does it consist, in part, and at the beginning, in a certain situation in which characters — however named, and irrespective of historical context — stand in certain relationships to one another?

An important clue to an answer is to be found in the allusion made by Racine to the gloss of a Greek scholiast on Sophocles:

> ...il ne faut point s'amuser à chicaner les poètes pour quelques changements qu'ils ont pu faire dans la fable, mais...il faut s'attacher à l'excellent usage qu'ils ont fait de ces changements, et la manière ingénieuse dont ils ont su accommoder la fable à leur sujet.[14]

Racine is of course defending those changes to the Greek and, more particularly, to the Virgilian story which he claims to be his source. Whereas in the first preface he had insisted on his fidelity to the source, now, in the second, after a lapse of eight or nine years, he is admitting his radical departures and justifying them. But what is important for the moment is to note that, as a general principle, such departures are accepted and that, above all, the dramatist evidently has a preconceived subject in mind when he writes a play, a subject to which the 'fable', the story of the source, must be accommodated in the most ingenious way possible. (One should note here that confusion exists in the use of the word 'fable' in Racine's writings: in the present instance it is clearly meant to indicate 'story' — Aristotle's 'traditional story' —, but in his own partial translation of the *Poetics*, he uses the word to translate the Greek 'mythos' which in that context, at any rate, means plot.[15] Likewise, Corneille frequently applies the word 'sujet' to the 'story'.[16] So one must beware of deducing firm conclusions from the dramatists' use of these terms.)

We need not leave the particular example of *Andromaque* to understand some of the implications and consequences of this differentiation of subject from traditional story. In the second preface, the opening quotation from Virgil is retained, together with the comments to the effect that in it the dramatist had found 'tout le sujet' of his play, the place, the action, the four principal

personages and their character, except for the jealousy and passion
of Hermione which is said to come from Euripides: 'C'est presque
la seule chose que j'emprunte ici de cet auteur. Car, quoique ma
tragédie porte le même nom que la sienne, le sujet en est pourtant
très différent.' It is, then, under cover of this reference to Euripides
that Racine proceeds to show some of the ways in which he has
departed from his sources: Andromaque is the virtuous widow of
Hector and mother only of his son, Astyanax, whose life he has
prolonged beyond the end of the Trojan war, and who has now
become the innocent and unseen pawn in a game of blackmail,
partly political, partly emotional; she is not the chattel and
concubine of Pyrrhus, nor the mother of his son, Molossus, but she
is the women he loves and tries to force into marriage with him —
while he is not above cruelty he is (see the first preface) less harsh
than in antiquity. But there are other changes, no less far-reaching.
Hermione is indeed jealous of Andromaque, but not because her
rival, the slave and captive, has borne Pyrrhus the son she, the
legitimate wife, is unable to give him, but because Pyrrhus, now
only betrothed to her, threatens to cast her aside in favour of
Hector's widow. Oreste, meanwhile, in love with Hermione,
arrives at the court of Pyrrhus bearing an invented embassy from
the Greeks, which is a pretext for his trying to elope with her. A
new situation is thus created, whose primary feature is the endless
chain of unreciprocated passion, Oreste loving Hermione, she
loving Pyrrhus, he loving Andromaque, she remaining faithful to
her dead husband. At the end, Pyrrhus is killed and Oreste is
instrumental in his assassination: that comes from the tradition;
but the place is changed and the motive is different. In Racine's
play, Oreste acts in order to win Hermione as a reward, not out of
jealousy of Pyrrhus who is already her husband. Such changes of
motive are compounded by the introduction of the political and
dynastic aspects of the situation which are made possible by the
prolongation of the life of Astyanax as a survivor of the royal house
of Troy and therefore a possible threat to the future security of the
Greeks should he return to his native city. It is difficult to believe
that Racine effected such radical changes merely in order not to
offend the proprieties. Andromaque's faithful widowhood, which
would have been impossible in the context of the ancient world,
may indeed allow her to conform to seventeenth-century French
morality, but in fact, though the dramatist does not admit it, it is
around her resistance to an unmarried Pyrrhus that the whole
situation is constructed, while he is able to consider himself free —
though not entirely — from his obligations to an Hermione
relentlessly driven by unslaked passion for him.

Although these facts are well known, it is as well to rehearse them because they so clearly allow us to see that Racine's concept of the subject of his tragedy is very remote from that of his classical sources, and that the primary aspect of the subject is the moral situation.

It is also well known that, as Voltaire long ago pointed out, that situation strikingly resembles that of Corneille's *Pertharite*,[17] the box-office failure of 1651–2, thanks to which he abandoned writing for the theatre until Foucquet's commission seven years later. The similarities, however, are circumstantial, and Corneille's play is not Racine's, partly because the supposedly dead hero reappears in Act III and causes the situation to be turned upside down and to require a series of new developments and decisions for its resolution, and partly because the passions driving the characters are by no means the same, nor of course is the emotional response from the audience.

What this appears to mean, then, is that Racine conceived of his subject in the form, not of the original story, but of a situation, probably drawn from Corneille and made up of a set of relationships which is exploited dramatically in emotional terms partly resembling those which feature in Euripides and Virgil. If we simplify matters by leaving aside the influence of contemporary interest in the psychology of love and in moral problems and that of a great body of literature, classical and modern, it is evident that Racine's subject demanded the most drastic alterations in the 'fable'. But therein, he implies in that second preface, lies a great part of his artistry. It is possible to grasp the situational parallel only when one is prepared to ignore the proper names and the historical context and to concentrate, in a rather abstract way, on relationships and motives. Perhaps it is not surprising that the *Pertharite–Andromaque* parallel should first have been mentioned by a writer of tragedies and the would-be successor of Racine.

That Racine should borrow from Corneille in this way is to be expected: the younger man learnt from and emulated the elder. But Corneille appears to have taken dramatic situations from others, too. In *Nicomède*, if we are to believe what he says in the *Examen*, he dramatized the account given in the thirty-fourth book of Justin's history. If, however, the proper names and context are ignored, the situation can be described as that of a weak old king who has two sons: the elder, by his first wife, is heir to the throne; the younger, by his second wife, has been brought up in a powerful country which now threatens his own. The king can choose resistance, as advocated by the loyal and patriotic elder son, or what we should call appeasement, as suggested by the younger

son, abetted by his mother who opposes the projected marriage of
her stepson to the queen of an allied power. Corneille, having
quoted his source, admits to having made certain far-reaching
modifications: he has eliminated the assassination of the king by
the elder son (and thus invested the latter with a heightened
nobility of character) and the king's intention to have the son
killed; he has made the sons rivals, not only politically, but
emotionally, by imagining them to be suitors of the foreign queen;
and he has invented the role of the ambassador of the strong and
hostile power. These are only the most obvious changes, but they
suffice to demonstrate that Corneille has in fact created an entirely
new disposition of the situation as it is to be found in Justin.

But is it new? Nothing could be less certain, for it has been
shown[18] that Corneille's situation is almost identical with that of
Rotrou's *Cosroès* (1648), a play set, not on the eastern edge of the
Roman Empire in the second century, but at Persepolis in the
sixth, and concerning an altogether less familiar historical episode,
which the dramatist develops in such a way as to exploit its
potential for pathos and to make of it a somewhat romanesque and
sentimental tragedy. Corneille, on the other hand, develops the
heroic potential, minimizing, as he says in the *Examen*, the
sentimental element. His procedure is in this respect no different
from that of Racine in adapting the situation of *Pertharite*, a play
concerning primarily heroic behaviour in characters drawn from
an obscure period in Lombard history, to an intensely passionate
treatment of a largely invented version of a group of celebrated
characters drawn from Greek legend. It is possible that Racine,
too, found some aspects of his situation in another play by Rotrou,
Hercule mourant (1634), though these were probably less important
than was once supposed.[19]

Then again, Racine in his turn appears to have adapted
important situational elements of *Nicomède* to his own *Mithridate*,
whose events occurred in partly similar circumstances and,
geographically, in a similar place. Racine exploited the situation in
his own way, accentuating rather than minimizing the love plot and
complicating it by unmarrying the king and making him the rival
of his sons for the hand of the foreign princess: as in *Nicomède*, she is
at his court. (There is also here, of course, an analogy with
Andromaque.) *Mithridate* becomes, not a play of grand if passive
heroism, but one of tempestuous love and jealousy, as well as of
treachery pitted against loyalty: these central aspects of it are
certainly suggested in the historical accounts[20] claimed by Racine
as his sources (see the preface), but their development is made
possible thanks, not to those sources, but to the adoption of the

situation of *Nicomède*. It is out of Andromaque's invented capacity to resist the advances of Pyrrhus that Racine virtually creates his play: in the same way, the virtue of Monime becomes the very centre of *Mithridate*, whether in the form of obedience to the king or of loyalty to Xipharès. The play is built round her, the one female character as against the three men. Her virtue, as presented by Racine, seems to owe more to Le Moyne's idealized portrayal[21] than to any of the strictly historical sources, and her situation — being wooed by a father and his two sons — to be derived from La Calprenède's *La Mort des enfants d'Hérode* (1639). (This would scarcely be surprising, since the same playwright had also written a *Mithridate*.) Some aspects of Racine's tragedy may also have come from *Rodogune*, in particular the brothers' both being suitors of the foreign princess at their ruling parent's court and the death of that parent at the dénouement.

Such comparisons of abstract situation could be multiplied: for example, Racine's debts in *Phèdre* not only to Euripides and Seneca, but to French writers who had dramatized the same story before him, from Garnier to Bidar, and even to Grenaille and Tristan who had used the analogous theme of Crispus, have been amply demonstrated. It could well be that the filiation goes back to Bernardino Stefonio's Latin *Crispus* of 1597, and that, if Racine did use dramatizations of this story, he found the suggestion in Corneille's second *Discours* (*Writings*, p. 42), where the comparison with the Hippolytus legend is explicitly made.[22]

Three other aspects of the problems concerning situation may be touched upon. First, it is evident that, as far in particular as circumstantial situations are concerned, Corneille and Racine sometimes imitate themselves. In *Sophonisbe*, Corneille takes up again from *Nicomède* the situation of the foreign prince or princess resisting the superior power of Rome. The heroine of the later play is the counterpart of the hero of the earlier in character and in patriotic idealism; both are also represented as being in love with an anti-Roman hero or heroine belonging to a different nation. A number of resemblances exist between the *Examen* of *Nicomède* and the preface to *Sophonisbe*: 'J'ai fait [Nicomède] amoureux de Laodice, reine d'Arménie' — 'je lui [Sophonisbe] prête un peu d'amour'; 'Ce héros de ma façon' — 'C'est une reine de ma façon'. To all intents and purposes, Corneille has invented the characters and their emotions and so created new situations which, however, in spite of parallels, develop and conclude in quite different ways. Other situational similarities may be noted in the unseen presence of Hannibal as the archetypal hero resisting the power of Rome, and in that power not being represented — as it is, for instance, in

Pompée — by a major historical figure (Scipio does not appear in *Sophonisbe*, though he had done so in earlier versions, notably Mairet's), but by Flaminius in one play (an invention) and by Lélius and Lépide in the other. Professor Stegmann has pointed out[23] the primary importance for Corneille both of the initial disposition of the characters and of his making use in later plays of aspects of it which he had already exploited in earlier ones. In the same way, Racine's *Bérénice*, in part of its situation, derives from *Andromaque*: Oreste, the luckless lover of Hermione, is transformed into Antiochus, rejected by Bérénice who loves Titus as Hermione loves Pyrrhus: where Pyrrhus abandons Hermione in favour of Andromaque, Titus repudiates Bérénice in favour of Rome.

The second point may be illustrated in terms of the use of a particular story with the rejection of the situation chosen or invented for an earlier treatment of it. Corneille's preface to *Sophonisbe*, while commending (tongue-in-cheek?) Mairet's treatment of the same story, emphasizes his own distinctive approach to it: Corneille has observed 'une scrupuleuse exactitude à [s] 'écarter de sa route [Mairet's] . . . par le seul dessein de faire autrement'. So Scipion, who played a part in Mairet's play (and in the historical narratives), does not appear, but the invented Éryxe does — this alone is also sufficient to set Corneille apart, not only from Mairet, but from Trissino and his French imitators (Melin de Saint-Gelais (1559), Mermet (1584), Montchrestien (1596), and Montreux (1601)), though, as he himself admits, he follows them (and history, which he does not mention) in not killing off either Syphax or Massinisse. His reason for inventing Éryxe is important: 'elle ajoute des motifs vraisemblables aux historiques, et sert tout ensemble d'aiguillon à Sophonisbe pour précipiter son mariage, et de prétexte aux Romains pour n'y point consentir.' (Exactly the same argument had been used by Le Vert, in 1646 (*Aricidie*, preface) to justify the addition of 'quelque épisode vraisemblable' to the essential material of his play, and it is interesting to note that in so doing he is following the example of his 'friend' and 'compatriot', Corneille, in *Cinna* and *Polyeucte*.) The interest of *Sophonisbe* is chiefly in the dramatic conflict of motives and passions, and this is reflected in Corneille's creation of a new situation and in his explicit rejection of Mairet's. In the same way, from the situational point of view, Racine in *Mithridate* alters history and avoids La Calprenède's version of the same story, preferring, it would seem, to make use of Corneille's situation in a play based on a quite different historical episode.

Thirdly, we may consider the *Bérénice* plays of 1670. I do not intend to reopen the question of priority as between Corneille and

Racine nor that of either having cognizance of the preparations of the other. What is of interest here is rather the quite different dramatic situations created by the two playwrights around the same historical story. Each of Corneille's main characters, Tite, Bérénice, Domitian, and Domitie, is historical, as are of course the final outcome (Bérénice's departure from Rome) and the fact that her presence there is not on the occasion of her first visit. But Domitie's relationship to Tite and to Domitian, her wish to marry the emperor out of ambition while admitting to being in love with his brother, and the ensuing triangular problem, complicated by the advent of Bérénice in Act II scene v, creates a highly dramatic situation unknown to the historians. Did Corneille find the germ of the situation in *Andromaque*, with Domitian (Oreste) in love with Domitie (Hermione), who seeks to marry Tite (Pyrrhus), the reigning monarch, who wishes to marry Bérénice (Andromaque), the foreign princess (a marriage frustrated, at least in part, by political forces), and a crisis provoked by the arrival of Bérénice, at the other end of the chain from Oreste? At all events, Corneille writes an essentially politically-centred play in which ambition in Domitie and 'gloire' in Bérénice stand opposite all-consuming passionate love in the male characters. From Racine's *Bérénice*, the political motives and the ambition are absent: although his tragedy is one of passionate love frustrated, which is what Suetonius's laconic account suggests, he has had to invent, too, in order to create a dramatic situation and not simply an elegy or a lament. In part, at least, that invention already existed for him, as we have seen, in *Andromaque*, and therein lies the clue to the existence of Antiochus, quite unhistorical in this context,[24] but essential to the development of a drama characterized — as is Corneille's very different play — in part by jealousy and rivalry and misunderstandings leading to a search for the truth, a search which is perhaps the real subject of the tragedy.

Whether Corneille and Racine make use of situations found in the plays of others based on stories different from their own, or reject situations created for stories the same as their own, or engage in self-imitation, or find each independently his own situation for the same story, such situations are, in relation to the original story, inventions.[25] In the ninth and fourteenth chapters of the *Poetics*, Aristotle states that although tragedies are based on traditional stories, characters, names, and episodes may be of the poet's invention, and that his aim should be to give his play a universal import in the action it represents. He is creative above all in being a plot-maker. And, as we have seen, the first stage in making a plot is the creation of a dramatic situation. What is sacrosanct is the nature of the catastrophe, to which the title of the play usually gives a clue from which the audience rightly derives certain expectations. So, when he discusses the Electra story, Corneille points out that in any acceptable version Aegisthus and Clytemnestra must die, but that, as Sophocles and Euripides show, there may be more than one way of reaching that conclusion.[26] (It is a little ironical and amusing to find him suggesting a rewriting of the story for modern audiences which involves an accidental killing of Clytemnestra of a kind advocated for Camille by D'Aubignac in a rewritten *Horace!*) It is in this context that Corneille distinguishes between the invented *acheminements* and the traditional *effet*: a notable example of the problems raised in this connexion by the exigencies of a *vraisemblance* directed for the seventeenth-century public precisely to a sense of universality is of course to be found in Racine's *Iphigénie*.

Taking precautions of a similar nature, for reasons of *bienséance* and in order to exculpate Antiochus, Corneille admits in his *Examen* to having made important changes to the historical story of *Rodogune*, although he claims as sources Justin, Josephus, the book of *Maccabees* and Appian ('Le *sujet* de cette tragédie est tiré d'Appien Alexandrin' — but in our sense 'sujet' is a misnomer). It is from Appian that he reckons to have drawn 'la narration que j'ai mise au premier acte' (i.e. a definition of the initial situation, but the parallel is only partial) and 'l'effet du cinquième' (i.e. the catastrophe, though he alters the means by which it is finally achieved). 'Le reste', he continues, 'sont des épisodes d'invention, qui ne sont pas incompatibles avec l'histoire.' In the next paragraph, Corneille explains his particular preference for this play by referring again to the 'incidents surprenants qui sont purement de mon invention et n'avaient jamais été vus au théâtre'. This tragedy is characterized first and foremost by 'la beauté du sujet, la nouveauté des fictions' (the juxtaposition is perhaps significant), and later in the same *Examen*, Corneille writes: 'J'ai déguisé quelque chose de la vérité historique...'. Indeed, in his critical writings, he frequently prides himself on his 'inventions', always suggesting that they do not actually contradict history and often that they are directed towards the arousal of particular emotions. Although he is less inclined to expatiate on this topic, it is clear that Racine, too, prides himself on his inventiveness, for example in the preface to *Bérénice* and, more defensively, in the second preface to *Andromaque* where he simply follows Aristotle in claiming to preserve 'le principal fondement' of the traditional story. It was through this process of innovation, at work first in the situation, that a dramatist made such a story his own and out of it created his subject. In so doing he attained that *nouveauté* for which the seventeenth-century public clamoured,[27] and which was a mark of the originality which, at certain times, Racine at any rate denied possessing. But such denials merely point to the paradox of adhering to the traditional stories (as the *doctes* and theorists demanded) and making them appear new (as the theatre-goers expected). Novelty, as Grenaille wrote in his preface to *L'Innocent malheureux* (1639), is directly linked to dramatic effect: 'Or devant que de parler plus avant de mon dessein [i.e. in this play], il faut que j'étale mon sujet, et que je décrive brièvement l'histoire qui lui sert de fonds pour mieux découvrir ce que j'ai ajouté de mon invention pour la rendre plus dramatique.'

The dramatic effects specific to tragedy are of course aimed at arousing certain emotions. In his second *Discours* (*Writings*, pp. 37 ff), Corneille comments at length on the fourteenth chapter of Aristotle's *Poetics*, in which the type of situation appropriate to tragedy is discussed: 'Whenever the tragic deed... is done within the family — when murder or the like is done or meditated by brother on brother, by son on father, by mother on son, or son on mother — these are the situations the poet should seek after.' Corneille's version of this takes some liberties with it: '... quand les choses arrivent entre des gens que la naissance *ou l'affection* attache aux intérêts l'un de l'autre, comme alors *qu'un mari tue ou est prêt de tuer sa femme, une mère ses enfants, un frère sa sœur*, c'est ce qui *convient merveilleusement* à la tragédie.' (My italics.) Although he remains much more faithful to the original text, Racine too (*Principes*, p. 22) makes one significant modification: 'Il reste... que ces événements se passent entre des personnes liées ensemble par les nœuds du sang *et de l'amitié*, comme, par exemple, lorsqu'un frère tue ou est prêt de tuer son frère, un fils son père, une mère son fils, ou un fils sa mère; et ce sont ces événements que le poète doit chercher.' (My italics.) It is clear that the extension of Aristotle's 'within the family' to include ties of love and friendship is a perhaps unconscious result of the traditions of seventeenth-century tragedy which was bound to include a love-interest; Corneille's alteration of the type of example given in the *Poetics* also probably arises in part out of the subjects of his own *Médée* and *Horace*. Eugène Vinaver points out (*Principes*, p. 6) that, in addition to current dramatic practice, Vettori's commentary ('inter personas aut sanguine aut benevolentia magna inter se coniunctas' — 'between persons bound together by ties of blood or of great friendship') could have influenced Racine's choice of formula. In his subsequent discussion, Corneille gives the reasons for the appropriateness of such alterations to tragedy: 'C'est...un grand avantage, pour exciter la commisération, que la proximité du sang et les liaisons d'amour ou d'amitié entre le persécutant et le persécuté, le poursuivant et le poursuivi, celui qui fait souffrir et celui qui souffre...'. Such situations give rise to 'ces actions tragiques qui se passent entre proches', and conflicts between and within the characters which constitute those actions arouse 'les grandes et fortes émotions qui renouvellent à tous moments et redoublent la commisération'. Although tragedy reaches perfection thanks to appropriately splendid language and spectacle, it is nevertheless in the subject thus conceived, in a particular kind of situation developed in action, that its essence consists.

Before going on to consider the 'grandes et fortes émotions' and their origins, as understood by the dramatists and their contemporaries, I wish to make two observations. The first is that just as Aristotle discusses the situations appropriate to tragedy in very general terms, thus reducing them to abstract concepts — such a procedure is hardly surprising in a philosopher —, so, when he elaborates on the text of the *Poetics*, does Corneille: 'Les oppositions des sentiments de la nature aux emportements de la passion ou à la sévérité du devoir, forment de puissantes agitations qui sont recues de l'auditeur avec plaisir...'. Other passages provide similar evidence — this one, for instance, from the *Examen* of *Nicomède*: '...la grandeur de courage y règne seule...Elle y est combattue par la politique, et n'oppose à ses artifices qu'une prudence généreuse, qui marche à visage découvert, qui prévoit le péril sans s'émouvoir, et qui ne veut point d'autre appui que celui de la vertu...'. It is striking that here — and we find something similar in the plays — Corneille defines the tragic situation in abstract nouns denoting, not characters, but characteristics, and that combined with them, and animating them, are verbs for which one would usually expect personal subjects. The situation, then, is an abstract conception, but it concerns persons whose passions are the source of their actions. In the same *Examen*, the playwright indicates that his reasons for modifying the historical story lie in the need to create a situation — which thereby becomes new, original, invented — characterized by conflicts of passions: jealousy, love, courage, pride, ambition, cowardice, 'gloire', 'générosité', 'fermeté'.

Such a way of looking at tragedy was not unusual in the seventeenth century. Grenaille, for example, writing of La Calprenède's *Édouard* in his preface to *L'Innocent malheureux* (1639), generalizes relationships and connects situation, action, and passions:

...nous voyons un roi qui devient esclave de sa sujette, un père qui confirme sa fille en ses bonnes résolutions..., une dame qui est soupçonnée d'être cruelle envers son roi parce qu'elle est trop fidèle à son honneur...La disposition correspond à la beauté de l'invention, les passions ont de beaux commencements et de très bonnes issues..., etc., etc.

La Mesnardière, in more prescriptive vein, writes in similar terms at about the same time: 'La poète ne produira point cet effet avantageux [the arousal of pity] s'il ne fait que les infortunes arrivent à son héros par l'insigne perfidie de ceux qui lui sont conjoints par les nœuds de la parenté, de l'amitié ou de l'amour.'[28] And he goes on to admit that the dramatist is at liberty to invent modifications 'en bâtissant son sujet'. Novelty, invention, the disposition of the subject are all connected, as was natural in an age where literary education was based on formal Rhetoric, in which *inventio* and *dispositio* loomed large. '*Inventio*', it has been said, 'was obviously useful in finding the characters something to say.'[29] But when Racine uses the word in the preface to *Bérénice*, he is referring specifically to the filling-out of his play with an appropriate number of episodes, and he doubtless has in mind the fact that in order to achieve this, as we have seen, it was necessary to invent the role of Antiochus and thus create a properly dramatic situation. 'Toute l'invention consiste à faire quelque chose de rien' may in fact be an exaggeration. After all, Antiochus is probably derived from Oreste, and *Andromaque* from *Pertharite*... Nevertheless, Antiochus is a pure invention as far as the Titus–Bérénice story is concerned. Furthermore, in his enumerations of the things in which invention consists, Racine places 'l'élégance des expressions' after 'action', 'passions', and 'sentiments'. The advantage, for the creative genius, of the truncated sentence from Suetonius which opens the preface was that it left Racine free virtually to invent the situation and the play in its entirety, not only in its words.

'*Dispositio* enables [the poet] to put together the tirades...; on a large scale the spirit of *dispositio* might influence him in the orderly overall construction of the play...'.[30] The implication of Professor Peter France's remark is of course that *dispositio* might indeed be applied to literary creation other than in the realm of language. That is precisely what we find in the work of D'Aubignac, for instance: 'La composition de la tragédie n'est autre chose que la disposition des actes et des scènes...', which follows 'la constitution de la fable', i.e. the formation of the subject. The inspiration comes from Aristotle and is echoed by Jean de la Taille in his *Art de la tragédie* in 1572: 'Or c'est le principal point d'une

Tragédie de la sçavoir bien disposer, bien bastir, et la deduire de sorte qu'elle change, transforme, manie et tourne l'esprit des escoutans de cà, de là...', and all the way down the following century until we come to Saint-Évremond's *Défense de Corneille* in 1677: 'J'ai soutenu que pour faire une belle comédie il fallait choisir un beau sujet, le bien disposer, le bien suivre, et le mener naturellement à sa fin...'. That we are faced with an aesthetic principle with applications beyond the realm of literature and drama is evident from the fact that Poussin, for instance, wrote in exactly parallel terms in a letter containing his definition of the art of painting: '[La matière] doit être prise noble [cf. 'le beau sujet'] ... Il la faut prendre capable de recevoir la plus excellente forme. Il faut commencer par la disposition, puis par l'ornement...'.[31]

However, this process of disposition is also one of the ways in which the demand for novelty could be reconciled with the use of the traditional stories. Again Poussin provides a valuable clue: 'La nouveauté dans la peinture ne consiste pas surtout dans un sujet non encore vu, mais dans la bonne et nouvelle disposition et expression, et ainsi de commun et vieux, le sujet devient singulier et neuf.' An almost exact parallel is to be found in Pascal:

Qu'on ne dise pas que je n'ai rien dit de nouveau: la disposition des matières est nouvelle. Quand on joue à la paume, c'est une même balle dont on joue l'un et l'autre, mais l'un la place mieux.

J'aimerais autant qu'on me dît que je me suis servi des mots anciens. Et comme si les pensées ne formaient pas un autre corps de discours par une disposition différente, aussi bien que les mêmes mots forment d'autres pensées par leur différente disposition.

The second sentence of this passage is particularly apposite to our discussion, because it provides an admirable commentary on the rivalries and borrowings of seventeenth-century dramatists and an indication that if Corneille and Racine differed from their lesser contemporaries it was in quality rather than in kind. Poussin's remark may be related to his own renewal of themes he had already treated. Chantelou having asked him to make copies for him of the *Seven Sacraments* painted for Cassiano del Pozzo, the painter at first tries to find some other copyist to do the work and then agrees to execute it himself. But this is repugnant to him and he decides to create an entirely new series:

Pensant en moi-même toutes ces choses, ... je souhaiterais être moi-même le copiste..., ou de tous les sept ou d'une partie, ou bien les faire d'une autre disposition.

[Pozzo] a été étonné de voir sur un même sujet une disposition si diverse et des actions de figures toutes contraires aux siens...

This possibility of diverse treatment of the same theme, of a re-creation amounting to a new subject, is the mark of 'tout l'artifice de la peinture', since in it lies 'une puissance d'induire l'âme des regardants à diverses passions'. The emotions aroused by Cor-neille's *Sophonisbe* (chiefly 'admiration' for superhuman heroism) are very different from those occasioned by Mairet's (pity above all), just as those awakened by Racine's and Corneille's Bérénice plays differ. The themes, the stories may be similar or identical, but the disposition of the material is the first clue to a difference of subject, in the dramatic and emotional sense.[32]

If, then, we take 'disposition' to mean the ordering and arrange-ment of the dramatic material within the play, and if it is there that novelty and originality are primarily to be found, it follows that the creation of a situation is the first and obvious step that the playwright must take. The fact that he may find that situation rather than create it *ex nihilo* (is such a thing possible?) does not detract from his originality and creativeness, because its novelty will lie in its application to and modification of a traditional story. I find it difficult, *pace* Raymond Picard, merely to dismiss the gloss — 'comptant le reste pour rien' — put by Louis Racine on the words attributed to his father: 'ma tragédie est faite'. And why should one despise the prose draft of Act I of the projected *Iphigénie en Tauride*? One of Racine's letters corroborates these indications that he did actually work in this way: 'J'ai refait et mis en sa der-nière perfection tout mon dessein ... Avec cela, j'ai commencé même quelques vers ...'. These words refer to a projected play, *Les Amours d'Ovide*. Where should the 'dessein' begin if not with the situation? An exact parallel can be found again in Poussin's letter to his patron Chantelou in March 1658, only three years earlier than Racine's to Le Vasseur: 'J'ai arrêté la disposition de la *Conver-sion de Saint Paul* et la dépeindrai en temps d'élection.'[33] In the letter to Chambray quoted earlier, we read that 'il faut commencer par la disposition'. This is not surprising, since it is clear that Poussin followed his own advice. His biographer, Félibien, writes of the artist's method of composing his great figure-paintings in terms which suggest theatricality:

Il étudiait sans cesse tout ce qui était nécessaire à sa profession [so did our dramatists], et ne commençait jamais un tableau sans avoir bien médité sur l'attitude de ses figures ... Aussi on pouvait, sur ses premières pensées et sur les simples esquisses qu'il en faisait [prose drafts?], connaître que son ouvrage serait conforme à ce qu'on attendait de lui. Il disposait sur une table de petits modèles qu'il couvrait de vêtements pour juger de l'effet et de la disposition de tous les corps ensemble, et cherchait si fort à imiter toujours la nature que je l'ai vu considérer jusques à des pierres, à des mottes de terre et à des morceaux de bois, pour imiter des rochers, des terrasses et des troncs d'arbres . . .[34]

It is clear that Poussin not only saw his figures on a kind of miniature stage (it has in fact been reconstructed[35]), but that he saw and set them in attitudes expressive of their relationship one to another. This came first, then individualization through costume, and lastly particularization of the setting or context through the judicious arrangement of details. The tragic poets, of whose methods of work we possess no complete accounts, seem also to be primarily concerned with a disposition involving attitudes and relationships in a general or abstract way first, and then particularized in individuals and set in a specific historical context. Both Corneille and Racine take great pains to give an impression of the appropriate setting — in an evocative and moral rather than in a picturesque way (one is very much aware of the contrast between them and, say, Rotrou in this respect) —, especially in the exposition (see, for example, the opening scenes of *Horace* or *Phèdre*), though it is usually sustained throughout (for instance in *Cinna*, II.i or *Bérénice*, I.v). Yet, suggestive as such passages are of a background, a background they remain. Like the features of landscape in Poussin's figure paintings, they are selected and arranged so as to give an appearance of being natural and realistic, but only in relation to the subject expressed in the figures. That the subordination of 'local colour', of details which particularize, to the action itself (or even their exclusion) was a feature of the art of Racine, of Poussin, and also of Le Brun, is clear from the comments and paintings of the latter, and from the remarks of Félibien, of the critical Jean-Baptiste de Champaigne, and of Perrault.[36]

Moreover, if one studies Poussin's preparatory drawings for, say, *The Seven Sacraments*, comparing them with the finished paintings, one is immediately struck by two things: that the background or setting is in some cases completely altered from drawing to painting, and that the figures are only indicated very sketchily and in blocks or groups without distinguishing features. With regard to the first point, it seems clear that the particularities of place suggested by the setting are not in themselves important: they contribute to the general shapeliness, atmosphere, and mood of the whole composition (but they may be of more than one form). They are subordinated to and are a function of the subject, as represented by the figures, to which they must be appropriate.[37]

'Finding' the subject was a matter for cogitation, as Racine wrote to Le Vasseur in July 1662 and Poussin to Chantelou in

December 1647 — in one particularly revealing remark: 'J'...ai trouvé la pensée, je veux dire la conception de l'idée, et l'ouvrage de l'esprit est conclu...' In his correspondence with Chantelou in 1653 concerning the painting of the Madonna which his patron had commissioned, this comment is echoed more than once: 'J'ai trouvé la pensée de la Vierge que je vous ai promise': 'Pour ce qui est de votre vierge, ... la pensée en est arrêtée, qui est le principal.' When another Madonna is commissioned by Mme de Montmort, in the following year, Poussin writes to Chantelou that he is about to begin work on it and that he 'étudie alentour de l'invention et disposition', and later says: 'Je travaille autour de la pensée et distribution de la Vierge en Egypte de Mme de Montmort.' The testimony of Bernini on the painter ('Vraiment cet homme a été un grand inventeur d'histoires et de fables!') and of Bellori ('Lisant les histoires grecques et latines, il annotait les sujets puis, à l'occasion, s'en servait') could well have applied to Corneille and Racine, as their critical writings show and as their adaptations and transpositions suggest. A passage from Félibien's *Entretiens*, in which he recounts how Poussin arrived at a subject, provides an exact parallel with what I have been arguing for the dramatists. Writing of competent critics, he says:

Ils examinent l'intention de l'auteur, la fin pour laquelle il a travaillé, le choix de son sujet, les moyens dont il s'est servi, les raisons qu'il a eues de se conduire d'une manière plutôt que d'une autre; et enfin ils jugent par l'exécution de son ouvrage s'il est parvenu à l'imitation parfaite de ce qu'il s'est proposé suivant la plus belle idée qu'il en pouvait concevoir.

Par exemple, quand le Poussin fit son tableau de Rébecca, quel fut, je vous prie, son dessein? J'étais alors à Rome lorsque la pensée lui en vient. L'Abbé Gavot avait envoyé au Cardinal Mazarin un tableau du Guide, où la Vierge est assise au milieu de plusieurs jeunes filles qui s'occupent à différents ouvrages ... Le Sieur Pointel, l'ayant vu, écrivit au Poussin, il lui témoigna qu'il l'obligerait s'il voulait lui faire un tableau comme celui-là, de plusieurs filles dans lesquelles on pût remarquer différentes beautés.

Le Poussin, pour satisfaire son ami, choisit cet endroit de l'Ecriture Sainte où il est rapporté comment le serviteur d'Abraham rencontre Rébecca qui tirait de l'eau pour abreuver les troupeaux de son père ...

Voilà quel est le sujet que le Poussin choisit pour faire ce qu'on désirait de lui ...[38]

This suggests a mode of thinking analogous to that of Corneille transposing the Persian situation of *Cosroès* to the Armenia of *Nicomède*, or of Racine enveloping the Lombard situation of *Pertharite* in the Epirus of *Andromaque*, and even at that he chooses the 'wrong' place, historically speaking, but, said Corneille, 'il faut placer les actions où il est plus facile et mieux séant qu'elles arrivent'.[39] It is no doubt partly a matter of thinking by analogy, as is also suggested, for instance, by Racine's conception of the subject of *Bérénice* being at least as Virgilian as it is Suetonian, by the situation of *Andromaque* probably owing much to the *Diana* of Montemayor and the pastoral tradition or to J.-B. Dupont's little novel, *L'Enfer d'amour* (1603), or that of *Bajazet* to Segrais's *Floridon* (1657).

Félibien expresses admiration for the 'belles dispositions de figures' in the works of Poussin, and comments on the way in which, even in his smaller compositions, the painter knew how to 'étaler dans de petits espaces de grandes et savantes dispositions'; he picks out as the first quality of *The Last Supper* done for the chapel at Saint-Germain-en-Laye 'la disposition du sujet'. An appropriate setting (*vraisemblable*) is consequential upon that disposition which indicates the subject, as it is also for Corneille and, apparently, for Racine. One notes also the analogy between Félibien's comment on Poussin's smaller pictures and Racine's frequent remarks, from his first preface (that to *Alexandre le Grand*, 1666) onwards, about his ability to write complete tragedies within the confines of the unities and a self-imposed simplicity of plot. One may infer that he, too, was capable within those limitations of 'de grandes et savantes dispositions'.[40]

We are clearly dealing with a whole mode of thought in which originality and inventiveness in the arts were associated with novel dispositions of well-known material, and, as far as tragedy is concerned, with dispositions in which situation came first because all else flowed from it. The evidence presented here may well make it difficult to accept M. Bénichou's criticism of Professor Knight's tentative suggestion that situation may have preceded story: 'J'ai peine à croire, quant à moi, à un cadre précédant la matière, on ne voit de telles choses que dans les paris légendaires qui jalonnent l'histoire littéraire. Et je crois que les auteurs tragiques de ce temps-là se mettaient d'abord en quête de sujets.'[41] Did something like the 'paris légendaires' not exist in the rivalries and *salon* gossip of the seventeenth century? And what if dramatic 'subjects' were first conceived of in the abstract and general terms of relationships and passions, as we have suggested?

It is to those relationships and passions that we must now turn. While both Corneille and Racine adopt, as occasion demands, the dual tactics of admitting to having altered their stories for reasons allegedly of *bienséance*, *vraisemblance*, observance of the unities or other technicalities, and of protesting fidelity to the sources (only those in antiquity being mentioned, for obvious polemical reasons), it is instructive to note some of the real reasons for and effects of the changes made. One of the clearest statements comes from Corneille, in his *Examen* of *Nicomède*. It has been shown that the web of relationships in the play probably came from Rotrou's *Cosroès*, but it is evident from the *Examen* that the motives which dictate the behaviour of the characters within those relationships are peculiar to Corneille. So, for political reasons, the projected marriage of Nicomède to Laodice will give 'plus d'ombrage aux Romains', whose fear and jealousy of the resulting alliance is the occasion for Flaminius's embassy. Nicomède is made the disciple of Hannibal 'pour lui prêter plus de valeur et de fierté contre les Romains'. Hannibal's death is attributed in part to the unscrupulous ambition and machinations of Arsinoé, who now manipulates her pusillanimous husband and her ultimately resistant son, Attale, in order to break down the marriage and alliance of Nicomède and Laodice and to have her stepson suspected of seeking to supplant his father. The play becomes a dramatic representation of the conflict between Arsinoé's Machiavellian tactics and the finally victorious 'grandeur de courage', 'prudence généreuse', 'générosité', 'vertu', and 'fermeté' of the other party, which are calculated to arouse the 'admiration' of the spectator.

What Corneille is saying, and what he has manifestly done in his play, is that he has devised a situation which derives from the unavowed *Cosroès* and to which he has adapted the historical narrative indicated by his title in order to create a dramatic conflict between certain emotions and passions, a conflict conducted and resolved in such a way as to excite a specific emotion or passion ('admiration') in the audience. This is a straightforward case, since the alterations to the 'source' material are admitted and explained.

But when a dramatist puts up a smoke-screen of plausible technicalities as his reasons for such alterations, the real reasons may be less easy to discern. We have observed that in *Andromaque* Racine claims to have altered the status of his heroine in order, he says in the second preface, to 'me conformer à l'idée que nous avons maintenant de cette princesse'. He has prolonged the life of Astyanax, and there is no Molossus, because 'Andromaque ne connaît point d'autre mari qu'Hector ou d'autre fils qu'Astyanax', and he goes on:

La plupart de ceux qui ont entendu parler d'Andromaque ne la connaissent guère que pour la veuve d'Hector et pour la mère d'Astyanax. On ne croit point qu'elle doive aimer ni un autre mari ni un autre fils. Et je doute que les larmes d'Andromaque eussent fait sur l'esprit de mes spectateurs l'impression qu'elles y ont faites si elles avaient coulé pour un autre fils que celui qu'elle avait d'Hector.

Such alterations, Racine then points out, were accepted in his time and had been practised by Homer and Euripides. He pleads, therefore, the need to observe *bienséance* and to accommodate the expectations (and ignorance?) of the audience, and his own wish to move that audience to pity. In the first preface, too, he had admitted making changes, particularly in the character of Pyrrhus, but there he had shown a much more cavalier attitude, especially to his critics: he was flushed with his first great success, no doubt, in 1668. However, while his motives as explained in the second preface may have played some part in his thinking, they seem less important than others, which must also have been at work in the adoption of changes we have already noted. The simple fact is that, without Andromaque's ability to resist her captor, without the presence of her and Hector's infant' son at the court of Pyrrhus, without also the unmarrying of Hermione and the arrival of Oreste, Racine's play would not have existed because there would have been no situation as he understood it. And that situation was indispensable to the conflicts which make up the plot and are occasioned precisely by the passions which drive the characters into those conflicts. Of course, the truth of what Racine has to say about 'ceux qui ont entendu parler d'Andromaque' depends on whether they have read only Homer, or Euripides as well: it is the Homeric Andromache of the Trojan war who is moved into the Virgilian or Euripidean accounts of events subsequent to the war, just as, a few years later, it will be the Homeric account of the *Iliad* which will so largely inform the Euripidean drama of *Iphigénie* whose events occurred before the siege of Troy.

That, however, is another question. What matters here is to observe that, although, unlike Corneille, he refrains from explicitly stating the creative reasons for his treatment of his 'sources', examination of the play itself reveals that those reasons are identical with his elder's: a situation giving rise to a conflict of passions, that being a dramatic action so conceived as to arouse specific emotions in the audience. The word 'passions' was used indiscriminately of those which drive the characters and of those experienced by the audience, and this lack of distinction was apparent in the arguments advanced in the controversies over the alleged immorality of the theatre; but the differentiation has to be made, and it is to the first kind of passions that I wish to refer here. Corneille made it at the beginning of his second *Discours*.

In general terms, it is clear that the theory of the passions was a universal preoccupation central to the moral and theological thought of the seventeenth century.[42] The dramatists cannot fail to have been affected by the ideas about man's nature and the sources of his behaviour as they were explored and analysed in the abundant moral and moralizing treatises from Du Vair to Senault. The word 'passion' was attached to a remarkably wide range of feelings or reactions: there is evidence enough of this in Descartes's enumeration in his *Les Passions de l'âme*, though it is by no means exhaustive.

I do not wish in any way to suggest that that work, published in mid-century, was directly influential in the formation of the psychological concepts of either Corneille or Racine, nor even that it may have been more important to them than others. It is quite possible, for example, that in its Longinian linking of passions characteristic of *personae* with the eloquence which communicates them to the audience, the work of Caussin may have been equally influential.[43] But some of Descartes's formulations seem to be very representative of contemporary modes of thought. It will be useful here to consider some of them in relation to dramatic situations such as those we have been examining.

I have already suggested that the nature of dramatic action is conflict and that, in Corneille and Racine, the conflict arises out of the convergence of opposing passions. At the outset of his book (Articles 1 and 2), Descartes virtually identifies action and passion and suggests, however discreetly, that passion involves suffering: that is of course the etymological sense of the word and it survives — and must have been particularly evident in a fundamentally Christian and Catholic society — in the Passion of Our Lord.

...Je considère que tout ce qui se fait ou qui arrive de nouveau est généralement appelé par les philosophes une passion au regard du sujet auquel il arrive, et une action au regard de celui qui fait qu'il arrive; en sorte que bien que l'agent et le patient soient souvent fort différents, l'action et la passion ne laissent pas d'être toujours une même chose qui a ces deux noms...

...Nous devons penser que ce qui est en [l'âme] une passion est communément en [le corps] une action...

To read, say, the *Examen* of *Nicomède* or the preface to *Phèdre* is to realize that their authors had a clear understanding of this relationship between action and passion. To read their plays is of course to be made aware that action springs from passion (Phèdre's declaration to Hippolyte in Act II scene v) and in its turn causes suffering. But the word 'passion' must be interpreted as broadly as it was in the seventeenth century.

It is evident from Descartes's treatise that the passions are very numerous and that their interplay is often complex, partly because they are closely linked together. Any given passion may engender a variety of frequently conflicting passions, especially if its activity is frustrated.[44] *Desire* to attain some object may lead either to *hope* of possessing it or to *fear* of not doing so; hope, in its turn, may lead to *courage* and *joy*, or, if it alternates with fear, to *irresolution*; while fear may bring in its train *cowardice* and *despair*. Descartes's analysis reveals that conflict may arise not only between opposing passions embodied in different dramatic characters ('gloire' in Horace against love in Camille) but between opposing passions within the same character ('gloire' and love in Curiace). This obviously goes a long way towards explaining the kind of situation and disposition we find in the tragedies of Corneille and Racine.

Descartes does not of course consider the passions to be bad in themselves, but only in excess or misuse. The greatest danger is that (see Articles 138, 160, 211), lacking self-control through the will or *générosité*, we may be led into hasty action, and that our passions may give us a false estimate of the good or evil inherent in their object. Again this provides explanation of behaviour and conflict in the plays of both our poets. What is more, Descartes points out the dangers (Article 145) of allowing our actions to be directed, thanks to uncontrolled passion, to false or unattainable ends: is not this the tragedy of Cléopâtre in *Rodogune*, of Suréna, of Agrippine, of Phèdre? The essence of seventeenth-century tragedy is that we are not capable, by the mere exercise of the will, of arousing or suppressing the passions (Articles 41, 45).

It is, of course, characterized by action, as Aristotle had observed in the sixth chapter of the *Poetics*. From that Corneille takes his cue at the beginning of the first *Discours* (*Writings*, pp. 8–9), not only in saying that 'la poésie dramatique... est une imitation des actions', but in finding the distinctions between comedy and tragedy in different types of action, 'illustre, extraordinaire, sérieuse' for tragedy, 'commune et enjouée' for comedy: these distinctions are 'les conditions du sujet'. It is perhaps not fortuitous that Racine begins his own partial translation of the *Poetics* at the same point: 'La tragédie est l'imitation d'une action grave et complète...'. Some of the prefaces make action and subject interchangeable, and in that to *Bérénice*, a play usually thought of as devoid of action or almost, action takes precedence over all else. Of *Mithridate*, the dramatist writes that the death of the king constitutes 'l'action de ma tragédie', and he is at pains to prove that the plan to invade Italy does not infringe the unity of action: 'J'ai... lié ce dessein de plus près à mon sujet', which, like that of *Phèdre* (see the opening sentences of the preface), is identified with action and can only be conceived of in terms of situation involving characters in particular circumstances which bring about the clash of passions, which in their turn cause them to act.

Here again, Poussin corroborates the evidence. Adapting a sentence from Tasso's *Discorso del poema eroico*, he writes that 'la peinture n'est rien autre que l'imitation des actions humaines, lesquelles sont proprement actions imitables...'. Both the painter's earliest biographers confirm this comment, Bellori saying that 'dans les sujets historiques, il recherchait l'action et disait que le peintre devait de lui-même choisir le sujet propre à être représenté', and Félibien that 'il fait consister l'excellence de cet art dans le beau choix des actions héroïques et extraordinaires' — exactly parallel to Corneille's definition and to part at least of Racine's: 'il suffit que l'action en soit grande, que les acteurs en soient héroïques, que les passions y soient excitées...'.[45]

Félibien's commentary emphasizes the importance for Poussin of arranging his figures in relation one to another and of giving to each an appropriate action and attitude expressive of those relationships and of 'les différentes passions de l'âme': 'il n'y a point de figure qui ne semble parler ou faire connaître ce qu'elle pense ou ce qu'elle sent... Combien de différentes actions représentées!' Writing to his fellow painter, Jacques Stella, of Lyons, concerning the *Manna* commissioned by Chantelou, Poussin himself says that he has found 'certaines attitudes naturelles' expressive of the feelings of the Jewish people, 'la joie et l'allégresse où il se trouve, le respect et la révérence qu'il a pour son législateur'.[46]

That Poussin and Félibien were aware of the analogies between painting and drama is clear from other passages. In the same commentary on the *Manna*, the biographer writes:

Comme une des premières obligations du peintre est de bien représenter l'action qu'il veut figurer, que cette action doit être unique, et les principales figures plus considérables que celles qui les doivent accompagner, afin qu'on connaisse d'abord le sujet qu'il traite, le Poussin a observé que les deux figures qui dominent dans son tableau sont si bien disposées, et s'expriment par des actions si intelligibles, que l'on comprend tout d'un coup l'histoire qu'il a voulu peindre.

There follows a passage in which Félibien shows how Poussin disposed his figures in groups according to their passions, ages, social status, etc., and in a letter to Chantelou the artist himself points out how the attitudes and disposition of the figures of his paintings must be 'read' (he uses the verb *lire*) as indicative of the passions. The subject of Poussin's pictures may be identified with action, which springs from a diversity of passions expressed in the disposition and attitudes of the figures.

Identifying action with subject, Corneille and Racine, as we
have seen, write in similar terms. In the two major quarrels arising
out of his plays (*Le Cid*, 1637; *Sophonisbe*, 1663), Corneille was
attacked precisely because his critics took exception to the passions
and consequent actions of his characters. The first two paragraphs
of the *Examen* of *Le Cid* show him still in 1660 unrepentant over the
nature of the passions which lie at the root of the admittedly
extraordinary action of the play; at about the same time, he was
defining in more general terms the nature of the tragic action and
subject (first *Discours*, *Writings*, p. 8): 'Sa dignité demande quelque
grand intérêt d'État, ou quelque passion plus noble et plus mâle
que l'amour, telles que sont l'ambition ou la vengeance... Il est à
propos d'y mêler l'amour...'. Racine's comments in, for example,
the preface to *Mithridate*, the second preface to *Alexandre le Grand*,
and the preface to *Phèdre* clearly indicate that he was thinking of
the subject (the action) as arising out of the passions. His
annotations on the Greek tragic poets often reveal the same
concept, though in the nature of things it is not developed.[47] The
most explicit and instructive reference is, however, in his
translation of the sixth chapter of the *Poetics* (*Principes*, p. 13):

La tragédie est l'imitation d'une action. Or toute action suppose des gens
qui agissent, et les gens qui agissent ont nécessairement un caractère,
c'est-à-dire des mœurs et des inclinations qui les font agir. Car ce sont les
mœurs et l'inclination, *la disposition de l'esprit*, qui rendent les actions telles
ou telles. Et par conséquent les mœurs et le sentiment, *ou la disposition de
l'esprit* sont les deux principes de l'action. Ajoutez que c'est par ces deux
choses que tous les hommes viennent ou ne viennent pas à bout de leurs
desseins *et de ce qu'ils souhaitent.*[48]

This passage, with its subordination of character to action and its
definition of character in terms of *mœurs*, is significant also for its
emphasis on the causal relationship of the latter to action. We
should be put on our guard against approaching the plays in terms
of character-study or modern psychology — yet it is on that that a
large part of criticism has concentrated without considering how
the dramatists conceived of the dramatic personages —, and
against reading them as repositories of ideas or simply as literary
works.

A terminological problem exists in English, because we use the
word *character* indiscriminately for the French *caractère* and *personnage*. We ought to use some such expression as *stage-figure* for the
French *personnage*, which is itself less appropriate than seventeenth-century usage of the word *acteur* (*comédien* meaning *actor*) to
indicate the *dramatis personae*, 'les gens qui agissent' as Racine puts

it. Having made this distinction, we are confronted with another, that between modern psychological concepts of character and what Racine, in the passage under review, calls *caractère* and what he then calls *mœurs*. With regard to Racine's use of these two words, for the word *caractère* there is no exact equivalent in the original text and certainly none in Vettori's Latin translations which accompany the text in the editions he annotated. The word most frequently used in the seventeenth century where today we should expect to find *caractère* is *mœurs*, by which Racine translates Vettori's *mores*, meaning habits of thought and feeling which give rise to certain fixed attitudes resulting in action. Pascal, Méré, La Rochefoucauld, and La Bruyère, among many others, are known as *moralistes*, not *psychologues*: they were interested in behaviour, i.e. a form of action, action which springs from passion. La Rochefoucauld's moral system, for example, is based on his observation of the effects of 'amour-propre', a passion. Descartes was aware, as the opening remarks of *Les Passions de l'âme* indicate, of the novelty of the interest shown by his contemporaries in these matters: 'il n'y a rien en quoi paraisse mieux combien les sciences que nous avons des anciens sont défectueuses qu'en ce qu'ils ont écrit des passions'. It is on that interest that the basis of seventeenth-century tragedy rests. In *Mithridate*, Racine tells us in his preface, he wished above all to depict 'les mœurs et les sentiments de ce prince' (one of the same phrases as in the passage from the *Poetics*), and these are characterized as 'sa haine violente contre les Romains, son grand courage, sa finesse, sa dissimulation, et enfin cette jalousie qui lui était si naturelle'. In the preface to *Sertorius*, Corneille defends the great interview between his hero and Pompée on the grounds of its being the manifestation of their passionate 'générosité'. The behaviour of Sophonisbe is similarly upheld because of her passion of 'gloire'. Corneille also discusses the question in more general terms, which are highly significant in our present context, in his first *Discours* (*Writings*, p. 19). First, he says, 'les mœurs ne sont pas seulement le principe des actions, mais du raisonnement'. This is closely parallel to what Racine wrote in the passage analysed above, particularly if, as Vinaver suggests in his note, it is understood that 'sentiments' can mean both the passions and emotions which are in conflict in tragedy (cf. 'inclination', 'disposition de l'esprit') and the thought which is often an attempt to fix or rationalize the passions: hence Corneille's 'raisonnement', which is related in its turn to Aristotle's statement towards the end of chapter VI of the *Poetics*: 'Thought...is shown in all they say when proving or disproving some particular point, or enunciating some universal proposition.'

Such a proposition may be couched in the form of a *sententia* or, as Corneille calls it, a 'maxim'. He does not approve of maxims if they are introduced in order to point a moral, and tragedy can dispense with them, even if they do result from habit-formed reasoning; but it cannot dispense with the habit itself: '... non pas de l'habitude même, puisqu'elle est le principe des actions, et que les actions sont l'âme de la tragédie, où l'on ne doit parler qu'en agissant et pour agir,' an echo, no doubt, of D'Aubignac's dictum that in the theatre, 'parler c'est agir' (*Pratique*, p. 282). The plays of Corneille and Racine are made up of speech, not static but active, and the expression of habits of thought in which opposing passions either bring two or more characters into conflict or involve them in what Corneille calls 'les combats intérieurs' (second *Discours, Writings*, p. 42). Descartes saw the passions as rooted in physiology, in the humours, which are what dispose men to certain passions.[49] The ineradicable nature of the passions is thus emphasized: it forms the very basis of the tragic inevitability of conflict and error. Reasoning itself is of course not objective: it is an attempt to justify action taken or contemplated, and is therefore frequently an occasion for deception or self-deception. The peculiar contribution, on one level, of Corneille and Racine, in their different ways, to the history of tragedy, lies in their dramatization of this process.

That dramatization — which we see also in Poussin — starts from the conception of a subject which requires an initial situation. Greek and Roman tragedy being little concerned with these problems of the passions, it is not surprising that Corneille and Racine had to look elsewhere for their subjects or situations. I should not wish at the conclusion to this chapter to suggest that my analysis results in a reconstruction of the genesis of any play by these authors. My purpose has been to put the emphasis on the aesthetic construction of a situation and mechanism of a particular kind, on methods and concepts which appear to be shared by Corneille and Racine and by at least one great artist working in a different medium, and on the perhaps inevitable affinity between their concepts and those of an important contemporary philosophical moralist, whose general mode of thought is by no means unique.

It may be thought that, taking my cue from such remarks as this: '... quoique ma tragédie porte le même nom que la sienne [that of Euripides], le sujet en est pourtant très différent' (Racine, second preface to *Andromaque*),[50] I have sought to dismiss or devalue the 'sources' as they are usually understood, and to ignore the importance of *vraisemblance* and *bienséance*. What is really argued here is that these things may be less significant at the outset than we usually think — and as the tragic dramatists themselves sometimes encourage us to believe — in the conception of the subject as initially defined in the situation. Their real importance, as I shall try to show, lies elsewhere.

CHAPTER II
Sources: Tragedy and history

La représentation n'étant que l'ombre de l'histoire, il n'est pas
requis que cette ombre soit égale au corps.

Discours à Cliton (1637)

Tragedie is to seyn a certayn storie,
As olde bokes maken us memorie,
Of him that stood in greet prosperitee
And is y-fallen out of heigh degree
Into miserie, and endeth wrechedly.

Chaucer, Prologue to *The Monk's Tale*

In no play does either Corneille or Racine follow his historical
source with any exactitude in the dramatic situation or in its development in the plot: the relationships between the characters do
not strictly belong to history, nor do the characters themselves, nor
does the sequence of episodes. None of this is surprising when one
considers that drama is a kind of fiction, related to but not repro-
ducing factual knowledge or observation or experience. None of it
is surprising when one considers the whole tradition of tragedy and
its first great theorist's analysis of the art. So, one might ask, why
did Corneille and Racine and so many of their contemporaries —
especially the critics — concern themselves so closely with the
problem of fidelity to historical sources? And why should the present-day reader interest himself in this question at all?

In the seventeenth century, the problem was less that of the way
in which history was to be treated by the dramatist than that of the
respect to be accorded to the 'sacred text', the *Poetics* of Aristotle,
which was regarded as authoritative in dramatic — or at any rate
tragic — theory. The quarrels over comedy took a quite different
turn because there was no exact equivalent of what survives of the
Poetics: only occasionally did they overlap with the quarrels about
tragedy, when strictly moral issues were raised. It is on moral
grounds that *Le Cid* was attacked, and *Horace* in some respects, and
Sophonisbe, and *Phèdre*, in spite of all Racine's precautions in the preface and in the play itself. Generally, however, critics took issue with
the tragic dramatists rather on aesthetic questions, and the playwrights were able to defend their productions by pointing, as the
theorists did, to the text of Aristotle and, sometimes, to those of his
interpreters. It was easy to make play with seemingly contradictory
statements in the ninth chapter of the *Poetics* alone, one claiming
that, unlike history, tragedy describes 'not the thing that has been'
but 'a kind of thing that might be' (hence the emphasis on verisimilitude rather than factual truth), the other that tragedians
'adhere to the historic names' because 'what convinces is the possible', yet 'one must not aim at a rigid adherence to the traditional
stories'. The two sides of the problem can be and were of course
resolved within the same chapter, but the hostile or pedantic critics, the 'doctes', could easily choose to attack a playwright with
one and ignore the other. And that is what often occurred. What
the tragic poets wrote in defence of their own plays must therefore
be read with this in mind, because they, too — as we have seen in
Racine's prefaces to *Andromaque* —, had to argue sometimes from
one premise, sometimes from the other, and sometimes from both
simultaneously. All this amounts to polemical tactics and has to be
understood as such.

Moral issues did also play their part in the discussions of tragedy: the playwrights took the liberty (or felt the obligation) of modifying the historical characters and their behaviour in order, they usually said, to bring them into line with contemporary moral values, in the same way as they turned Greek marriage ceremonies into church services by having them solemnized at an altar. One may however question the reasons they give for these alterations. When Corneille wipes out the sensuality of Cléopâtre in *Pompée* or the parricide of Nicomède, or when Racine obliterates the enforced concubinage of Andromaque or the unnatural and barbaric (to the seventeenth-century spectator) chastity of Hippolyte, are they simply, as they put it, 'correcting' their character and behaviour? Such correction can scarcely have been a consistent purpose when the examples of Cléopâtre in *Rodogune* or of Néron in *Britannicus* are brought to mind. The simple fact is that when one sees or reads the plays, one becomes aware that the corrections, if such they be in any moral sense, are calculated to produce certain dramatic effects and emotional responses, and to set the characters in certain relationships, by similarity or contrast, with others — which is, of course, the same purpose as is served by blackening in other cases. The problem is of course largely one of the relative importance of *vraisemblance* and *bienséance* on the one hand and of historical truth on the other.

It is not my concern to discover the extent of the historical knowledge of Corneille and Racine: that question may have its interest, but it is largely beside the point when one is studying their tragedies as artefacts. The important matters here are to try to discover how history was used and the extent to which the plays suggest historical episodes and characters and produce a sense of what is tragic in them. Put another way, this entails an attempt to determine the dramatic and poetic function of history in the tragedies.

Since we are dealing with a matter in which detail is significant, the most fruitful approach is likely to be a comparison of one play of each author in which certain broad analogies in subject-matter are to be found: *Pompée* and *Bérénice*. In each of these two plays, the master of the Roman Empire is in love with a foreign, oriental princess. This situation is responsible in part for provoking a crisis which features a conflict between one aspect of the greatness of Rome and moral values which are alien and even, to some extent, hostile to it, particularly in terms of political traditions and a sense of honour. The dramatic crisis is in each play provoked by the death of the emperor or his would-be equivalent: that death is the historical point of departure for the tragedy and the resolution of the immediate consequences is its historical outcome. The dramatic conflict which characterizes the plot, the 'middle' between the 'beginning' and the 'end', has at its root an incompatibility between Rome and the Barbarians.

So in *Pompée* the peril, in Corneille's sense of that term,[1] to which César is exposed arises out of the threat to his authority and his life represented by the activities of the Egyptian king, Ptolomée, and his evil counsellors. As the dramatist points out, that peril constitutes the 'nœud' of the play.[2] It is the consequence of the battle of Pharsalus (48 BC) and, with paradoxical irony, of the victory of César over Pompée. But it is not the conflict of the civil war, nor the threat to the victor from a Roman rival, whether directly from Pompée or indirectly through his widow, Cornélie (in spite of her menaces and her determination to pursue César — partly to restore Roman honour, as she sees it, and partly to ensure her personal revenge on him — in a renewed campaign), which Corneille has chosen to dramatize. The title perhaps suggested that it was: we never see Pompée, whose death is reported early in the play (II.ii), but it is the plot to kill him, the motives of those involved in it, and its immediate consequences, dramatically arranged, which form the substance of the tragedy. Corneille can say that at the dénouement César is 'hors du péril', but only in the sense that those consequences threatened him with a death as treacherous as Pompée's and that, thanks to the action of Cornélie, he escapes that particular danger. Cornélie, however, intervenes as she does simply in the hope of saving Rome from further dishonour and of inflicting on him a worthier punishment, for a different crime, than that plotted by Ptolomée.

In conceiving of his material in this way, Corneille has made use of many historical sources, whether works of history proper (like those of Appian or Dio Cassius) or biographical sketches (like those of Plutarch) or imaginative literature (like Lucan's epic poem, *Pharsalia*). Corneille does not distinguish between them nor does he, in his *Examen* or elsewhere, mention all of them. 'A bien considérer cette pièce,' he writes in that *Examen*, 'je ne crois pas qu'il y en ait sur le théâtre où l'histoire soit plus conservée et plus falsifiée tout ensemble. Elle est si connue que je n'ai osé en changer les événements, mais il s'y en trouvera peu qui soient arrivés comme je les fais arriver.' The one source which in the *Examen* Corneille discusses in any detail is Lucan's poem (which is already a largely fictitious work, to which Robert Graves, in his translation,[3] has given the significant subtitle of 'Dramatic Episodes of the Civil Wars', and which he has described as 'melodramatic'); the only other sources mentioned there are Caesar's *Commentaries* — negatively — and Plutarch, merely in passing. The major events of the latter part of the *Pharsalia* — the defeat of Pompey by Julius Caesar in Thessaly, Pompey's flight to Egypt and his assassination at the hands of Ptolemy's henchmen — all feature in some form in Corneille's play, but in a strictly dramatic sense they really constitute its prelude. In his edition of the play, Marty-Laveaux[4] follows up the lead of Corneille's own editions of 1648, 1652, and 1655, by indicating some of the most obvious direct verbal borrowings and adding others, less conscious, in the form of echoes and adaptations. The most important borrowings occur in Ptolomée's initial deliberation (I.i) and amount to some fifty of the Latin poet's lines. In the short address *Au Lecteur* of the original (1644) edition, Corneille declares, however, that the extent of his direct borrowing amounts to 'cent ou deux cents vers' in all.

He also states there that, had he referred to all his sources 'dont cette histoire est tirée', his preface would have been ten times longer. He contents himself with quoting in Latin some twenty-five lines from Lucan's account of Cato's obituary speech on Pompey, only the last seven of which refer to his death, the rest being an idealized character-portrait; and two short passages from Velleius Paterculus, both flattering portraits, one of Pompey and the other of Caesar, and neither alluding even indirectly to the events of the play. None of these can truly be considered a source, except in so far as the first two contain perhaps the germ of Cornélie's and César's respect and admiration for the dead hero. Although Corneille claims to have taken so much from Lucan, and although it is true, as he says, that the rhetoric of the tragedy owes something to that of the epic, it is by no means as flamboyant or sensational as the Latin poet's. Besides, the dramatist has no axe to grind, whereas Lucan makes a transparent attempt to vilify Julius Caesar (because of his jealous hatred for Nero, another Caesar) and to glorify Pompey as the wronged but true son of Rome. Corneille does not take sides: some of his characters — César, Cornélie and Philippe — do honour to Pompey's memory, but César is far from being represented as in any way his inferior.

Moreover, what Corneille called the dramatic necessities involve him in ruthlessly abridging and condensing Lucan's account: the rhetorical digressions disappear, and the events which were spread over a period of months are reduced to a scale of about thirty hours (concentration resulting from an application of the unities). The episodes of the play cover a period extending beyond the end of Lucan's never-completed narrative to include part of the Alexandrian War, and they are arranged in a relentlessly logical and inescapable sequence absent from the epic. Another aspect of the

same preoccupation is to be found in Corneille's elimination of the
kind of inconsistency to be discerned in Lucan's episodic treatment
of Pompey's behaviour — in Book VII he leaves the battlefield at
Pharsalus quietly and deliberately when the struggle seems to be
going in Caesar's favour, but in Book VIII he is fleeing towards
Tempe in a state of panic. Such inconsistencies might pass in an
epic poem, but not in a tragedy in which consistency of character
was essential.[5]

Corneille's remarks in the *Examen* reveal a concern for unification of his plot — a purely technical matter, but one involving a
fundamental characteristic of tragedy, the arrangement of the historical episodes in a logical and close-knit pattern. Pompée's flight
to Egypt, after his decisive defeat at Pharsalus, and César's pursuit
provide the starting-point both in history and in the play. The impending arrival in Egypt of victor and vanquished is the occasion
for Ptolomée's indecisiveness and his subsequent determination to
have Pompée killed in the hope of winning César's favour. In the
play, as in history, the enmity between Ptolomée and his sister
Cléopâtre introduces a serious complication, since the king has
usurped the queen's share of royal power, a share guaranteed by
their father's will, of which Pompée, representing the Roman people, is the guardian.[6] This provides an additional motive for Ptolomée's decision. At this point, however, Corneille abandons the historical sequence of events and the complex intrigues and
manœuvres resulting historically in the Alexandrian War, which
lasted many months and brought about the death of Ptolomée as
he fought against the Roman troops,[7] and in the victory of César
who was extricated from the most serious military difficulties only
by the arrival of an army from Syria.[8] Instead, Corneille turns the
war into a civil commotion directed by Ptolomée against César in
retaliation against the latter's alleged ingratitude at being rid of
Pompée and because of his anger at finding Cléopâtre's rights as
coregnant recognized.[9] Again, while in history it was true that Cornélie accompanied her husband to Eygpt,[10] it was not true that she
was captured as she fled after his assassination.[11] Once more Corneille alters the facts in order partly to put César in a position of
greater peril and to subject him to a conflict of passions, partly to
provide him with an additional incentive to avenge Pompée, and
partly to ensure a means of having Ptolomée's plot revealed to him.
The initial situation, the catastrophe (for Pompée and then for Ptolomée), the restoration of her rights to Cleopâtre, the safety of
César, and Cornélie's escape are all historical 'events', as Corneille
calls them, which must be safeguarded; but the way in which they
are arranged dramatically is the poet's own invention.

The arrangement of the facts of history finds a parallel in the treatment of time and place. Since the episodes are tightly organized in terms of immediate cause and effect and concern a small group of people directly involved in them, it is natural (i.e. *vraisemblable* in the circumstances of the plot) that they should occur in the one place. In history, Pompée landed at Pelusium, where Cléopâtre's army was facing Ptolomée's on the Eygptian frontier with Syria, while César disembarked at Alexandria, a hundred and fifty miles to the west.[12] Corneille situates his play at Alexandria, in the royal palace, leaving it to be assumed that it is there that Pompée is killed, and refraining from naming either city in his text. He is right to suggest that the audience is unlikely to concern itself with the precise place once the play has begun. At the same time the arrangement created by Corneille necessarily excludes from the play the fact that Cléopâtre had, in history, reigned with her brother from 51 to 48 BC, when he had driven her out, and that she had gathered an army in Syria and was now on the point of invading Egypt. Corneille strengthens Cléopâtre's resolve and will to have her rights restored because she has never enjoyed them, a circumstance paralleled, with similar consequences, by Racine's restoration of her virginity to Bérénice, as we shall see. César is already in love with Cléopâtre[13] and she relies on this in order to assure herself of his aid. The Alexandrian War being transformed to a revolt confined to the city, it is there also that Ptolomée[14] meets his death. The episodes are thus brought close together in time and space and the tension arising from them is heightened.

The most obvious aspect of this observance of the unity of time is of course the reduction of the long months of the Alexandrian War to a few hours of rebellion. This cannot, however, be separated from the logical organization of the episodes. The passions to which the characters are subject in the play are imperious: they drive them to immediate action..César's anger with the Egyptians takes effect at once and has the consequence of causing Ptolomée to plot against him at once. Since the characters are brought together in a particular situation and in one place, the conflicts between them are thus bound to erupt in this way: any prolongation of time beyond the limit imposed by a generous application of the rule would diminish the dramatic power of the passions.

Corneille's treatment of the Alexandrian War allows him moreover to reduce the number of important characters involved in his play and to concentrate attention on the conflict of passions and moral values. Cléopâtre's sister Arsinoe (who ultimately ordered the death of Achillas) and her henchman Ganymedes, both of whom played a conspicuous part in the war, are omitted. Similarly, one of Ptolomée's advisers, Theodotus, does not appear in the tragedy: in reality, the plot against Pompée was determined by a discussion involving the king and his three guardians (he was then only seventeen), Theodotus (who was responsible for his education), Photin (in charge of finance and general administration), and Achillas (in command of the army).[15] In the play, Photin appears to advance some of Theodotus's arguments, so that the two characters have become one,[16] while Septime is the third adviser.[17] In history, too, he played the traitor's part at the actual assassination. Since he had served in Pompée's army,[18] the death of his old commander assumes an additional and ironical tragic dimension.

Now if, in order to tighten dramatically the organization of the plot, Corneille omits certain historical characters, why does he include Antoine and Lépide? Both, according to all the historians, were with César at Pharsalus, but it can be argued that the presence of neither is necessary in the play. Simply from the point of view of plot, this may well be so, and it has been suggested that, like the opening description of the battle of Pharsalus and the introduction of César's magnanimity towards his defeated enemies, Corneille simply took it over from Scudéry's *La Mort de César*.[19] It could well be that the suggestion came from that quarter, but Corneille's use of it requires examination. In terms of plot, Antoine plays the part of messenger from César to Cléopâtre: this enables Corneille to keep the audience in suspense awaiting the meeting between the lovers and to show that Cléopâtre is not entirely secure in César's love. More important, César is enabled to put affairs of state first (the interview with Ptolomée — III.ii) while Antoine's good offices allow him not to delay greeting the queen, at least through an intermediary. Corneille is also able (scene iv) to introduce Cornélie just at the moment when César thinks the way is clear — since he has for the time being disposed of the king — for him to see Cléopâtre. César is in fact constantly being frustrated (see ll. 977–80) — in his desire for reconciliation with Pompée, in his hope of seeing Cléopâtre, in his attempt to save Ptolomée. But by his almost constant presence at his commander's side, Antoine is a reminder that Cléopâtre was finally to be partly responsible for the death of both the Romans. He is, then, of some importance both in the contrivance of the plot in the middle of the play and in underlining one aspect of the inner meaning of the tragedy. Lépide, who does not speak, performs a similar function. After the assassination of César in 44 he became a member of the ruling triumvirate with Augustus and Antoine. Augustus eventually had him deprived of all his offices, but not before he had attacked Pompée's son Sextus who, continuing the civil war, had taken possession of Sicily. Lépide is a reminder of Antoine's ultimate discomfiture and of the final defeat of the Pompeian party.

In his conception of the subject, Corneille differs quite radically from Chaulmer whose play, *La Mort de Pompée* (also Corneille's original title), was completed in 1637. It is possible that the idea of dramatizing this particular episode came from this source, and signs are visible that Corneille knew it. The deliberation scene of *Pompée* (I.i) follows Chaulmer's corresponding scene (IV.v) fairly closely in construction, though important differences also exist: Photin, who speaks first in Chaulmer's play, advises sparing Pompée; Achillas suggests simply refusing him asylum; and Théodote, one of Ptolomée's closest advisers in historical fact, believes that safety can be ensured only by assassinating Pompée and presenting his head to César; Septime does not appear. The spellings Ptolomée (where Ptolémée might be expected) and Photin (for Pothin) are found in Chaulmer and repeated by Corneille, though the form Photinus does occur in certain manuscripts of Caesar's *Commentaries*, cited in the *Examen*. Chaulmer, however, made his play out of the way in which Pompée's death is decided upon: the Roman general lands safely in Egypt (where he is well received by the Queen Mother) accompanied by Cornélie and his son Sexte, with whom Cléopâtre falls in love. She promises to try to save Pompée until she discovers that Sexte is unwilling to abandon in her favour the girl he has promised to marry: Cléopâtre's jealousy causes her to order Théodote to advise Pompée's assassination. Cornélie appears in the last act to witness the catastrophe. Chaulmer clearly departs much further than does Corneille from the historical accounts, chiefly, it would seem, in order to create a conventional love-and-jealousy plot as the immediate cause of Pompée's death. Corneille disposes speedily of the assassination, and although he exploits the dramatic interest of the motives of Ptolomée and his counsellors, he shifts the emphasis to the contrast between their behaviour and that of the Romans after the killing.

It has also been suggested[20] that Corneille may have been indebted to Benserade (*Cléopâtre*, 1635) and Guérin de Bouscal (*Cléomène*, 1639) for the choice of his Roman–Egyptian theme; but no detailed similarities are discernible. Other plays[21] bear in part on the same or kindred themes (Mairet, *Sophonisbe*, 1634, *Marc-Antoine*, 1635; Scudéry, *La Mort de César*, 1635; Tristan, *Mariane*, 1636): clearly this cluster in the middle to late 'thirties betokens popular interest in such themes at that time. Corneille's play comes several years later, however, and treats some of the materials

used in the earlier plays differently — with less deliberate pathos
and (in spite of Lucan) less epic quality, that is with less specific
and picturesque detail in description and narrative, including a
notable absence of particularity about Cléopâtre's physical beauty.
These distinctions reveal the essentially dramatic nature of Corneille's play, the drama being centred on the conflict of values passionately held and expressed by the characters.

In this dramatic struggle the extremes are represented by Cornélie and, presumably, the dead Pompée on the one hand, in their
absolute devotion to what they conceive to be Roman honour,
dignity, and rectitude, and by Ptolomée and his Egyptian advisers
on the other, in their clinging unscrupulously to power and in their
base and dishonest opportunism. Between the extremes are set,
first, César, a 'généreux' in true Roman style, but subject to an
illicit, and from the Roman standpoint, dishonourable passion for
a foreign queen; and second, Cléopâtre herself, ambitious, self-
centred, but not underhanded, yet an opportunist, too, though
obviously touched by Roman values while still willing to try to
save her brother and even his henchman. Septime, although a
traitor, is at least more candid than Photin and Achillas. It is this
situation and struggle which Corneille has created out of the historical narratives or which he has seen in them.

Part, at least, of the motivation, and almost certainly the suggestion for telescoping the historical events, could have come from
Florus's history (II.iv. 51 ff.). There, Ptolemy is characterized as
the 'most contemptible of kings' and is not excused, as he is in
some sense by Dio Cassius and by Dupleix after him, by his extreme youth. The idea that destiny intervened to bring about
vengeance for the illustrious Pompey is also in Florus and echoed
in Ptolomée's first words in the play: 'Le Destin se déclare...'.
Caesar is captivated by Cleopatra's beauty (Florus is not critical of
him nor of her) as well as moved by hatred for her brother, and he
thinks her wronged. Caesar believes that Pompey's death arose out
of the ambitions of a faction, not out of consideration for himself,
and that, if the opportunity had arisen, he also might have been
their victim. The sequence of events, with Caesar falling out of one
peril into another immediately after the death of Pompey, is also in
Florus: after he had ordered that Cleopatra be restored to the
throne he was immediately surrounded in the palace by Pompey's
assassins. The Alexandrian War is dealt with in a few lines, Caesar
exacting vengeance for Pompey on the Egyptians ('cowardly and
treacherous') and in particular on Pothinus, Theodotus, and
Ganymedes, who are characterized as 'monsters'. Of all the historical accounts, Florus's seems to be the most likely source for Cor-

neille's conception of the actual sequence of events and perhaps
also for his favourable characterization of César and Cléopâtre. It
could be significant that in his *Examen* he does not mention him.

In *Bérénice*, as in *Pompée*, the conflict is engaged in part on a
political and moral problem. The historical episode, although its
centre as presented by Racine appears to be the unavoidable parting of hapless lovers, turns nevertheless on the irreconcilable political and moral attitudes of Rome and the East. Bérénice and Antiochus, monarchs of barbarian countries, represent a personal des-
potism entirely at variance with the authority of the Roman senate and people which is embodied, but only embodied, in the person of the emperor. For Bérénice, the whole drama consists in her
incomprehension of this reality and in her consequent resistance to
the Roman concept of sovereignty. Far from freeing the sovereign
to act in accordance with his personal wishes, Roman political
tradition chains him to responsibility for the real authority of senate and people. (That most of the emperors abused or tried to
abuse the authority vested in them does not alter the situation as
far as Titus is concerned.) It is of her failure to understand that
fact that Bérénice's stubborn delusion is born. Her confidant, Phénice, understands it well enough — but then she is not emotionally
involved — and says so, early in the play, but the queen replies:

> Le temps n'est plus, Phénice, où je pouvais trembler.
> Titus m'aime, il peut tout, il n'a plus qu'à parler.   (ll. 297–8)

Ptolomée, trying to ingratiate himself with César, expresses the
same incomprehension:

> Seigneur, montez au trône et commandez ici,

to which César retorts:

> Connaissez-vous César, de lui parler ainsi?   (ll. 807–8)

For Titus, the whole drama consists in his endeavours to bring
Bérénice to an understanding of the fact that he is not all-powerful,
that he is responsible to senate and people for the guardianship of
their political and anti-royalist traditions as for all else, that, in
short, he is not a barbarian despot. That is the dramatic problem
which the play must resolve, compounded as it is by Titus's ardent
love for Bérénice, his reluctance to see her depart, his certainty
that the cruelty of her dismissal will make him inhuman. For he is
a man before he is an emperor and it is the man, and not the
emperor, whom Bérénice loves; and whereas his firmness in dismissing her may enhance his reputation as a ruler among his people, it
can only diminish his stature as a man in the eyes of his beloved
(and in his own). This dramatic problem, with its oscillations between groundless hope and bleak despair, between firm resolution
and wavering cowardice, is to all intents and purposes invented by

Racine out of the suggestion of the simple words 'invitus invitam'. Nothing however in his play is incompatible with history, as Corneille expresses it.[22] Titus puts himself 'hors de péril' by overcoming a foreign threat to his realm: he does so in a purely personal situation which demands moral courage, while César achieves the same end by physical courage and a military victory. Both heroes thereby enhance their glory, one reluctantly, the other eagerly. In one sense, the separation of the lovers, the only historical fact represented as a fact in Racine's play, demonstrates the irreconcilability of the two moral codes which are its cause: their tragic function is to make us aware of the abyss which separates intention from achievement. This is a strictly poetic interpretation of the historical fact.

If Racine's imaginative powers took wing on those two words 'invitus invitam', they probably did not do so unaided. The clue is in the preface, in the reference to the separation of Dido and Aeneas in the fourth book of Virgil's epic. That episode ends, of course, very differently from Racine's: Dido destroys herself out of despair while Bérénice does not, though she has every intention of doing so until, in the very last scene, she is confronted with the possibility of the deaths of both the men who love her. Aeneas, responding like Titus to the call of duty, runs away in an act of evasion if not of cowardice. Titus, too, is tempted to run away, first from his imperial duty into marriage with Bérénice and into abdication, then into suicide. In a complex way, the Virgilian narrative seems to have been one at least of the sources of Racine's inspiration for investing the simple words of Suetonius with so much drama: he makes use of the 'historical' episode from the *Aeneid*, but partly by reversing it.

Virgil's narrative includes, of course, the famous hunt and storm scene, in which Aeneas and Dido consummate their love, following which they live openly as man and wife. This resembles the historical accounts of Titus's relationship with Bérénice. As we shall see, Racine will have none of it. While the Dido and Aeneas story had been dramatized many times, and the rivalry and jealousy theme had been incorporated in it as early as 1525,[23] it is interesting to note that, at D'Aubignac's suggestion, so Corneille says in his preface to *Sophonisbe* (*Writings*, p. 164), Boisrobert had, in 1642, written a play, *La Vraie Didon*, with the subtitle *La Didon chaste*. Although this play has nothing to do with Dido's love for Aeneas, but concerns her resistance to Hyarbas, king of Getulia, and her faithfulness to the memory of her dead husband (the Pyrrhus–Andromaque–Hector situation, as it will become in Racine's hands), Racine could well have found the suggestion here for Bérénice's chastity. While Boisrobert's version does have Didon, out of despair, die by her own hand, as in the *Aeneid*, the motive is quite different — the preservation of her virtue. Here is another of Racine's reversals: his heroine does not resist her suitor's advances — she desires nothing so much as union with him — and in spite of despair and the threat of suicide she has to go on living in eternal separation from him.

Boisrobert's play was presumably a morally 'corrected' reply to Scudéry's *Didon* of 1636, which, with its very large number of characters, includes the hunt and storm and all the pastoral and romantic features that might be expected, with the lovers ensconced in the shelter of their cave. While Racine reduces the number of characters to the minimum and avoids all picturesque details, he does occasionally seem to recall Scudéry's tragedy. Bérénice's ecstatic picture of Titus, painted for her confidant, Phénice:

> . . . Ce port majestueux, cette douce présence.
> Ciel! avec quel respect et quelle complaisance
> Tous les cœurs en secret l' assuraient de leur foi [etc.],
>
> (ll. 311–13)

recalls Didon's confidences to her sister, Anne, in Scudéry's play:

> Quelle mine! quel port! et qu'il fait bien connaître
> En ses nobles discours sa valeur et son être!   (I.i)
> Qu'on voit à travers ce port respectueux
> De rayons éclatants d'un air majestueux!   (I.iv)

Like Titus after him, Énée reluctantly faces the call of duty (III.iij):

> Et pour pouvoir donner un coup si rigoureux,
> Je suis trop peu barbare, et trop bien amoureux.

Titus addresses himself in similar terms in his great soliloquy (IV.iv), and like Didon (IV.vii) before her, Bérénice in the following scene takes up the word 'barbare'. Titus's failure (II.ii) to bring himself to confront Bérénice with the fact of their impending and inevitable separation parallels Énée's inability to acquaint Didon with the truth (III.vi), and Didon's reproaches to Énée (IV.i) are echoed by Bérénice's to Titus (IV.v). Scudéry also has Énée depart once without taking leave and then return (IV.ii) to do so, pleading the force of his destiny and swearing eternal love. Titus, too, does not finally yield to the temptation to evade the crucial issue, but he makes the same plea and swears the same love.

Some features of *Bérénice* seem to make it a Dido-and-Aeneas tragedy, as the preface hints, set in a new context, whether the theme or the historical episode came first to Racine. Suggestions, whether directly followed, or reversed, or in part contradicted, appear to have come to him, not only straight from Virgil but from dramatic adaptations of the fourth book of the *Aeneid* — in characterization and motive, in some aspects of inner conflict, and in verbal formulas. But a number of other features, mostly matters of detail, find their way from Virgil's narrative to Racine's play. Bérénice had, in her own country, given aid to Titus, as Dido succours Aeneas on his arrival at Carthage: he assists her in the fortification of her city, as Titus had consolidated the queen's realm. The splendours of the banquet offered to the Trojans on their coming to Dido's city (Book I, ll. 719 ff.), splendours which dazzle Aeneas, are paralleled by those of the apotheosis of Vespasian (I.v) which call forth the admiration of Bérénice. The roles are reversed, the man's being taken by the woman, a process we have just seen and shall see again. The stern calls of duty, personified in the visitations of Mercury to Aeneas, are imperative for Titus, too. Dido's suspicions of Énée's intentions have their echo in Bérénice's: they are countered in each case by the appeal to unavoidable but unpalatable duty. Titus, like Aeneas, is reduced to silence by a lover's reproaches, and longs to prevent her suffering. Mercury's second visit to Aeneas comes at a point of possible delay, like the Senate's message which Titus awaits; and Dido's incomprehension of the hero's departure foreshadows that of Bérénice. All these poetic details, which seem to nourish Racine's imagination, helped him no doubt to fill out into his five acts the laconic accounts found in the historians, and certainly in the elegaic and fatalistic mood the two works have much in common.

But Racine, like Corneille in *Pompée*, needed to find means of dramatizing an epic or historical account, and his first requisite was a third major character. Working within the conventions and traditions of tragic drama as he knew it in his own time, he resorted to the introduction — in reality the creation — of Antiochus and, with him, of a triangular love-plot. This device allowed him also to take on the challenge of the ultimate simplicity of dramatic structure and to simplify that of his own *Andromaque* while keeping some of its characteristics, including something of the personality and misfortune in fidelity of Oreste, now embodied in Antiochus. It is not necessary to repeat the arguments about his doubtful historical credentials:[24] the part he is called upon to play certainly has no counterpart in history. Louis Racine saw well enough that that was of no consequence: '... qu'importe au spectateur? Le poète qui avait besoin dans sa pièce d'un prince étranger, pouvait-il mieux choisir que cet Antiochus qui avait accompagné Titus au siège de Jérusalem?'[25] The idea of introducing Antiochus as the ally and friend of Titus could well have come from another play, on the marriage of this emperor, long after Bérénice's first visit to Rome and before the death of Vespasian. This play is Le Vert's *Aricidie* (1646), in which the alliance and friendship between Vespasian and Vologeses I, king of the Parthians, is to be sealed by the marriage of the king's daughter to Titus. Although Vologeses does not appear in person, the germ of Antiochus's relationship with Titus is suggested in Le Vert's preface, where he refers to Tacitus's account (*Histories*, iv): '... sur cette amitié effective qui était entre ces deux princes, j'ai fait une alliance apparente qui peut y avoir été'. This suggestion is set in the context of Le Vert's discussion of the real nature of his subject and of his belief that, although essentially fictitious, it is not incompatible with historical fact. His play contains references to Bérénice's relationship with Titus and a number of passages which have echoes, if no more, in Racine's play (and indeed in *Britannicus*), e.g.

> Je viens vous visiter pour la dernière fois
> Avecques la contrainte où m'obligent vos lois,
> Seul, de nuit, augmentant dedans des lieux si sombres
> Sans suite, sans éclat, le silence et les ombres.
> Mais il est temps enfin que je vous fasse voir
> Ce qu'exigent de moi l'amour et la devoir ...
>   ... Je veux apprendre à l'univers
> Que c'est vous seulement que j'aime et que je sers ...   (I.iii)

The nearest, however, that the historical Antiochus comes to being a rejected suitor of Bérénice is in Josephus's mention of his having been betrothed to her sister Drusilla and of his breaking off the engagement.[26] But that suggestion (with the rejection reversed), together with those of the historians about his alliance, if not his friendship, with Bérénice and Titus, may have been one of the initial sources, however tenuous, of Racine's creation whose real purpose was, as his son suggests, to fill a dramatic need. The poet's development of the theme of friendship and the addition of that of the lovers' rivalry — as in *Andromaque* — provides him with the occasion for dramatic conflict and misunderstanding 'entre proches', as Corneille puts it, with one of the most poignant ironies of the play on the emotional level — that ardent love and faithful friendship can end only in separation —, and on the dramatic level — that Titus should ask Antiochus to convey to Bérénice the news that she must leave Rome. This act of unconscious cruelty is made to mask the cruelty of which the emperor is only too conscious and too ready to accuse himself. Finally, after his disappearances (III.iv; V.i and iv), he reappears (V.vii — from where?) at Titus's behest (l. 1362) to perform perhaps his most essential dramatic role in revealing the truth about his own love and friendship[27] and thus sealing the separation which in a way prevents what Racine calls unnecessary bloodshed and death.

Like Corneille, he has obviously elaborated considerably on the historical accounts in creating a situation and a plot, and above all in imagining a web of relationships between people who might conceivably have been (but were not) involved in them in this particular way and have met together in these circumstances and in this place. In characterization, too, the two dramatists share certain fundamental concepts. For both, it is clear that the destiny of humanity depends on the decisions of men not entirely good nor entirely bad, in the Aristotelian formula so much discussed. The decisions must be taken and the choices made in circumstances where rational argument inevitably fails to reconcile opposing passions and moral values or to find a preference for one rather than another. Corneille and Racine introduce into their plays a feature common to most of the tragedy and tragi-comedy of their time, the dilemma, and make highly dramatic use of it; but they give it a moral charge missing from the works of most of their contemporaries. One of the means of doing this is to be found in the way in which their characters are presented in terms of the much-debated Aristotelian 'goodness'. Any reading of the tragic drama of antiquity precluded interpretation of this concept in terms of moral virtue as that was understood in a Christian society. Corneille comes to a definition of the 'caractère brillant et élevé d'une habitude vertueuse et criminelle, selon qu'elle est propre et convenable à la personne qu'on introduit', and he makes use of this interpretation in order to justify his presentation of Cléopâtre, in *Rodogune*.[28] Racine could doubtless have used it had his rivalry with his elder allowed it, in order to prove the orthodoxy of his picture of Néron. Like Corneille, too, Racine includes in his version of Aristotle's text a reference to 'l'habitude...au vice et à la vertu'[29] absent from the relevant part of the fifteenth chapter of the *Poetics*. It is clear that both playwrights see 'goodness' in terms of a preconceived idea of what is appropriate to a character in relation to the part he has to play: it is the dramatic conception as a whole which determines characterization. Here again, therefore, that conception determines also the presentation of historical personages.

Let us see the consequences of this. In *Pompée*, Cléopâtre has none of the lasciviousness and sensuality attributed to her by most of the historians, who were usually prejudiced against her because of her disastrous influence on Roman policy. If she loves César, it is for his 'gloire' and her own, and she defines her love in terms worthy of a great lady, a 'grande précieuse' conscious of her dignity. Her overriding passion is ambition and her desire to achieve her rights which have been usurped by her brother. Her love for

César may be real enough, but she uses it to promote her ambition, yet without adopting the suppliant's stance such as we find it in Lucan and Suetonius, but with calculation and confidence in her own power over César as well as in her rights over the kingdom. The Cléopâtre of Benserade and of Mairet (in his *Marc-Antoine*) is a quite different character from Corneille's; she is given to self-pity, yet makes use of guile and deception to escape captivity and to commit suicide. The episode dramatized in the two earlier plays concerns of course the death of Cléopâtre (and of Antoine), and the historians' account of her attempt to seduce Octavius in order to secure her fortunes is not forgotten. Nothing of all this appears in Corneille's portrayal, in spite of Cléopâtre's all-consuming ambition.

The emphasis on her ambition and on her confidence in pursuing it is to be found in Scipion Dupleix's account (XXIX, 9, pp. 37–8):

... Voici comment César, victorieux de tant de grands capitaines, demeure captif et vaincu d'une fille.

Cléopâtre était une princesse douée d'autant de *grâces d'esprit* et de corps, naturelles et acquises, que pas une des siècles passés; les attraits de sa rare beauté en la fleur de son âge étaient capables de ravir les cœurs des hommes les plus continents, et *la gentillesse de son esprit avec l'afféterie de son langage pouvaient charmer les plus fortes âmes* ...

Elle, ayant ... appris que César etait de complexion amoureuse, *ne voulut plus traiter par ambassadeurs ni par députés, mais elle-même s'en vint présenter à lui* sans lui avoir donné connaissance de son dessein, de sorte que, l'ayant surpris, il en demeura tout interdit, et après l'avoir vue et ouïe, il admira autant *l'excellence de son esprit et son éloquence* que sa beauté et bonne grâce.

Dès lors, il ne songea qu'à lui plaire et, après avoir joui de ses impudiques amours, la justice s'éteignant en lui, *toutes ses inclinations penchèrent du côté de Cléopâtre au préjudice de Ptolomée* son frère et cohéritier, *si bien que de juge des parties il devint l'avocat de Cléopâtre*, dont les Alexandrins furent si outrés qu'ils commencèrent à conspirer contre lui ...

This brief account, derived as it obviously is from more than one of the ancient historians, could well have been one of the sources on which Corneille drew — see the italicized phrases — for his portrayal of Cléopâtre and for her influence over César. What Corneille omits is also noteworthy, however — her shamelessness and his lack of a sense of justice: neither character is presented in this way in *Pompée*. The cause of the Alexandrian War is attributed by Corneille to Ptolomée's jealousy and anger, aroused by his counsellors, whereas for Dupleix it was to be found in the displeasure of the people of Alexandria.

When the play opens, Cléopâtre's love, like Bérénice's, is already

established, and Corneille avoids romantic stories about her, such
as her immediate conquest of the Roman commander when she
secretly arrives before him rolled up in a carpet, or her 'honey-
moon trip' up the Nile with him: the first would detract from her
dignity, the second from her (and his) moral stature.[30]

The presentation of Cornélie undergoes similar transformations.
The importance accorded to her in Corneille's play may well have
been suggested by Garnier's tragedy of 1574 of which she is the
eponymous heroine. Her apostrophe to Pompée's ashes owes something to a parallel scene in the sixteenth-century play, but what is
there a passive lamentation over the dead hero is transformed and
contemplation of the funerary urn becomes a source of courage, of
determination to act by pursuing César. That ability to act, even in
the deepest sorrow and harshest adversity, has characterized her
already in her will to punish César and yet to save him from an
ignominious end at the hands of the Egyptians — for she is fired
above all by a zealous, ardent patriotism of which the dead Pompée had been, for her, the epitome —, and in the steadfast courage
with which she faces the conqueror of Pharsalus. None of this is in
Garnier, nor in Lucan, nor again in Chaulmer, where Cornélie's
role is limited to that, in the last act, of helpless witness of her
husband's assassination. Something, at least, of Cornélie's de-
fiance may well be derived from the heroine of Tristan l'Hermite's
Mariane (1636).

If Corneille strengthens and in a sense idealizes Cléopâtre and
Cornélie, does he, as Voltaire among others suggests, weaken
César by comparison with the historical character? In spite of Cornélie's accusations, César's patriotism is not in doubt. In part —
and it is one of the paradoxes of the play — César and Cornélie,
for all their mutual hostility, strive for the same Roman ideal,
which each recognizes in the other. A curious prejudice seems to
exist that all Corneille's heroes must somehow be monolithic characters pursuing a single aim single-mindedly: if true at all, it is true
of only a very small number, among whom César cannot be
placed. Even in relation to the historical character, such a portrayal would have been false. César's devotion to Rome is indeed
less than single-minded: Corneille presents him as admitting that
his love for Cléopâtre — and she is an Egyptian and a Queen, a
member of a suspect barbarian race, living in part under Roman
suzerainty — has inspired his generalship. This puts him in the
tradition of the courtly lover, the warrior who lays his trophies at
his lady's feet. Yet, although his amorous propensities are evident
in the historians and would be familiar to Corneille's audiences
(and therefore expected in the play) through Montaigne and

through Amyot's Plutarch,[31] these are in fact much attenuated in spite of the values, emotions, and language which ally him to the 'galant' and the 'soupirant' of the seventeenth century. Plutarch's portrait, in Amyot's version,[32] still lies at the root of Corneille's — the portrait of a man whose greatest satisfaction comes from being (or striving to be) in his own words in Corneille's play, 'le premier et de Rome et du monde' (1. 1278), whose restless ambition will drive him away even from Cléopâtre (1. 1330) until he has only the gods above him (1. 1258). Love may be the spur to glorious action, but it is that glory which comes first in his mind: the sequence of scenes in the middle of the play shows him putting both affairs of state (III.ii) and the Roman widow (III.iv) before his meeting with Cléopâtre (not until IV.iii), and even when he does meet her, he greets her with another of his triumphs: 'Reine, tout est paisible, et la ville calmée...' (ll. 1241 ff.). There is no real conflict in his mind between his (and Rome's) honour and glory and his love for Cléopâtre.[33] She may foreshadow his ultimate fate (the play is not without discreet allusions to it), but César is by no means the weak-willed man portrayed by Suetonius and Appian nor simply the seventeenth-century lover. Corneille departs from some of the historical accounts also in making his grief over the death of Pompée genuine and in giving him magnanimity towards his defeated foes, including Cornélie.[34] César is, however, by no means fault-less: his love for Cléopâtre, the barbarian queen, and his admission that she is the inspiration for his military exploits, do diminish his stature, at least by ideal Roman standards (and Cornélie taunts him with them). Yet he is ennobled by the nature of his love, at least by comparison with some of the historical portrayals.

This ennoblement was perhaps to some extent inevitable if César, Cornélie, and Cléopâtre were to be fitting characters for tragedy. For all that, they retain a considerable likeness to their historical counterparts. On the other hand, the 'Machiavellian' characters, while also showing clear affinities with those depicted by the historians, are, if not exactly blackened, certainly presented in an unsympathetic light, so that all are in fact 'good' of their kind, appropriate to the roles they have to play in a drama which is, in all its essentials, of Corneille's own conception, and in which what takes place is a struggle between conflicting values which are held with passion by those who subscribe to them. The characters are the incarnation of those values, animated, vigorous, active: this, and not the slavish copying of their historical originals, is what gives them life and reality and, paradoxically, recreates them as recognizable historical personages.

In part, the characters — and episodes — of the tragedy give an

impression of authenticity because of the many references in the
text to historical events which took place before, during, and even
after the supposed duration of the play. It is easy to dismiss such
references as a kind of scholarly window-dressing provided to impress the 'doctes' among Corneille's critics. In fact, however, it has
to be remembered that ancient history was not only an important
feature of education in his day, but formed, through translators
and adapters and popularizers, a considerable source of general
reading matter for the educated public. The significance of the historical allusions would not be lost on Corneille's audiences, and it
should be noted that they are usually allusions rather than parts of
set-pieces of narrative or description, and that they are introduced
only to the extent that they bear some relation to the episodes and
characters of the play. Thus Cléopâtre, for instance, introduces
(I.iii) references to the kindnesses of Pompée as executor of her
father's will at the precise moment when she is involved in her
altercation with her brother over the fate of the Roman general;
and this is directly connected with their joint succession to the
Egyptian throne (the cause of conflict between brother and sister)
and to Cléopâtre's influence over César who would in any event
see justice done should Pompée no longer be alive to do so.
Cléopâtre's long speech (ll. 283–320) sets these references within
the context of her motives and emotions: they subserve the human
drama, but also raise it to the level of history, to that of 'quelque
grand intérêt d'État' which, for Corneille, is the hallmark of
tragedy.[35]

In the same way, Pompée's reported command to his wife, at the
moment of his assassination (ll. 469 ff.), to continue the war contains references to the future campaigns of the Pompeian party
which Cornélie herself takes up in her last defiant speech
(ll. 1701 ff.).[36] The obvious immediate effect of these allusions is,
first, to show Cornélie intent on carrying out her dead husband's
wishes — she is inspired to do so by César's promise to free her
and by her impassioned contemplation of the urn containing Pompée's ashes — and, second, to make it clear that at the end of the
play César may be free of danger for the moment but that threats
still hang over him, including that of repudiation by the Romans
should he in fact marry Cléopâtre (ll. 1743–4). But, for the spectator familiar with subsequent history, it is also clear that the ral-
lying of the Pompeian troops is doomed to failure: the allusions to
the battles in North Africa and Spain and to Pompée's successors
and supporters are all, unconsciously and ironically, connected
with defeat and death. So their second effect is to suggest the ultimate frustration of Cornélie's hopes. César, then, may be momen-

tarily 'hors de péril' after his victory over Ptolomée, but he is still threatened; at the same time, although in that sense he is not secure, there is no final triumph for Cornélie, either, nor yet, for that matter, for Cléopâtre whose destiny is prefigured by Antoine's presence in the play and the suggestion (ll. 951–2) that he has already fallen love with her.

In *Bérénice* the same creative processes may be discerned. It has frequently been remarked that Racine rejuvenates his heroine and obliterates her marriages and those of Titus. It has almost as frequently been suggested that the playwright did not wish to offend his public's sense of propriety; but even supposing that to be so the need is turned to good effect: by attributing to his characters innocent and unsatisfied passions, he made them more sympathetic and their passions more imperious.[37] At the same time, we have here an example of the kind of idealization, of the 'goodness' of character, which we have seen in *Pompée*. In the case of Titus, that 'goodness' does take on a moral sense and the suggestion for it is, as we shall see, already in Suetonius, as Louis Racine and Paul Mesnard have observed. And if the suggestion was developed in Racine's characterization of the emperor, it also seems to have affected by contamination his portrayal of Bérénice.

Here too, however, some suggestions come from historians. Racine makes his heroine attractive, morally, and in doing so transposes the physical attractiveness mentioned by Tacitus, for example. Even at the age when she first met Titus (she was over forty, he under thirty), she was 'florens aetate formaque' ('in the flower of her age and beauty') and sufficiently so to attract Titus and his ageing father.[38] What Racine does not refer to — it is alien to the kind of character his dramatic conception required — is the moral defects mentioned by Josephus.[39] The separation in the play is that of young and innocent lovers driven apart by forces over which they have no control, and it is there that the formula 'invitus invitam' begins to be charged with its peculiar emotional power. We have already noticed that Bérénice's resistance to the strength of Roman tradition arises out of her own illusory conception of the emperor as an Oriental despot. That illusion is nourished not only by her passionate love, which blinds her to the truth, but also by her youth, which makes her naïve, inexperienced, and capable of self-delusion. It was essential, within the dramatic framework as created by Racine, that his heroine should be young, innocent, and passionate — and he had to endow her with the first two of these qualities and to enhance the third.

The point is illustrated by the way in which she allows herself to be lulled into a world of unreality when at the end of Act I she

imaginatively recalls the apotheosis of Vespasian. That apotheosis
was a precise historical event, and for his description of it Racine
seems to have gone to the detailed accounts given by Dio Cassius
of that of Pertinax and by Herodian of that of Severus.[40] To details
from those accounts, he has added from Suetonius that of the way
in which Titus was acclaimed by his soldiers at the end of the
Palestinian campaign.[41] But Bérénice's account is far from precise:
it is evocative, lyrical, allusive, characterized by the use of the de-
monstrative adjective which reveals not so much the event itself as
its life in her own imagination. Is she vaguely aware, already, of
the powerful yet illusory nature of her dreams? 'Mais, Phénice, où
m'emporte un souvenir charmant?' (l. 317.) Still, however, in Act
V, her incomprehension or refusal of the hard truth stands in the
way of its acceptance (ll. 1103, 1137–8, 1175 ff.).

In fact, the apotheosis which Bérénice recalls is not for her
Vespasian's but Titus's. All the splendour of the ceremony is
attached, not to the man whose reign it ostensibly celebrates, but
to that of his son which has not yet begun:

> Ces flambeaux, ce bûcher, cette nuit enflammée,
> Ces aigles, ces faisceaux, ce peuple, cette armée,
> Cette foule de rois, ces consuls, ce sénat,
> ... Tous de mon amant empruntaient leur éclat.   (ll. 303–6)

The accumulation arises out of adoration of the beloved and wonder at his magnificence. It is true that Dio Cassius and Herodian
make mention of the central part played towards the end of the
ceremony by the new emperor, and of the joy with which his funer-
al oration is acclaimed; but, for Bérénice, the real purpose of the
obsequies is effaced by the sole centre of her attention. Her imaginative recollection is hallucinatory and consistent with her
dreams of love with Titus, alone, withdrawn from the public gaze,
idyllically happy. It already foreshadows the hallucination in
which Phèdre in recollection substitutes herself for Ariane and
Hippolyte for Thésée in the Cretan labyrinth. In *Bérénice*, the irony
is that at that very moment when the queen is carried away on
dreams of love and marriage she is witnessing, not the liberation of
Titus from his mourning for his father, but his imprisonment in
imperial duty. Like Corneille, Racine chooses to omit or generalize
the picturesque details, alluding to them only in order to suggest
the power of Bérénice's illusions and passion. The many references
in the play to Roman history, tradition, and customs contribute of
course to the creation of an authentic-seeming background to the
tragedy, but they are not a form of local colour because they represent to the characters the reality of the forces against which their
desires will not prevail. It is there, perhaps, more than anywhere

that the playwright's knowledge of history is combined with his powers of poetic evocation to serve a properly dramatic function. All the historical references have a direct significance within the events of the play: the force of the allusions, made largely through the use of proper names, cannot fail to strike the spectator who is familiar with Roman history. Nowhere is this clearer than in the scenes in which Paulin confronts the emperor with the realities of Roman tradition (II.ii) and Titus himself finally comes face to face with the fact that they can no longer be evaded (IV.iv), scenes which are Racine's equivalents of the deliberations in *Cinna, Pompée, Sertorius,* and several other of Corneille's plays.

The idealization of Bérénice and Titus into a pair of guiltless lovers (though not perfect because weak) is one of the major features through which Racine invests their separation with a tragic sense. Dio Cassius states that the queen had, before her departure from Rome, lived there as Titus's wife, and Aurelius Victor that her influence over him was entirely bad.[42] Racine not only sets aside the first statement but has Titus claim that Bérénice has inspired him to goodness and greatness (II.ii, to Paulin; IV.v and V.vi to Bérénice herself). For those were qualities he had not always possessed: the historians inform us that they were notably absent until, after his accession, Titus gradually became what Bérénice sees in him, 'amor et deliciae generis humani' ('the darling and delight of all mankind').[43] Only the briefest and most discreet reference is made in Racine's play to Titus's debauched past (ll. 504–8), and even at that it is set in the context of the moral benefits wrought by Bérénice when he first knew her: it belongs to a dead and already distant period of his life. This is another example of Racine's reversing the received data: whereas for the historians it was the assumption of imperial office which caused Titus's reform, for Racine the change had taken place at least five years earlier, and was due to Bérénice's good influence, something entirely fictitious. Racine rejects the accounts of the emperor's cruelty before his accession, of his two marriages, of his dissoluteness:[44] he is already what he was to be at the end of his reign. It is in the context of his debauchery that Suetonius writes of Bérénice ('insignem Berenices amorem, cui etiam nuptias pollicitus ferebatur' — 'he had an extraordinary passion for Berenice, to whom he was said already to have promised marriage') and of the resemblance contemporaries saw between Titus and Nero ('denique propalam alium Neronem et opinabantur et praedicabant' — 'indeed people were openly thinking and prophesying that he would be a second Nero'). That was all left behind on the death of Vespasian ('At illi ea fama pro bono cessit conversaque est in maximas laudes, neque

vitio ullo reperto et contra virtutibus summis' — 'But that low
opinion stood him in good stead and was turned to the highest
praise; no vice was discovered in him but, on the contrary, the
highest virtues') and it is precisely in this context that we find the
statement: 'Berenicem statim ab urbe dimisit invitus invitam'
('against his wishes and against hers, he immediately sent Berenice
away from Rome').[45] So, even in the preface, Racine has arranged
the historian's account by combining parts of two of his sentences,
thus fixing attention, not on the scarcely virtuous side of Titus's
nature and behaviour, but on his goodness and unhappiness.
Again this represents a transposition: Bérénice in the play has
known Titus for several years, and long before his accession, and
the Titus she has known is the virtuous man whom he became in
reality only afterwards. Indeed, the first sign in historical reality of
that virtue was perhaps his dismissal of Bérénice. Titus is idealized
for the same reasons as the queen, but Racine does not in his case
depart entirely from history: rather he uses suggestions found in
history and transposes and combines them in such a way as to
make of the new emperor the tragic and sympathetic figure required to play a particular dramatic part. His struggle with himself
is not between the demands of love and imperial duty, but between
his humanity and the cruel necessity to state the truth to Bérénice.
Suggestions for all this came particularly from Suetonius, as Louis
Racine observed,[46] but the poet does idealize, even suppressing
Titus's alleged monarchist ambitions.[47] For Titus is not, after all, a
Cornelian hero seeking fulfilment on the stage of public affairs:
they merely expose him to finding his deepest desires unfulfilled.

    Corneille's play, *Tite et Bérénice*, reveals a quite different form of
idealization. Both he and Racine draw on Suetonius, Corneille concentrating on the *Life* of Domitian and Racine on that of Titus.
The elder dramatist places at the head of his play two short passages, without comment, from Xiphilinus's digest of Dio Cassius:
unlike Racine, he does not omit from his source or from his play
the fact that Bérénice had lived with Tite at Rome as his mistress
and that after her departure, prior to her secret return in the
course of the drama, he had been 'unfaithful' to her. On the other
hand, while not particularly sympathetic characters, Domitian and
Domitie are cleared of the misdeeds attributed to them by all the
historians. The earlier period of Tite's and Bérénice's cohabitation
is of course essential to Corneille's conception of the play.
Although they are still in love, they must part for reasons of state,
but Bérénice demands that her 'gloire' be first satisfied, that no
shame attach to her departure and that she leave of her own free
will, and that Tite assure her that he loves her still and will not

substitute Domitie for her. It is in their agreement to these conditions and in the motives of self-sacrifice (for Tite does love Bérénice and, unlike Racine's Titus, is ready to abdicate for her sake; and Bérénice does love him and sacrifices her love to his imperial duty) that Corneille idealizes them. Domitian's passionate ardour for Domitie is pure — an idealization no doubt of the fact that, although in history he divorced her, he loved her well enough to take her back in spite of her earlier unfaithfulness. Domitie does genuinely love him in the play, but is governed by her passion for the throne, hence her wish to marry Tite; that passion is ultimately satisfied when the emperor, proving his love for Bérénice, refuses to go through with the marriage arranged with Domitie, but allows her to marry his brother with whom he will share the throne. Whereas Racine's tragedy is that of passionate and guiltless love frustrated by an inescapable but repugnant duty, Corneille's play concerns the willing acceptance of a duty which brings its own satisfactions in its train and, through renunciation of marriage, allows the characters to achieve their 'gloire'. This distinction being clear, one may see how the dramatic conception in each case demanded its own form of idealization.

Both Corneille and Racine appear to owe something to Magnon's *Tite* of 1660; the modern editor of that text[48] has drawn attention in his introduction to a number of parallels. From the point of view of situation it seems to have contributed more to Corneille than to Racine, but its plot turns on a *quiproquo* and a disguise absent from the later plays. These do, on the other hand, have close affinities with the portrait of Bérénice we find in Scudéry's *Les Femmes illustres*. (As far as the idealization of Titus is concerned, that is suggested by Scudéry and may owe something to Dupleix's exclusion of criticism and — important for Corneille's play — to the emphasis he (i.e. Dupleix) places on the emperor's generous behaviour towards his brother.[49]) Scudéry may well have been the first to attribute to Bérénice the passionate tenderness, the grief, and the ultimate self-sacrifice to be found in varying proportions in the 1670 plays. The absence of political ambition in her love for Titus, and her awareness of its absence in him, is one of the major themes of her address to him in Scudéry's work:

Je sais que quoique l'ambition soit une passion aussi forte que l'amour, elle ne la surmonte pas en votre âme; et je veux même croire, pour me consoler dans ma disgrâce, que si vous étiez en état de disposer absolument de vous, vous préféreriez la possession de Bérénice à l'empire de tout le monde. Mais la raison d'état . . . ne peut souffrir que l'invincible Titus, après avoir tant de fois hasardé sa vie pour assurer la félicité des Romains, puisse songer à la sienne particulière.   (I. 142)

The recognition of the inescapability of imperial duty, and of the reality of love, is to be found in different forms in both the plays. In this recognition, ambition has no place:

Non, Titus, la magnificence de Rome ne m'éblouit point. Le trône qui vous attend [in Scudéry's version, Titus is not yet emperor: in the plays it is his accession which precipitates the drama] n'a rien contribué à l'affection que j'ai pour vous, et les vertus de votre âme et l'amour que vous avez eue pour moi, ont été les seules choses que j'ai considérées quand j'ai formé la résolution de vous aimer.   (I. 145–6)

Racine's Bérénice follows this line exactly:

> Depuis quand croyez-vous que ma grandeur me touche?
> Un soupir, un regard, un mot de votre bouche,
> Voilà l'ambition d'un cœur comme le mien.   (ll. 575–7)
> Mon cœur vous est connu, Seigneur, et je puis dire
> Qu'on ne l'a jamais vu soupirer pour l'Empire.
> La grandeur des Romains, la pourpre des Césars
> N'a point, vous le savez, attiré mes regards.   (ll. 1475–8)

Speaking of Tite's fortune, Corneille's heroine also echoes Scudéry's:

> Si j'eusse eu moins pour elle ou de zèle ou de foi,
> Vous seriez moins puissant, mais vous seriez à moi,
> Vous n'auriez que le nom de général d'armée,
> Mais j'aurais pour époux l'amant qui m'a charmée...
>   (ll. 1021–4)

Indeed, Corneille's Bérénice goes on:

> Et je posséderais dans ma cour, en repos,
> Au lieu d'un Empereur, le plus grand des héros.   (ll. 1025–6)

This is almost exactly what Scudéry's says a little later on:

Mais s'il m'est permis de vous dire tout ce que je pense, je voudrais qu'étant nés sans couronne, sans royaume et sans empire, nous puissions vivre ensemble en quelque lieu où la vertu seule regnât avec nous.

Racine's heroine dreams also, as does Scudéry's Cléopâtre (I. 55), of love fulfilled without the constraints of high office. In both the plays, as in Scudéry, Bérénice can accept separation provided that Titus still loves her, but Corneille's, unlike Racine's, sees in that a need to satisfy her 'gloire', which is also to be found in Scudéry, where she likewise urges the emperor to do his duty even to the point of marrying someone acceptable to the Roman people — with the exception, in Corneille's play, of Domitie. The question of rivalry between the heroine and another woman does not arise in Racine's play: his Bérénice wishes to be assured that Titus's refusal

to marry her is not the result of love discontinued. Corneille's Bérénice, on the other hand, wishes to be satisfied that Tite will not
marry anyone other than herself *for love*: such an assurance safeguards her 'gloire'.[50] If the two dramatists draw on Scudéry's
work — and many other similarities could be cited —, exploiting it
in partly parallel, partly diverging directions, the same observation
can be made about Magnon's *Tite*. In that play, Bérénice has all
the single-minded passion of Racine's heroine (though she is much
more active in its pursuit, like Corneille's, and not resigned, as she
is also in Scudéry), and she is contrasted with the ambitious Mucie
whose counterpart in Corneille is Domitie. What this illustrates is
of course the fact that even when drawing on common non-
historical sources, Corneille and Racine select from them and use
and develop them for their own specific dramatic purposes. A preconceived dramatic subject draws out from the sources those features which it requires for its realization in a play.

The dramatic preconception involves not only selection and rejection, and a particular emphasis affected by creative development — plausible both historically and dramatically —, but, as we
have seen, rearrangement and transposition of the data, whether historical or fictional. Two small points about Racine's *Bérénice*, not
noticed, I think, by other commentators, will add a little further
evidence of the effect of transposition. The first concerns the human goodness and kindness of Titus which are at the root of the
conflict and drama to which he is subject. These qualities are of
course suggested by the word 'invitus' when it is taken together
with other details we have observed in Suetonius's account. The
same historian[51] refers to the emperor's public conduct at the time
of the several great catastrophes which marked his reign: 'non
modo principis sollicitudinem sed et parentis affectum unicum
praestitit' (— 'he showed not only an emperor's concern, but even a
father's feelings for his children'). Moreover, the phrase, 'sed peri-
turum se potius quam perditurum adiurans' (— 'swearing that he
would rather die than take life'), is also transposed by Racine from
the public sphere to that of Titus's personal relations with Bérénice.

The second point concerns the way in which the new emperor
accepts his destiny (ll. 715, 1394) and resigns himself to carrying
the burden it imposes on him (ll. 462, 719–22, 736). Suetonius
recounts[52] that when two patricians were accused of having designs
on the imperial throne, Titus simply told them: 'principatum fato
dari' (— 'that the empire was bestowed by Destiny'). When he mentions the emperor's love for Bérénice, Tacitus states:[53] 'Neque
abhorrebat a Berenice iuvenilis animus, sed gerendis rebus nullum

ex eo impedimentum' (— 'nor was he unattracted to Berenice, but
that was no hindrance to his conduct of affairs of state'). Do not
these phrases suggest Racine's representation of Titus resigned to
the inevitable and putting public duty before all else, before even
his love?

It is by a whole series of small touches like those we have examined that, in spite of fundamental changes to both event and
character, Racine, like Corneille, is yet able to give to his plays an
air of being genuinely historical. Such details are not independent
of the dramatic aspects of the plays but are woven into their tex-
ture in support of motives, feelings, and arguments, and woven in
not in their historical order, but where the dramatic conception
requires them to be.

History, moreover, often itself imposes a duty on the dramatic
characters or acts as a spur to them in their self-projection on its
stage. Paul Mesnard remarked[54] that in the scene in which Titus at
last confronts Bérénice with the inescapable truth (IV.v), Racine
seems to have remembered passages of the sixth book of the
Aeneid. The same might in part be said of Act II scene ii, with
Paulin playing the role of Anchises. A radical difference exists,
however, between Virgil's and Racine's use of the allusions to the
sweep of Roman history. For Aeneas, the future history of Rome is
unfolded in a vision inspiring him for the heroic actions he must
undertake and kindling him with a passion for the glory to be. The
direct parallel to this episode seems to me to be in Sabine's prophetic words in the first scene of *Horace*, in le Vieil Horace's at the end
of Act III scene v, and in the theme of the play as a whole, in
which the hero is drawn magnetically into an awesome and inhuman duty to the future glory of his city and his own. In *Bérénice*, on
the other hand, we have another example of Racine imitating in
reverse, because in the play history is not in front of Titus, but
behind him, not drawing him on with a vision of glory, but pursuing him with a burden of disasters whose repetition he must try
to avoid. His concept of glory is negative and itself a burden to be
shouldered (ll. 452–4).[55] In part it consists merely in the prevention of further disaster. He may say to Bérénice:

> ... Si nous ne pouvons commander à nos pleurs,
> Que la gloire du moins soutienne nos douleurs,
> Et que tout l'univers reconnaisse sans peine
> Les pleurs d'un Empereur et les pleurs d'une Reine.

>   (ll. 1057–60)

That pathetic glory is, however, small consolation, for all that Titus
can really hope for is the reputation of not being

Un indigne empereur, sans empire, sans cour,
Vil spectacle aux humains des faiblesses d'amour. (ll. 1405–6)

The visions of history, no less than the reality behind Bérénice's vision of the apotheosis of Vespasian, are simply the sign of Titus's imprisonment in empire. Cornélie, in *Pompée*, may, as we have seen, be inspired to threaten César by her vision of the future, but she will prove to be no less imprisoned by that future than is Titus when the imagined victories turn out to be defeats.

Of course, the irony is quite differently treated in the two plays. In *Bérénice*, Titus's real imprisonment in what the queen takes to be the moment of his liberation is the crux of the whole action, whereas in *Pompée* the triumph of César is real and not illusory within the confines of the play but, for all that, the allusions to the future conduct of the Civil War do play on the spectator's imagination and, ultimately, attenuate that triumph.

The processes which we observe Corneille and Racine adopting in the plays just analysed are at work in all their plays. They make use of historical materials, whether drawn from history proper or from adaptations of history in the form of moral portraits or novels or epic poems or plays, arranging those materials freely in order to accommodate them to situations and subjects which may be conceived independently of them. It is not without interest to note that in *Pompée* Corneille's principal model is Lucan's *Pharsalia*, a work not of history, but already of imagination, and that in this same way in *Bérénice* Racine made extensive use of the *Aeneid*. As in the matter of situation, so in that of sources, the playwrights seem to pillage the work of other creative writers, if only because these have already made an attempt to simplify complex historical data or to select from them such features as could be conveniently dramatized. This does not mean that Corneille and Racine follow in the same direction as their predecessors: on the contrary, as we have seen, the lesson learnt may be negative or the new situation require its own peculiar treatment. As D'Aubignac (*Pratique*, p. 88) saw, it was necessary sometimes to practise rigorous selection, sometimes to fill out a sketchy series of episodes:

Il ne sera peut-être pas inutile d'avertir notre poète que si l'action principale ... était chargée dans l'histoire de trop d'incidents, il doit rejeter les moins importants et surtout les moins pathétiques; mais s'il trouve qu'il y en ait trop peu, son imagination y doit suppléer ..., ou bien en inventant quelques intrigues qui pouvaient raisonnablement faire partie de l'action principale, ou bien rechercher dans l'histoire des choses arrivées devant ou après l'action dont il fait le sujet de son poème ...

If we substitute for incidents arousing pathos something a little
more general — incidents arousing emotion — D'Aubignac's statement is clearly applicable to both Corneille and Racine: they select
from history what their subject requires; they add to it in the same
spirit; they relate the incidents to the principal action; they arrange
the data of history into their own particular pattern. It would be
misleading to suppose that the process of simplification or amplification of the data is the prerogative of either dramatist as
opposed to the other. Radical simplification in the relationships
between the episodes is evident in *Pompée*, but motivation, particularly in *César*, is complicated, and Cornélie is introduced into episodes in which she had no part. D'Aubignac accused Corneille of
having failed to simplify in some of his later plays, notably in *Sertorius*, and it is true that in some he appears to do the reverse. So,
for example, in *Horace*, as we have seen, he creates the part of
Sabine; but this has to be set against the exclusion of two of the
brothers on either side as well as of the negotiations preceding the
truce and the battle, and against the simplification of the judicial
proceedings following the death of Camille. In much the same way,
Éryxe is an innovation in *Sophonisbe*, complicating the knot of the
politico-matrimonial intrigue, but large tracts of the historical
accounts are omitted, as is the appearance of Scipio. Racine's practice differs little from Corneille's: in *Andromaque* the situation, which
draws together the four principal characters in a way unknown to
the ancient legend, is thereby complicated in terms of the new relationships and of the conflicts which arise from them; in *Britannicus*,
Bérénice, *Iphigénie*, and *Phèdre*, the invention of an important character complicates the incidents and the emotions which they arouse.
If, in *Mithridate*, Racine excludes some of the characters found in
the historical accounts and in La Calprenède's play, he prolongs
the life of Monime and imagines the complex relationship between
the king and his sons that her survival entails. Whether Racine
practises greater simplification than Corneille in terms of the number of characters or of relationships or of incidents is another question which will be dealt with in its proper place; but in terms of
historical sources, it is clear that both playwrights feel free to adapt
them to their needs and that this involves sometimes complication
and sometimes simplification, but usually both together.

The public attitude adopted by the dramatists towards history is
ambiguous because they were involved in polemics. Both claim at
times to adhere closely to their sources, at times to depart from
them, occasionally both simultaneously. In creating situation and
plot, they often owe little to history. It has been remarked, however, that they claim more often to be faithful to it than do most of

their contemporaries, for whom all that mattered was conformity to the tastes and prejudices and ignorance of their public. It is also true that Corneille and Racine, for all the modifications they introduced, did show considerably greater respect, in their dramatic practice, for their sources than did their rivals. That respect applies more generally to characterization than to situation and plot, but even there it was limited by regard for *bienséance* and *vraisemblance*.

Yet to a remarkable degree their plays do convey an air of historical authenticity even today. The characters may be involved in unhistorical relationships and episodes, but the invented or partially invented situation does develop through the plot to a historically authentic catastrophe which is the first guarantee of their credentials. The plot at times bears some resemblance to historical developments, as does the behaviour of the characters. Frequently departures from historical fact take the form of a rearrangement of the episodes rather than that of their complete falsification, or the creation of episodes not historically attested but bearing some resemblance to historical events connected with the characters concerned. The general tendency to idealize may involve a diminution of factual accuracy, but it often results in the impression of a character true in his essentials to his historical counterpart: such idealization usually takes the form of exclusion of certain features rather than accentuation of others, but the effect of such exclusion is of course to emphasize the remaining essentials. Guez de Balzac praised Corneille for idealization: he and his contemporaries saw no necessary contradiction between historical truth and its embellishment:

Vous nous faites voir Rome tout ce qu'elle peut être à Paris, et vous ne l'avez point brisée en la remuant... Vous avez même trouvé [in *Cinna*] ce qu'elle avait perdu dans les ruines de la république, cette noble et magnanime fierté... Vous êtes le vrai et fidèle interprète de son esprit et de son courage...

Balzac clearly attaches fidelity to history not to its facts as such but to the interpretation of what he and his contemporaries, brought up on a diet of Latin moralists and historians, understood by the spirit of Rome. This was not incompatible with their view of Roman history:

Vous êtes le réformateur du vieux temps, s'il a besoin d'embellissement ou d'appui. Aux endroits où Rome est de brique, vous la rebâtissez de marbre; quand vous trouvez du vide, vous le remplissez d'un chef d'œuvre; et je prends garde que ce que vous prêtez à l'histoire est toujours meilleur que ce que vous empruntez d'elle.[56]

We have seen that this process is not peculiar to Corneille. If Andromaque is released from concubinage with Pyrrhus, it is in part, as Racine says in his second preface, so that she may conform 'à l'idée que nous avons maintenant de cette princesse', that is to say that she becomes the embodiment of perfect widowhood and perfect motherhood which represents an idealization already suggested by Homer and Virgil, but one which for Racine becomes a dramatic and psychological necessity in the subject as he has conceived of it. He will make the same kind of change, for similar reasons, in Bérénice, as we have seen, and again in Monime. For Bérénice, the idealization had at least been begun by Scudéry. For Monime, it is already clearly present in Le Moyne's portrait,[57] in which she receives the message that she is to die by Mithridate's order 'avec sa mine de fête et son visage de réjouissance', and her resignation is described as 'cette action orgueilleuse et bienséante, mêlée de fierté et de modestie'. It would be a mistake, I think, merely to attribute Racine's idealized representations to a desire to make moral 'improvements', introduced in order to avoid offending the susceptibilities of a public governed by *bienséance*. In Corneille, we can observe the contrary, in the Cléopâtre of *Rodogune* for instance, and Racine does little, if anything, to relieve the blackness of Néron's character or of his mother's. On the other hand, Corneille does not allow Antiochus to commit Orestes's crime of matricide, because, he says in the second *Discours* (*Writings*, p. 48), that would deprive him of the audience's sympathy: the idealization of Antiochus, even in relation to his own history (ibid., p. 47), not only to that of Orestes, is essential to Corneille's concept of his subject which consists in the conflict between unspoilt goodness and unmitigated evil. And if Racine introduces Junie into *Britannicus* the effect is not that of partially exculpating Néron, but of setting her purity against his sins and, by involving her in them, of adding to the motive of political jealousy towards his step-brother that of amorous jealousy as well. The characters become, at least in part, moral exemplars, as many of them had already become, not least in contemporary translations of the historians of antiquity[58] whose interpretations were developed by the moral portraitists. For all that, in the discreet yet telling use of allusions and proper names, introduced neither gratuitously nor necessarily in their historical place, but in direct relation to the characters' passions and motives, Corneille and Racine were able to give to their dramas an authentic historical colouring. That colouring was essential, no less than was the idealization of the characters, to the suggestion of the tragic pitch of the drama. The characters were really involved in the creation of history: unless the audience were given some sense of that fact, the plays would have remained, as did those of the lesser dramatists of the period, on the level of their abstract situations, that of domestic feud and intrigue, without responsibility save on an immediately personal basis.

The expectations of the public are not without interest, as they point to an interpretation of the dramatists' practice and of the ways in which they justify it. D'Aubignac, for example, writes (*Pratique*, p. 225): '[Les poètes] prennent de l'histoire ce qui leur est propre, et changent le reste pour en faire leurs poèmes, et c'est une pensée bien ridicule d'aller au théâtre pour apprendre l'histoire.' How far this critic was prepared to go in this direction is of course evident from Corneille's assertion, already referred to, that it was he who suggested the 'disposition' of *La Didon chaste*, in which little of the traditional story remains other than the actual suicide of the heroine. Most of the 'historical' tragedies of the 1650s and early 1660s, particularly those of the really popular authors, Thomas Corneille (*Timocrate, Bérénice, Darius, Pyrrhus*) and Quinault (*Cyrus, Astrate*), are based on romanticized history or simply on novels. It is the taste for this kind of drama that is reflected in D'Aubignac's statement. Almost twenty years after it, Villiers had the speakers in his dialogue on tragedy echo it. What they say shows how little regard the public had for historical accuracy:

Cléarque: Vous me faites faire une réflexion que je ne veux pas laisser échapper, c'est qu'il est difficile qu'une pièce de théâtre réussisse quand tout ce qu'elle représente est inconnu. Car ce qui est inconnu semble fabuleux; et quoiqu'une tragédie puisse être toute fabuleuse, néanmoins on se plaît bien plus à voir sur le théâtre un nom illustre, et des aventures dont on a déjà quelque légère connaissance, qu'un nom barbare et des incidents romanesques . . .

Timante: . . . Quand le titre d'une tragédie est connu, cela prépare mieux les esprits, et je ne voudrais pas qu'un auteur qui n'a point encore travaillé pour le théâtre commençât par un sujet et un nom caché.[59]

In spite of the reference in Cléarque's statement to the possibility, suggested by Aristotle in his reference (*Poetics*, chapter IX) to the *Antheus* of Agathon, a tragedy 'in which both the incidents and names are of the poet's invention', of entirely 'fabulous' or fictitious subjects, it is clear that Villiers is reflecting a preference for plays which have some connexion with history. He is criticizing by implication, not plays which are drawn only indirectly from historical accounts and in which the author's inventiveness may have been stimulated by romanticized history (as we have seen — Lucan, Virgil, Scudéry, Le Moyne, and many others), but those like

Quinault's *Astrate* (1664–5) in which only the hero's name has historical authority, the rest — names, plot, episodes — being, in their historical context, purely imaginary. Corneille admits to having travelled some distance in this direction twenty years before (*Rodogune, Héraclius*), and the kind of process we have observed in one or two of his plays and of Racine's reveals that their arguments about historicity have to be taken with at least a grain of salt, and Corneille, at any rate, more than once refers to the possibility of entirely fictitious tragedy.

Yet in spite of all this and of the polemical nature of many of their justifications — often quite specious — of their own departures from historical fact, and notwithstanding the clear evidence that in terms of situation, relationships, and plot their plays owe little or nothing to the traditional stories, Corneille and Racine do create something of an authentic atmosphere and endow their characters with some recognizably true qualities. Although the bitter sarcasm of Corneille's attack on the criticisms of D'Aubignac is obviously due to its polemical context, his preface to *Sophonisbe* should not be dismissed merely for that reason. Some of his remarks provide us with valuable clues to his respect — when compared with most of his rivals, the notable exception being Racine — for history and to his understanding of his art as a dramatist:

Quoi qu'il en soit, comme je ne sais que les règles d'Aristote et d'Horace, et ne les sais pas même trop bien, je ne hasarde pas volontiers en dépit d'elles ces agréments surnaturels et miraculeux qui défigurent quelquefois nos personnages autant qu'ils les embellissent, et détruisent l'histoire au lieu de la corriger. Ces grands coups de maître passent ma portée; je les laisse à ceux qui en savent plus que moi, et j'aime mieux qu'on me reproche d'avoir fait mes femmes trop héroïnes, par une ignorante et basse affectation de les faire ressembler aux originaux qui en sont venus jusqu'à nous, que de m'entendre louer d'avoir efféminé mes héros par une docte et sublime complaisance au goût de nos délicats, qui veulent de l'amour partout... (*Writing*, p. 167.)

Embellishment, idealization, 'correction' of history: this is acceptable, but not disfigurement or destruction. In his turn, Racine defends (first preface to *Britannicus*) the moral correction of the alleged historical counterpart of Junie: '...je n'ai pas ouï dire qu'il était défendu de rectifier les mœurs d'un personnage'; and at the same time treats ironically those who criticized his excessively dark portrayal of Néron: '...peut-être qu'ils raffinent sur son histoire.' This is a justification of idealization in both directions, as it were: the play required an innocent and virtuous heroine to be contrasted with a vicious and cruel hero and to be the victim of his passion. Each needed to be superlatively good of his or her kind, partly for dramatic reasons and partly so that the spectator's imagination be vividly struck by their exemplary nature. Almost every preface by Racine, like many of the *Examens* of Corneille, contains a reference to the dramatist's right — indeed duty — to idealize in this way. Sometimes, even in the critical writings 'corrections' are discernible, if only by omission: in the second preface to *Alexandre le Grand*, for example, the 'degrading' phrase, 'concubitu redemptum', is omitted from the quotation from Justin's account of Cléofile's love for Alexandre.

Such changes as need to be made for dramatic reasons must obviously be compatible with history and not simply calculated to produce effects incompatible with *la vraisemblance*: nothing miraculous or supernatural, but also nothing to detract from the heroism of historical heroes and heroines. The insistence on heroism is of course a characteristic Cornelian touch, but Andromaque's devotion to Hector, Néron's cruelty, Mithridate's resistance to Roman power, Iphigénie's purity, Phèdre's passion are all historical features which Racine safeguards with a similar regard for historical truth. And of course all these characteristics are idealized by both dramatists, history being 'corrected' so that it may, in its turn, be made compatible with the emotional and aesthetic function of the tragedies in which they feature. So, for example, Corneille tells us that the virtually inverted character of Nicomède is conceived in such a way as to arouse a specific emotion, which he calls 'admiration' (*Examen: Writings*, p. 152), and that in *Sophonisbe*, that of Éryxe is a new creation, not gratuitous because 'elle ajoute des motifs vraisemblables aux historiques, et sert tout ensemble d'aiguillon à Sophonisbe pour précipiter son mariage, et de prétexte aux Romains pour n'y point consentir' (Preface: *Writings*, p. 167). It is, then, the needs of the subject, understood as an aesthetically complete and satisfying organism calculated to arouse specific emotions, which lead to the choice of materials drawn from history, to the creation of new ones, and to the arrangement of the whole in a particular combination and order. Only in that context can the way in which Corneille and Racine treat *la vraisemblance* and *la bienséance* be understood. Their regard for those criteria is relative and flexible as is their regard for history, but in both cases it is real.

It may be that the attraction for Racine of the truncated quotation from the already laconic account of Suetonius which stands at the head of the preface to *Bérénice* was that it allowed him to show that he could 'faire quelque chose de rien' — but not exactly of nothing, with Virgil in important respects an inspiration (perhaps the chief one). He set himself the opposite problem to the one Corneille usually faced — that of inventing ways of reducing complex historical events to the exigencies of tragedy — yet he, too, could claim to invent characters, motives, episodes. Racine's version of Aristotle's distinction between history and poetry (*Principes*, p. 16, *Poetics*, chapter IX) is revealing: here it is, followed by Bywater's translation:

La poésie est quelque chose de plus philosophique et de plus parfait que l'histoire. La poésie est occupée autour du général, et l'histoire ne regarde que le détail. J'appelle le général ce qu'il est convenable qu'un tel homme dise ou fasse vraisemblablement ou nécessairement. Et c'est là ce que traite la poésie, *jetant son idée sur les noms qui lui plaisent, c'est-à-dire empruntant les noms de tels ou de tels pour les faire agir ou parler selon son idée.*

L'histoire, au contraire, ne traite que le détail; par exemple, ce qu'a fait Alcibiade, ou ce qui lui est arrivé.

... Poetry is something more philosophic and of graver import than history, since its statements are those of the nature rather of universals, whereas those of history are singulars. By a universal statement I mean one as to which such or such a kind of man will probably or necessarily say or do — which is the aim of poetry, *though it affixes proper names to the characters*; by a singular statement, one as to what, say, Alcibiades did or had done to him.

Comparison of the italicized passages reveals the extent to which Racine understood that it is the 'idea' in the mind of the poet, his conception of the subject, which on the one hand controls his use of historical material and on the other indicates what, within the framework of the subject, is probable or necessary. The characters and their names are borrowed from history at the choice of the poet in order to be presented in accordance with his conception of the dramatic subject and with the exigencies of his art.

Professor Henri Gouhier has drawn attention to the distinction between history as a source of inspiration and as 'un répertoire d'histoires', pointing out that if it is treated as the former it leaves the poet free in the conduct of his plot, but if as the latter it chains him to a greater measure of conformity to historical accounts. So, he writes,

si l'histoire inspire le poète, ce n'est point pour le transformer en conteur d'histoires vraies, ce n'est pas pour l'inviter à mettre en scène ce qui a été réellement vécu, mais pour émouvoir le visionnaire et le prophète qui croit découvrir une histoire invisible dont l'histoire des historiens est l'apparence figurative. La transformation du fait en signe, telle est sans doute la métamorphose propre au théâtre dont l'action trouve ses personnages et ses situations fondamentales dans l'histoire. Que la signification l'emporte sur le signe, qu'elle refoule le signe, qu'elle en arrive à créer d'autres signes pour se mieux manifester, c'est ce que montrent les plus grandes œuvres:

l'auteur de *Bérénice* crée Antiochus, celui de *Marie-Stuart* crée Mortimer; ici et là, le poète invente des personnages qui joueront un rôle important dans les péripéties où l'action prend forme et où les personnages se révèlent tels qu'eux-mêmes enfin le poète les a vus.[60]

If Corneille and Racine seem, as we have seen, to make as much use of imaginative as of purely historical sources, this is surely because their predecessors had already, at least to some extent, shown the way to the extraction of the significant or universal aspects of history. Even historians themselves occasionally provide insights into the ironies of history and into the apparently irresistible force of destiny, chance, coincidence, and their tragic consequences.[61] That history is seen as a force in human affairs, impinging particularly sharply on those on whom those affairs seem to depend, and acting through them in an exemplary way, turning them into exemplars, is evident in the way in which it is treated by both Corneille and Racine. It is precisely that ability to penetrate beyond the particular to the universal and yet authentically to relate the universal — the significant — to the historical and the particular — the sign, which distinguishes their tragedies from those of Quinault, Boyer, Thomas Corneille, Gilbert, and a host of others.

CHAPTER III

'La Vraisemblance':
its dramatic function and significance

> ... Toutes les vérités sont recevables dans la poésie, quoi-
> qu'elle ne soit pas obligée à les suivre. La liberté qu'elle a de s'en
> écarter n'est pas une nécessité, et la vraisemblance n'est
> qu'une condition nécessaire à la disposition, et non pas au
> choix du sujet, ni des incidents qui sont appuyés de l'histoire.
> ... Quoique peut-être on voudra prendre cette proposition
> pour un paradoxe, je ne craindrai pas d'avancer que le sujet
> d'une belle tragédie doit n'être pas vraisemblable.
>
> Corneille, *Héraclius, Au Lecteur* (1647)

It is unnecessary, after René Bray and many others, to summarize
here the history and meaning of *la vraisemblance* and *la bienséance*: it
can be taken for granted that anyone interested in the subject of
this study will be familiar with them. They certainly form the
background to the plays of Corneille and Racine. It has generally
been accepted that the elder playwright simply disregarded them if
their observance was dramatically inconvenient, whereas the younger never experienced any difficulty in adhering to them. Their
own statements, as well as their practice, invite such a clear distinction, but can it be expressed in such black-and-white terms?

Discussion and controversy revolved around the relative importance of *le vrai* (historical fact), *le vraisemblable*, and *le nécessaire*, and
the degree to which these were to be influenced by *la bienséance* (the
audience's sense of what was proper in a moral sense and
appropriate in an aesthetic sense). All these concepts were ultimately derived from Aristotle but were interpreted in the light of
seventeenth-century ideas about logic and morals. It can be shown
without difficulty that in his critical self-justification Corneille
stood apart from the majority of his contemporaries and, in particular, from Racine — hence the usual contrast between a Corneille rebelling against the literary orthodoxy of his day and claiming the right to create freely the original works of his imagination,
and a Racine conforming to that orthodoxy and claiming to be the
faithful follower and imitator of the great writers of antiquity. The
fact that Corneille's rebelliousness finds expression in a body of
critical writing of which Racine produced no equivalent is often
provided as evidence in support of this thesis, as is Corneille's
attempt to reduce the importance of the central doctrinal tenet of *la
vraisemblance* contrasted with Racine's acceptance of it. But some
attention must here be paid to the objects to which that term is
applied and to the purpose it was meant to serve.

Corneille's examination of the problems of *la vraisemblance*, conducted at length in the second *Discours* as its title testifies (*Discours
de la tragédie et des moyens de la traiter selon le vraisemblable ou le nécessaire*), and in their ramifications in the third, on the unities, stems
from the critical views expressed by his adversaries on *Le Cid* and
on his tragedies more generally. The immediate occasion for the
elaboration of his ideas was, of course, the publication in 1657 of
D'Aubignac's *La Pratique du théâtre*, which had been written, at least
in part, within a year or two of the great quarrel, that is, fifteen to
twenty years earlier. This work belongs, therefore, to the same
period as do the *Sentiments de l'Académie sur le 'Cid'*, La Mesnardière's *Poétique*, Sarasin's *Discours de la tragédie*, and a considerable
number of critical prefaces all of which reveal similar preoccupa-

tions and preconceptions. It is clear from a letter dated 25 August 1660 (*Intégrale*, pp. 859–60), in which he describes his work on the *Examens* and *Discours* and appeals for the support of the recipient, the Abbé de Pure, that Corneille sees himself in opposition to D'Aubignac, whom he names, and the anonymous 'Messieurs de l'Académie'. It was with their views on *la vraisemblance* that he found himself principally at variance.

As early as 1623, Chapelain had stated that 'l'objet immuable de la poésie' must be 'la fable vraisemblable', while 'la vérité, considérée comme vraie' is the province of the historian.[1] Then, in the *Sentiments de l'Académie sur le 'Cid'*, he wrote:

... nous maintenons que toutes les vérités ne sont pas bonnes pour le théâtre, et qu'il en est quelques-unes comme de ces crimes énormes dont les juges font brûler les procès avec les criminels. Il y a des vérités monstrueuses, ou qu'il faut supprimer pour le bien de la société, ou que si l'on ne les peut tenir cachées il faut se contenter de remarquer comme des choses étranges. C'est principalement en ces rencontres que le poète a droit de préférer la vraisemblance à la vérité, et de travailler plutôt sur un sujet feint et raisonnable que sur un véritable qui ne fût pas conforme à la raison.[2]

This kind of comment is of course directed specifically at *Le Cid* which Chapelain would like to have seen modified so radically, as we have already observed, as to make it unrecognizable. The possibility of entirely invented subjects, as suggested here by Chapelain, is not excluded by Aristotle, but the reason the French critic puts forward is not to be found in the *Poetics*: what he objects to is the immorality ('qu'il faut supprimer pour le bien de la société') of certain kinds of subject — including that of *Le Cid* —, and the appeal to reason is, in this context, an appeal to moral sense and *la bienséance externe*, as Bray called it, acceptability in terms of the moral and social preconceptions of the audience.

D'Aubignac's views run on the same lines (*Pratique*, pp. 76–7):

... la vraisemblance est ... l'essence du poème dramatique, et sans laquelle il ne se peut rien faire ni dire de raisonnable sur la scène. C'est une maxime générale que le *vrai* n'est pas le sujet du théâtre, parce qu'il y a bien des choses véritables qui n'y doivent pas être vues et beaucoup qui n'y peuvent pas être représentées ...

Il n'y a que le *vraisemblable* qui puisse raisonnablement fonder, soutenir et terminer un poème dramatique: ce n'est pas que les choses véritables et possibles soient bannies du théâtre; mais elles n'y sont recues qu'en tant qu'elles ont de la vraisemblance ...

This critic makes the same appeal to reason, but does not confine it to moral sense, though that is clearly in his mind ('des choses véritables qui n'y *doivent* pas etre vues'): he is also concerned about the practical possibilities, with which he was so much preoccupied, of staging the play ('qui n'y *peuvent* pas être représentées') — that is to say, he was conscious of the need to overcome the spectator's disbelief, to present something credible on the stage. This has partly to do with rationality and partly with prejudice and ignorance, and in the latter context again, by a different route, brings us to the criteria of *la bienséance externe*. What, however, is crucial here is the idea that the passport for admission to the theatre is that the subject of a play must be *vraisemblable* and that historical truth does not of itself guarantee this. In this light we may see how D'Aubignac's criticism of *Théodore*, directed solely at the subject, arises out of the offensiveness of that subject in itself. In the same way, if it is true that he suggested the 'disposition' of *La Didon chaste* to Boisrobert, he will have done so in order that the supposed moral susceptibilities of the public might not be offended by the Virgilian account of Dido and Aeneas. Likewise, Chapelain's suggested rewriting of *Le Cid* in no way corresponds to Corneille's concept of the subject of that play — but at least it was morally safe. The same could scarcely be said of the subject of *Phèdre*. Racine, like Corneille, retains the essential element of the story, the incestuous love of the heroine for her stepson, where Gilbert (*Hippolyte*, 1645) and Bidar (*Hippolyte*, 1675) had — as no doubt Chapelain and D'Aubignac preferred — not only to make Hippolyte sensitive to love (in Gilbert, he returns Phèdre's passion for him; in Bidar, he loves Cyane, the equivalent of Racine's Aricie), but to unmarry Thésée and Phèdre and so to avoid the incest theme. It is small wonder that Racine devoted the last paragraph of his preface to a defence of the morality of his play. The other aspect of *la vraisemblance*, connected with *la bienséance externe*, did however compel him to show Hippolyte in love. If this was done partly in order to render him sympathetic to the audience (so that he might arouse pity), it was also to make him 'un peu coupable' (so that his fate might not arouse indignation) by treating his love as a 'weakness' and as a cause of guilt in the eyes of Thésée, for the invented Aricie is the sister and daughter of his sworn enemies. On the central and moral issue, however, Racine did not yield any more than did Corneille, and the moralizing views of the preface look very much like special pleading.

Here, it is the question of the *vraisemblance* of the subject that I wish to develop. What Chapelain's criticism of *Le Cid*, and D'Aubignac's of *Théodore*, amounts to is that certain kinds of truth are not appropriate for drama because they offend against *la bienséance*. But it is also clear from what both critics say that such subjects and indeed many others are inadmissible because they fail to satisfy the requirements of *la vraisemblance*, which, says D'Aubignac, is 'la première et la fondamentale de toutes les règles' (*Pratique*, p. 232). If we spell the word as they did — 'la vray-semblance' — we see that they were looking for something which, to their contemporaries, had the semblance of truth judged by the sum of their ordinary experience. Since that experience was of a world governed by *la bienséance*, that concept and *la vraisemblance* were interconnected. As I have suggested elsewhere,[3] however, these views were not universally accepted: those who gave Corneille their support at the time of *Le Cid* did so chiefly because they did not accept these criteria absolutely. It was possibly from their arguments that he derived encouragement to defend his own practice with vigour. His first fully developed statement came in the preface to *Héraclius* (1647: *Writings*, p. 190):

...l'action étant vraie..., il ne faut plus s'informer si elle est vraisemblable, étant certain que toutes les vérités sont recevables dans la poésie, quoiqu'elle ne soit pas obligée à les suivre. La liberté qu'elle a de s'en écarter n'est pas une nécessité, et la vraisemblance n'est qu'une condition nécessaire à la disposition, et non pas au choix du sujet, ni des incidents qui sont appuyés de l'histoire. Tout ce qui entre dans le poème doit être croyable; et il l'est, selon Aristotle, par l'un de ces trois moyens, la vérité, la vraisemblance, ou l'opinion commune. J'irai plus outre; et quoique peut-être on voudra prendre cette proposition pour un paradoxe, je ne craindrai pas d'avancer que le sujet d'une belle tragédie doit n'être pas vraisemblable. La preuve en est aisée par le même Aristote, qui ne veut pas qu'on en compose une d'un ennemi qui tue son ennemi, parce que, bien que cela soit fort vraisemblable, il n'excite dans l'âme des spectateurs ni pitié ni crainte, qui sont les deux passions de la tragédie; mais il nous la renvoie choisir dans les événements extraordinaires qui se passent entre personnes proches, comme d'un père qui tue son fils, une femme son mari, un frère sa sœur; ce qui, n'étant jamais vraisemblable, doit avoir l'autorité de l'histoire ou de l'opinion commune pour être cru: si bien qu'il n'est pas permis d'inventer un sujet de cette nature. C'est la raison qu'il donne de ce que les anciens traitaient presque mêmes sujets, d'autant qu'ils rencontraient peu de familles où fussent arrivés de pareils désordres, qui font les belles et puissantes oppositions du devoir et de la passion.

This crucially important passage, which opens with a flat contradiction of the statement just quoted from Chapelain's *Préface à l'Adonis*, contains a summary of Corneille's central ideas about the tragic subject. Although it revolves around an apparent differentiation of truth from verisimilitude, it is not there that its main interest lies. The references to the twenty-fifth and fourteenth chapters of the *Poetics* are perfectly orthodox. In the fourteenth, Aristotle discusses the various types of tragic situation in relation to the

emotions of pity and fear which their representation is calculated
to arouse in the audience. In the twenty-fifth, he implicitly makes
room for the extraordinary and the improbable, provided that they
serve a specific dramatic purpose: 'there is no possible apology for
improbability of plot or depravity of character, when they are not
necessary and no use is made of them'.

Corneille's assertion that 'le sujet d'une belle tragédie doit n'être
pas vraisemblable' appears to be perfectly logical when related to
the kinds of subject Aristotle suggests as suitable for tragedy. *La
vraisemblance* is reduced in importance to being necessary only for
the 'disposition' of the play: it is, that is to say, a logical and aesthetic requirement for the ordering and arrangement of the episodes. This is made even clearer when Corneille reiterates his view,
boldly and at the very beginning of his first *Discours*, thirteen years
later (*Writings*, pp. 1–2):

> ... on en est venu jusqu'á établir une maxime très fausse, qu'*il faut que le
> sujet d'une tragédie soit vraisemblable*, appliquant aussi aux conditions du
> sujet la moitié de ce qu'il [Aristotle] a dit de la manière de le traiter.

And it is of course true that Aristotle's references to the 'probable
and necessary' concern, not the subject, but the construction of the
plot. This does not, therefore, mean that Corneille's many justifications of his plays on the grounds that, however unlikely they may
seem, they are historically attested and so acceptable, are to be
taken as referring to anything more than the basic situation (but
that, as we have seen, may be modified) and the actual catastrophe. Racine, in similar circumstances, makes use of similar
arguments. In the first preface to *Andromaque*, for example, he justifies his portrayal of Pyrrhus ('Pyrrhus n'avait pas lu nos romans,
il était violent de son naturel, et tous les héros ne sont pas faits
pour être des Céladons') implicitly on the grounds that it may not
appear *vraisemblable* to Racine's polite spectators, yet it is historically attested and so acceptable. Exactly the same argument is repeated in regard to the cruelty of Néron in *Britannicus* (first preface).
Racine does, however, in these same prefaces admit that he has
made concessions to the sensibilities of his spectators. 'Il n'a pas
encore tué sa mère, sa femme, ses gouverneurs; mais il a en lui les
semences de tous ces crimes...' (second preface). Néron, the 'mon-
stre naissant', is cruel enough in Racine's portrayal, he says, 'pour
empêcher que personne ne le méconnaisse'. The argument, then, is
conducted in two directions at once — it claims fidelity to history
to counter the criticisms of the 'doctes', and points to *bienséance* to
mollify the ordinary, polite member of the audience. The statements, as such, need not, therefore, be taken too seriously or too

literally, but those which make pretensions to historical accuracy are based on the idea that Pyrrhus and Néron are not *vraisemblables* by the standards of seventeenth-century French society. Racine takes a similar line with regard to the behaviour of his characters in *Bajazet* — see the preface — and there maintains that the very unfamiliarity of the social and moral context, as far removed from the audience, in its way, as the classical world of antiquity, makes the *invraisemblance* (judging by the criteria of ordinary experience) acceptable. As far as they go, these arguments differ little in significance from Corneille's. But then, in that same first preface to *Britannicus*, we find the famous diatribe against Corneille, with allusions to several of his plays. In its contrast between 'le naturel' and 'l'extraordinaire', and dramatic effects 'surprenants' and 'vraisemblables', it may look as though Racine comes down on the side of *la vraisemblance* as heavily as do Chapelain and D'Aubignac. In fact, however, these things refer, not to the subjects chosen by Corneille, but to their treatment — plots too complex, unities observed with obvious difficulty, contorted language, characters behaving in an extraordinary manner. Racine's remarks may indeed suggest important differences between his conception of his art and Corneille's, but those differences do not concern the *vraisemblance* of the subject any more than they do the conception of the subject or the treatment of history. Racine does, however, find it necessary, on the grounds both of internal and external *vraisemblance*, to avoid the traditionally miraculous dénouement in *Iphigénie*, 'qui pouvait bien trouver quelque créance au temps d'Euripide, mais qui serait trop absurde et trop incroyable parmi nous', and at the same time would not have appeared to be 'tiré du fond même de la pièce'.

One need scarcely try to demonstrate that Corneille's practice in choosing and disposing his subjects conforms with the terms of his own apology for it: his subjects are indeed extraordinary and of the kind suggested in the preface to *Héraclius* and of course by Aristotle — mortal enmity between members of the same family, or friends, or lovers; domestic strife occasioned by and affecting the destinies of nations. The important political dimension in Corneille's plays, inextricably bound up with personal desires and emotions, is so obvious as not to require comment beyond drawing attention to the fact that these subjects are, once thus defined, in themselves extraordinary and *invraisemblables*: they do not arise for ordinary people every day and are alien to their normal experience of life. Racine's subjects have, on the other hand, been defined as 'faits divers',[4] rather sordid family squabbles among ordinary people — who may, it is true, have the names of kings and queens — and giving rise to the 'crimes passionnels' beloved of the popular news-papers and their readers. But neglect of the status of these people changes the whole character of the plays. *Andromaque* is of course a tragedy in which the passions bring the characters into collision one with another and their passions are those of love, jealousy, and hate between individuals; but they are involved in a dynastic prob-lem and inseparable from it — the future of the Greek alliance and the possible resurgence of Troy. *Bérénice* does concern the separa-tion of a man and a woman who love one another, but that separa-tion is occasioned by political forces beyond their control and is required by the maintenance of Titus's imperial rule. The dynastic problem arises in *Britannicus, Bajazet, Mithridate, Phèdre,* and *Athalie.* The future of Troy is at stake in *Iphigénie.*[5] In this sense, Racine's plays do not differ from Corneille's — though their emphasis does differ — and they do not involve ordinary men and women but those who shoulder great responsibilities.

At the same time, if we return to Corneille's definition and ex-amples of tragic subjects, these certainly apply to Racine's plays. The situation and outcome of *Andromaque* — the chain of unrecipro-cated passion binding the four principal characters, the death of two of them, the insanity of a third and the almost miraculous survival of the fourth and of her infant son — are scarcely *vraisem-blables* by everyday standards of experience, in the seventeenth cen-tury or now. The renunciation of his love by a seemingly all-powerful ruler is surely a rare event, and at any time extraordi-nary, not to say sensational. A father rival of his own sons for the hand of a captive princess is *invraisemblable* in itself, without the military and political rivalries in which Mithridate is involved. A great fleet becalmed until the leader of the expedition it is sup-posed to carry to the land of its enemies allows his daughter to die in a blood sacrifice is not and never has been within the verisimili-tude of day-to-day life. A stepmother in love with a stepson who is dynastically her rival and then, out of jealousy, allowing him to be sent to his death by her husband, his father, is not a subject to which one would apply the epithet *vraisemblable.*

Corneille's distinction between the credible and *le vraisemblable* is crucial here. *Le vraisemblable* is only one of the three means of achieving credibility, and it clearly is not the means by which to render the subject credible. That is achieved by the fact that the subject is historical or at least generally accepted to be so. But it would scarcely be memorable had it been commonplace. It is of the nature of tragedy that its subjects be extraordinary in the stric-test sense, and Racine must have been as keenly aware of this as was Corneille, not only when taking up themes like those of *Béré-*

nice, Iphigénie, or *Phèdre,* but also when to all intents and purposes inventing the subject of *Andromaque* into which he moulded themes connected with the aftermath of the Trojan war: the relationships between his four characters have no greater verisimilitude — less perhaps — than the stories found in Homer, Euripides, or Virgil. *La vraisemblance,* then, applies not to the subject itself but to its treatment. That that treatment is concerned with the probable and necessary relationships between the episodes and the probable and necessary connexions between passion and action was clear to Racine from the time that he wrote his very first preface (that to *Alexandre,* 1666), and it is for the sake of achieving that kind of internal *vraisemblance* that rules such as the *liaison des scènes* had been evolved.

Professor Eva Schaper, in her brilliant commentary on the *Poetics,* points out that it has to do with the achievement of coherent and logical relations distinct from the arbitrariness of life:

By forming an internally structured whole, the work presents a cosmos. We can assent to the possibilities in such a cosmos only because, though autonomous, the cosmos is in a world to which it is mimetically related. Strictly speaking a work therefore constitutes a microcosm. It formulates its statements in its own way — that of necessity or probability of connections — but the terms which are shown as related are recognisable because we can also find them incoherently and purely factually related elsewhere, namely in life as lived . . .[6]

Once the nature of the microcosm is established in the spectator's mind, he will judge the verisimilitude of its characters and events, not by relating them directly to his experience of actual people and events, but by imagining how the dramatic characters would behave if confronted with similar circumstances in real life. This is what lies behind Corneille's explanation, in the second *Discours* (*Writings,* pp. 57–8), of Aristotle's discussion (*Poetics,* chapter XXIV) of a likely impossibility being preferable to a possible improbability:

. . . il y a des choses impossibles en elles-mêmes qui paraissent aisément possibles, et par conséquent croyables, quand on les envisage d'une autre manière . . . Elles paraissent manifestement possibles quand elles sont dans la vraisemblance générale, pourvu qu'on les regarde détachées de l'histoire . . .

The appeal to 'la vraisemblance générale' relates to Aristotle's 'universal statement' (chapter IX), that is, 'one as to which such or such a kind of man will probably or necessarily do'. Characters in effect invented, like Nicomède or Oreste, are *vraisemblables* or otherwise in respect of their behaviour being or not being in conformity with the kind of person the dramatist has created in them. In the fifteenth chapter, Aristotle adds that 'the necessary or the probable' ought also to be the characteristic of the sequence of the incidents. This is one of the passages translated by Racine (*Principes*, p. 16). The nature of the dramatic microcosm is inseparable from the kinds of character who inhabit it. No matter, then, how improbable, when judged by the criteria of ordinary experience, the behaviour of Nicomède or of Oreste may appear, it bears all the marks of *vraisemblance* when one considers that the subject of the play demands the existence of a man of extraordinary courage or of extraordinary passion, and that his action will therefore necessarily spring from that courage or that passion. It will remain extraordinary when judged by everyday experience, but not when situated within the world of the play as imagined by the dramatist. To this extent Racine does not differ from Corneille.

The world of the play may include supernatural elements. In *Iphigénie*, for instance, it hovers between the realm of the gods and that of men: the very situation is divinely created, Calchas is the unseen but ever-present oracle of the gods, the characters refer to their divine ancestry and are aware that it is only by supernatural means that the fleet can be enabled to sail to Troy. Similar comments may be made about *Phèdre*, particularly in regard to Neptune's answer to Thésée's prayer. Racine indicates in his preface how his portrayal of Thésée is a combination of 'historical' and 'fabulous' elements:

C'est dans cet historien [Plutarch] que j'ai trouvé ce qui avait donné occasion de croire que Thésée fût descendu dans les enfers pour enlever Proserpine était un voyage que ce prince avait fait en Épire, vers la source de l'Achéron, chez un roi dont Pirithoüs voulait enlever la femme, et qui arrêta Thésée prisonnier, après avoir fait mourir Pirithoüs. Ainsi j'ai tâché de conserver la vraisemblance de l'histoire, sans rien perdre des ornements de la fable, qui fournit extrêmement à la poésie; et le bruit de la mort de Thésée, fondé sur un voyage fabuleux, donne lieu à Phèdre de faire une déclaration d'amour qui devient une des principales causes de son malheur, et qu'elle n'aurait jamais osé faire tant qu'elle aurait cru que son mari était vivant.

Thésée's fabulous career is established in the very first scene, in the conversation between Hippolyte and Théramène who has been searching for the missing king. The multiplicity of allusions and proper names puts the audience from the outset in a world of which the gods form part and in which the human characters move. The imaginative creation of that world persists when in the following scene Phèdre addresses Venus and the Sun, her own ancestor. This world is sustained and enriched as the play develops. Racine's reference to the mythological dimension as a source of poetry relates of course to the 'poetic' aspects of the play in the narrow sense, but also surely to the whole imaginative concept, which includes 'history', that is, the purely human elements, without losing its *vraisemblance*. As the passage quoted suggests, Thésée's fabulous disappearance is the foundation on which the whole human drama depends. That, once it begins, is entirely self-consistent, whether one considers it in terms of the *vraisemblance* of the behaviour of the characters and their passions (treating the supernatural features as poetically symbolic), or in those of the autonomous microcosm which rests on an interplay of natural and supernatural. Racine's respect for the scepticism of his audience accounts perhaps for the form of Ulysse's final *récit* in *Iphigénie*, where belief in the supernatural nature of the dénouement is placed, not in the speaker's mind, but in that of the watching soldiers: this device makes for greater *vraisemblance* by external and modern criteria. In *Phèdre*, Racine dares to go further, being more confident perhaps in his creation of the poetic world of his tragedy, because the death of Hippolyte is recounted directly by an eyewitness: even there, however, the intervention of the god is introduced indirectly (ll. 1539–40). The safeguarding of *la vraisemblance* on the purely human level should not, however, blind us to Racine's creation of a poetic world into which we are drawn from the beginning and which is, in some ways, far removed from the prosaic world of every day but which is *vraisemblable* in terms of its own inner coherence. Corneille, when he moved along this road, wrote his machine-plays, *Andromède* and *La Toison d'or*, whose conventions and staging almost preclude *la vraisemblance* in their deliberate search for the marvellous, that quality which, from the psychological and moral point of view, was never absent from his tragedies, and which calls forth its own emotional response.

It is of course obvious that in creating the microcosm of the play and all that it contains the dramatist aims at arousing certain emotions.[7] Corneille develops his discussion of *la vraisemblance* at the beginning of the first *Discours* by relating it directly to the emotional effect: '...les grands sujets qui remuent fortement les passions, et en opposent l'impétuosité aux lois du devoir et aux tendresses du sang, doivent toujours aller au-delà du vraisemblable ...,' which is why they require the guarantee of history. This statement is clearly connected with what we have seen of the preface to *Héraclius* and with the need to have the audience prepared, by the very title of the play, to enter an extraordinary world and experience the emotions it is created to arouse. Now here Corneille is suggesting that it is only subjects which are 'au-delà du vraisemblable' which produce these emotions. This idea, manifestly based on his own practice as a playwright, is often contrasted with that expressed by Racine in the preface to *Bérénice*, where he seeks to answer criticisms that his tragedy was too simple:

> ... il ne faut point çroire que cette règle [the principle of simplicity] ne soit fondée que sur la fantaisie de ceux qui l'ont faite: il n'y a que le vraisemblable qui touche dans la tragédie; et quelle vraisemblance y a-t-il qu'il arrive en un jour une multitude de choses qui pourraient à peine arriver en plusieurs semaines?

Racine is continuing the argument — and the polemics — of the preface to *Alexandre le Grand* and the first preface to *Britannicus*. It does not, however, concern the *vraisemblance* of the subject as such, but, as we have already noted, that of its treatment. That the action of *Bérénice* is particularly simple does not alter the fact that it concerns an extraordinary subject. No one could deny that improbabilities in the conduct of the plot or in the behaviour of the characters would inevitably hinder the emotional effect. This is the argument used in both prefaces for the portrayal of the youthful Britannicus, which, says Racine, is 'très capable d'exciter la compassion'. In the second preface to *Andromaque* and in the preface to *Iphigénie*, he refers to moving his audience to tears. And the pleasure of experiencing emotion is already implicit in the first preface of all, that to *Alexandre le Grand*. The subject of *Bérénice* appealed to the dramatist particularly for 'la violence des passions qu'elle ... pouvait exciter', as 'les grands sujets' appealed to Corneille because they 'remuent fortement les passions'. These views are very far removed from those of Chapelain, for instance, who, in *Les sentiments de l'Académie sur le 'Cid'*, argues that *la vraisemblance* is the equivalent of credibility and that without credibility a play cannot achieve a moral purpose. D'Aubignac also (*Pratique*, pp. 7–10) requires drama to fulfil a moral function, which is presumably part of the pleasure (p. 38) the poet seeks to give. Corneille and Racine see *la vraisemblance* as a source of aesthetic pleasure and as a condition for the arousal of emotion: their views on this matter are exactly parallel to their treatment of historical sources. Corneille certainly puts pleasure first, as the preface to *La Suite du Menteur* (1645) shows (*Writings*, p. 183): he bases himself explicitly on Aristotle and Horace. Fifteen years later, in the first *Discours*, he makes some concession to the moralizers (*Writings*, pp. 3–4), but the whole tenor of his argument at that time suggests that he still sees his art as giving delight. From the first preface to *Alexandre le Grand* onwards, Racine repeatedly adopts the same line, either explicitly or by implication, the only real exception being part of the preface to *Phèdre*, a play whose subject perhaps called for some moral justification.

Does all this mean that, in the light of their views on la *vraisemblance* and of its place in their dramatic practice, Corneille and Racine are indistinguishable? I have suggested in passing that verisimilitude is connected for Racine with a kind of simplicity of plot of which Corneille says virtually nothing. This question, though I shall have occasion briefly to return to it here, is the subject of a later chapter. I have also suggested that, although Racine's plays possess a political dimension, the emphasis placed on it is different from Corneille's. This emphasis is closely connected with the kind of emotions to be aroused and the kind of characters involved in the dramatic action.

The political dimension forms part of Corneille's definition of tragedy as elaborated in the first *Discours* (*Writings*, pp. 8–10). He begins by distinguishing it from comedy:

Lorsqu'on met sur la scène un simple intrigue d'amour entre des rois, et qu'ils ne courent aucun péril, ni de leur vie, ni de leur État, je ne crois pas que, bien que les personnes soient illustres, l'action le soit assez pour s'élever jusqu'à la tragédie. Sa dignité demande quelque grand intérêt d'État, ou quelque passion plus noble et plus mâle que l'amour, telles que sont l'ambition ou la vengeance, et veut donner à craindre des malheurs plus grands que la perte d'une maîtresse...

...S'il ne s'y rencontre point de péril de vie, de pertes d'États, ou de bannissement, je ne pense pas que [le poème] ait droit de prendre un nom plus élevé que celui de comédie...

...La comédie diffère donc en cela de la tragédie, que celle-ci veut pour son sujet une action illustre, extraordinaire, sérieuse: celle-là s'arrête à une action commune et enjouée; celle-ci demande de grands périls pour ses héros: celle-là se contente de l'inquiétude et des déplaisirs de ceux à qui elle donne le premier rang parmi ses acteurs.

Although the love interest is never absent from Corneille's tragedies, it is always subordinated to 'quelque grand intérêt d'État'. It may, says Corneille, 'servir de fondement à ces intérêts et à ces autres passions dont je parle', but must always be content 'du second rang'. And of course it gives rise to the 'combats intérieurs' which are the real interest of the plays: without Horace's love for Sabine and hers for him, without Camille's and Curiace's passion one for another, the combat in which the men are involved would scarcely be tragic or even particularly moving. Cinna's love for Émilie is the spur to his conspiracy against the emperor, but it is put at the service of her desire for vengeance: the political plot is dominant even if its motives lie in the love plot. The battle and the conspiracy are of the nature of 'une action illustre, extraordinaire, sérieuse' such as should characterize tragedy. In the same way, Cléopâtre may be in love with César, but she makes use of his love for her in order to satisfy her ambition, and is quite explicit about this. Nicomède and Laodice may be genuinely in love, but the alliance which that love represents is put at the service of resistance to Roman imperialism. In the later tragedies, Corneille's characters always find themselves having to sacrifice genuine love to political necessity — ambition, vengeance, patriotism — even sometimes to the point of entering or seeing their lovers entering into marriages which are, from a sentimental viewpoint, entirely undesirable. The alternative is 'péril de vie, pertes d'États, bannissement'.

In the preface to *Bérénice* we find the nearest statement by Racine to those of Corneille. When he says that 'ce n'est point une nécessité qu'il y ait du sang et des morts dans une tragédie', Corneille would doubtless agree — *Cinna* and *Nicomède* are there to prove the point. As Racine continues, Corneille would continue — in part — to agree with him: 'il suffit que l'action en soit grande' (Corneille's 'action illustre'), 'que les acteurs en soient héroïques' ('des rois', 'les personnes illustres'), 'que les passions y soient excitées' ('sujets qui remuent fortement les passions'). But when it comes to 'que tout s'y ressente de cette tristesse majestueuse qui fait tout le plaisir de la tragédie', although Corneille accepts the need to arouse pity and fear and writes of the 'commiseration' excited by characters torn between contrary imperatives such as passion and duty, his definition of tragedy leads one to suppose that the 'admiration' which is added to these emotions in the *Examen* of *Nicomède* is a permanent feature of his art as a tragedian. This addition to the Aristotelian emotions (though by no means contrary to them) is to be found in Vossius (in 1647) and in Mambrun (in 1652), and has at least as much to do with astonishment, awe, and wonder as with ethical admiration.[8] One may pity Nicomède and fear for him in his state of powerlessness (in every sense but moral), but he excites wonder (and admiration) for his courage and steadfastness. One may pity Horace in his being, as he sees it, obliged to fight his kith and kin, both on the battlefield and within the circle of his family life, and one may fear for him in his great enterprise, but one wonders at (and admires) his bravery and his willingness to sacrifice all to what he sees as his duty. No equivalent of this wonder and admiration exists in Racine's tragedies, chiefly because their subjects are not conceived in the same way as are Corneille's. Presumably, Corneille conceived of his subjects as he did, and gave to the political dimension the major importance, because his object was to arouse this emotion as well as pity and fear. It may be true that Racine's plays involve an action which is 'grande', important above the purely personal and domestic level, and in which the personal and domestic impinge on great public issues (the dynastic problems, for example), but ambition (Roxane) and vengeance (Hermione) are inspired, not by motives of politics (Cléopâtre) or honour (Émilie), but by passionate love, thwarted and jealous. If Agamemnon were dominated by ambition, patriotism, and 'gloire', he would either find some way of circumventing the oracle's demand or go forward with the sacrifice. If Hippolyte is dispossessed of his father's kingdom, it is because he rebuffs Phèdre's advances, and if he dies, it is because of her jealousy of Aricie. What we see here, then, is that although Racine's subjects as such are as *invraisemblables* as those of Corneille, they are different in kind, because Corneille creates them in such a way — and thereby further removes them from *la vraisemblance* — as to arouse 'admiration' at least as much as pity and fear.

This in turn affects characterization. Although Racine's characters know themselves to be involved in what Corneille in the second *Discours* (*Writings*, p. 61) calls 'les affaires publiques' which are usually 'mêlées avec les intérêts particuliers des personnes illustres', they do not see their problems primarily as being concerned with public issues. While it contains a degree of exaggeration, Racine's statement in the first preface to *Britannicus* does reveal his particular emphasis: 'Il ne s'agit point dans ma tragédie des affaires du dehors: Néron est ici dans sa famille et dans son particulier.' This may be contrasted with Corneille's assertion in the *Examen* of *Nicomède*: 'Mon principal but a été de peindre la politique des Romains au dehors...'. For Pyrrhus, the Greek alliance may be in jeopardy, but he has as little regard for that as for the need to honour his pledge to Hermione: his problem, as he sees it, is to overcome Andromaque's resistance to his advances and quietly to be rid of the presence of Hermione. For Agamemnon, the real problem is not that of releasing the Greek fleet from Aulis but that of saving a daughter whom he loves (and thus not incurring the wrath of Clytemnestre and of Achille) without losing face before his allies, Ulysse in particular. Such characters may refer to 'la gloire' as often as do Corneille's, but for Pyrrhus it consists in victory over Andromaque (and haughtiness towards Hermione and Oreste), and for Agamemnon in avoiding humiliation before family and allies. It is little wonder that Racine insists so strongly in his prefaces on his having created characters 'ni tout à fait bons, ni tout à fait méchants': characters who by their nature are *vraisemblables* in the sense that they are of 'middling virtue', 'a little better than the average', 'men like ourselves' as Aristotle puts it, who arouse our sympathy a condition of which is their very *vraisemblance*. Yet, by their status if not by their nature they are involved — though their passions are such as to allow them sometimes to forget this, whereas those of Corneille's characters make them acutely aware of it and cause them to exploit it — in what he calls, in the same passage from the second *Discours*, 'des batailles' (*Mithridate*), 'des prises de villes' (*Bajazet*), 'de grands périls' (*Britannicus*), 'des révolutions d'États' (*Bérénice*). It is precisely these things which, in Corneille's plays, embody what he calls 'le péril d'un héros qui constitue [la tragédie]': 'lorsqu'il en est sorti, l'action est terminée' (first *Discours: Writings*, p. 10). They are the tests through which the hero must pass in order to prove himself. Racine's characters generally evade them: they are concerned to possess the object of their passion, while Corneille's in their passionate ardour desire to possess and to realize their projected and idealized selves by passing through the ordeal, whether they survive or perish.

The two playwrights' comments on Aristotle's 'goodness' of character (*Poetics*, chapter XV) are of some importance here. They are connected with what we have seen in the way of idealization. Both Corneille and Racine seem to be justified in their interpretations, but their emphases differ in such a way as to throw light on their individual concepts of characterization which in their turn may account for the *vraisemblance* of the action of their plays. After stating that 'there should be nothing improbable among the actual incidents', Aristotle returns to the question of goodness of character with which he had opened this chapter. It is in the context of verisimilitude that one reads what follows, here in Racine's version (*Principes*, p. 31):

La tragédie étant une imitation des mœurs et des personnes les plus excellentes, il faut que nous fassions comme les bons peintres qui, en gardant la ressemblance dans leurs portraits, peignent en beau ceux qu'ils font ressembler. Ainsi le poète, en représentant un homme colère ou un homme patient, ou de quelque autre caractère que ce puisse être, doit non seulement les représenter tels qu'ils étaient, mais il les doit représenter dans un tel degré d'excellence qu'ils puissent servir de modèle, ou de colère, ou de douceur, ou d'autre chose. C'est ainsi qu'Agathon et Homère ont su représenter Achille.

These lines call for some comment. Racine adds the words 'des mœurs' to 'des personnes', thus confirming what we have observed about the importance of this word and the concept it denotes in the creation of tragedy in the seventeenth century: it forms the basis for characterization, referring to abstract and general qualities and not simply to individuals. Then the phrase 'les plus excellentes' represents a degree of idealization absent from Aristotle's text, which simply has 'better than ordinary men'. This interpretation evidently seemed so right and natural to Racine that he repeated it in 'dans un tel degré d'excellence', which follows also from 'peignent en beau' and leads to 'qu'ils puissent servir de modèle, ou de colère, ou de douceur' and — another small addition — 'ou d'autre chose'. The idealization, as understood by Racine, is calculated to produce a representation of exemplary humanity, exemplary, that is, in its characteristics, whatever these may be, good or bad.

Corneille translates the same passage thus, in his first *Discours* (*Writings*, p. 15):

La poésie . . . est une imitation de gens meilleurs qu'ils n'ont été, et comme les peintres font souvent des portraits flattés, qui sont plus beaux que l'original, et conservent toutefois la ressemblance, ainsi les poètes, représentant des hommes colères ou fainéants, doivent tirer une haute idée de ces qualités qu'ils leur attribuent, en sorte qu'il s'y trouve un bel exemplaire d'équité ou de dureté; et c'est ainsi qu'Homère a fait Achille bon.

Then Corneille comments:

Ce dernier mot est à remarquer, pour faire voir qu'Homère a donné aux emportements de la colère d'Achille cette bonté nécessaire aux mœurs, que je fais consister en cette élévation de leur caractère, et dont Robortel parle ainsi: *Unumquodque genus per se supremos quosdam habet decoris gradus, et absolutissimam recipit formam, non tamen degenerans a sua natura, et effigie pristina.*

Whereas Aristotle has 'better than ordinary men' and Racine 'les plus excellentes', Corneille has 'des gens meilleurs qu'ils n'ont été': the idealization he sees is one which is compatible with the original character, but specifically an embellishment of it and not of ordinary men. The type of character portrayed in history is already the exceptional man worthy of remembrance: for Corneille his exceptional nature is to be enhanced further by the tragic poet, so that he becomes akin to Racine's conception. The 'haute idée' which results is that of an exemplary humanity again ('un bel exemplaire'), and, as Corneille's gloss (particularly 'cette élévation de leur caractère') reveals, exemplary of its kind, whether morally good or bad. In this connexion, the phrase 'absolutissimam recipit formam' from Robortello's treatise of 1548, which Racine also would know, if only from this quotation, may have been the starting-point, with its superlative, for the interpretation given by both French poets.

Now Corneille's 'cette bonté nécessaire aux mœurs, que je fais consister en cette élévation de leur caractère', refers back to a passage (p. 14) in the same paragraph of the first *Discours*, in which he is grappling with the idea of 'goodness', particularly in the light of the character of such 'personnages méchants, ou vicieux' as Medea, Ixion, and Achilles. Goodness, he suggests, is 'le caractère brillant et élevé d'une habitude vertueuse ou criminelle, selon qu'elle est propre et convenable à la personne qu'on introduit'. It is on these grounds that he then proceeds to justify his portrayal of Cléopâtre in *Rodogune* and of Dorante in *Le Menteur*. Corneille's use of the word 'habitude' corresponds to Racine's addition of 'mœurs', but 'brillant et élevé', when taken together with the comment on Cléopâtre ('tous ses crimes sont accompagnés d'une grandeur d'âme qui a quelque chose de si haut, qu'en même temps qu'on déteste ses actions, on admire la source dont elles partent') introduces a concept not to be encountered in Racine, that 'grandeur d'âme' which, when seen in action, gives rise to the wonder, the admiration we have already noted. (The only place in which one finds Racine using the expression is in the preface to *Phèdre*, where he attributes the quality to Hippolyte in his attempt to save Phèdre's honour by not accusing her.) The formula 'une habitude vertueuse ou criminelle' suggests in its context, and with the example of Cléopâtre, that the exemplary and superlative characteristics take an extreme form, so that, at the opposite pole from Cléopâtre, we find Polyeucte and Nicomède. It is to that 'grandeur d'âme' that D'Aubignac takes exception in *Sertorius* and *Sophonisbe*, and to which Corneille appeals in order to justify the conduct of the action and the behaviour of the characters in these plays. That quality is clearly an essential aspect of what he calls 'le beau sujet', and in the *Examen* of *Le Cid* (*Writings*, p. 102) it is explicitly set against 'cette médiocre bonté' of the tragedy of antiquity to which Aristotle refers in the *Poetics*. One of the characteristics of 'la grandeur d'âme', if one is to judge from the plays, is of course resolution and will-power, and the plots are so constructed as to allow the characters to demonstrate it in action by facing ordeals, tests, challenges. Closely connected with this is Corneille's idea, which features so prominently at the beginning of his first *Discours* (*Writings*, pp. 8–10) in particular, of the peril which besets the hero and which is the equivalent of the unity of action.

As Corneille's analysis of *Pompée* shows, in this same passage, his conception of the plot and of its function includes those 'batailles', 'prises de villes', 'grands périls', 'révolutions d'États', referred to in the second *Discours* as means of testing and challenging the hero's 'grandeur d'âme'. It is significant that in the same passage (*Writings*, p. 60) Corneille should defend the multiplication of the tests and challenges in terms which are manifestly those of pride in his achievement rather than of repentance:

Il est de beaux sujets où on ne peut éviter [la violence à l'ordre commun des choses], et un auteur scrupuleux se priverait d'une belle occasion de gloire, et le public de beaucoup de satisfaction s'il n'osait s'enhardir à les mettre sur le théâtre, de peur de se voir forcé à les faire aller plus vite que la vraisemblance ne le permet.

If the unities of time and place necessitate some 'violence à l'ordre commun des choses' and some *invraisemblance*, that may be unfortunate and, as the preface to *Sertorius* (*Writings*, p. 163) puts it, due to 'l'incommodité de la règle', but it ought not to involve the dramatist in abandoning his subject. The 'beau sujet', conceived as productive of wonder, may occasion certain absurdities, as Corneille freely adapting a passage from the *Poetics* (chapter XXIV) in the *Examen* of *Le Cid* (*Writings*, p. 104) admits, but 'il est du devoir du poète... de les couvrir de tant de brillants qu'elles puissent éblouir'. Exactly the same view was already expressed in the preface to Corneille's collected edition of 1648 (*Writings*, p. 194), where he wrote thus of the unity of time: 'Je crois que nous devons toujours faire notre possible en sa faveur, jusqu'à forcer un peu les événements que nous traitons, pour les y accommoder; mais si je n'en pouvais venir à bout, je les négligerais même sans scrupule, et ne voudrais pas perdre un beau sujet pour ne l'y pouvoir réduire.' For Corneille, the 'beau sujet', for all the technical problems it incurs, some of them virtually insoluble, is inviolable, because it arouses, in the ordeals through which the exceptional hero is made to pass, the emotions expressed in the terms 'admiration' and 'éblouissement', emotions which, if powerful enough, will not allow the spectator to perceive technical defects. In such a context, compression of events for the sake of observing the unities may appear as a logical absurdity but will not, in the imaginative and emotional experience of the theatre, strike the audience as *invraisemblable*.

That *invraisemblance* is precisely, however, the stick with which Racine chooses to beat Corneille in the first preface to *Britannicus* and the preface to *Bérénice*. As we read these attacks, however, we have to bear in mind the fact that they form part of their author's justification, in the face of adverse criticism, of his own very different practice in adopting the greatest possible simplicity, and that he singles out — rather unjustly — some of Corneille's more recent and less successful plays (*Sertorius*, 1662; *Agésilas* and *Attila*, both 1666). In order to satisfy his critics, says Racine, 'il ne faudrait que s'écarter du naturel pour se jeter dans l'extraordinaire', and abandon a simple action 'soutenue par les intérêts, les sentiments et les passions des personnages' in favour of 'quantité d'incidents qui ne se pouvaient passer qu'en un mois' (first preface to *Britannicus*), a formula repeated in the preface to *Bérénice*, in which he asserts that 'tout ce grand nombre d'incidents a toujours été le refuge des poètes qui ne sentaient pas dans leur génie ni assez d'abondance ni assez de force pour attacher durant cinq actes leurs spectateurs par une action simple...'. The relatively poor reception given to Corneille's more recent plays can of course be attributed to their failure to evoke the emotional response their author hoped for, and so to the inefficacity of the 'beau sujet' as far as they were concerned. But Racine's tone in the first preface to *Britannicus* shows that he was still smarting from the effects of the *cabale* mounted against the first performances of the play.

All this is not, however, merely a question of polemics. Corneille's 'beaux sujets' may well preclude all thought of *vraisemblance* in the mind of the spectator provided that they really do succeed in arousing the emotions he specifically aims to produce. That does not necessarily mean that the conduct of the plot, with its multiple episodes and ordeals, appears to lack *vraisemblance* when seen as the testing ground for the resolute hero endowed with 'la grandeur d'âme'. When Racine says that 'il n'y a que le vraisemblable qui touche dans la tragédie', in the same context as his remarks on 'cette tristesse majestueuse qui fait tout le plaisir de la tragédie' and on the fact that *Bérénice* 'a été honorée de tant de larmes', it is clear that the emotional effect he wished to produce was quite different from Corneille's: pity and fear, but pity above all. 'La principale règle est de plaire et de toucher', he writes, and 'toucher' does here imply pity. The fact that the statement is couched in these terms and is subsequently developed indicates that for Racine, as for Corneille, tragedy must above all arouse specific emotions. It must not, however, arouse others, as Aristotle observed in chapter XIII of the *Poetics*, which is why Racine tells us in the preface that he could not countenance 'le meurtre horrible d'une personne aussi vertueuse et aussi aimable qu'il fallait représenter Iphigénie'. This is also of course a case of *la bienséance externe*, in which the moral sense of the audience would have been outraged by the sacrifice.

We have seen that 'goodness' of character means, for Racine, excellence of its own kind and an exemplary quality. The plays, however, reveal that that quality does not, in his case, call forth wonder or admiration. His heroes are not paragons of vice or virtue, but men precisely of 'une bonté médiocre'. In his translation of chapter XIII of the *Poetics*, he defines the tragic character (*Principes*, pp. 19–20):

Il faut... que ce soit un homme qui soit entre les deux, c'est-à-dire qui ne soit ni extrêmement juste et vertueux, et qui ne mérite point aussi son malheur par un excès de méchanceté et d'injustice. Mais il faut que ce soit un homme qui, par sa faute, devienne malheureux, et tombe d'une grande félicité et d'un rang très considérable dans une grande misère.

Only such a character can arouse pity and fear, 'car on n'a pitié que d'un malheureux qui ne mérite point son malheur, et on ne craint que pour ses semblables'. The 'excellence' of the tragic character is not that of superlative 'grandeur d'âme', but of the passions which are common to all men. Heroic ambition is not one of these, but love, hate, and jealousy are. Racine's characters are subject to these to an exceptional and exemplary degree — we may be thankful for ourselves that this is so! — to a degree which is catastrophic in its effects.

It is of the nature of such passions that they do not require great events to challenge or frustrate them and to reveal their capacity for self-contradiction: all that is needed is a particular situation in which close personal relationships produce emotional conflicts between and within the characters. Instead of a succession of ordeals and tests occasioned by great events, Racine's characters must undergo a crisis (often 'political' in origin) which occasions an intensification of passion. They are not challenged by some power greater than themselves to exercise their will in órder to achieve self-fulfilment, but are frustrated by the nature of their relationships with others in their efforts to achieve control or possession of them. Because the passions, although constant in their force, are unstable in their nature — love turning to hate or jealousy, for instance — the situation in which the characters find themselves is self-renewing from within. Given the nature of the passions, this is *vraisemblable* and, given a self-renewing situation in which the relationships are the sole constant, the succession and multiplicity of incidents to which Racine took exception is unnecessary. The actual situation, however, is in itself no more *vraisemblable* than it is in Corneille, and the force of the passions — Pyrrhus's love for Andromaque, Oreste's for Hermione, her love–hate relationship with Pyrrhus; Néron's lust for power and for Junie; Bérénice's passion for Titus, Roxane's for Bajazet, Mithridate's for Monime, Phèdre's for Hippolyte; Thésée's anger; Joad's religious zeal — is always carried to a pitch of 'excellence' which, although in itself it remains a universal characteristic, still raises it and its effects above the level of the *vraisemblable*. But Racine's much vaunted simplicity, which allows him to observe the unities with *vraisemblance* in a tragedy with a single crisis, is made possible by the nature of the emotions he wishes to arouse for characters driven by certain kinds of passion. The greater complexity of Cor-

neille's plays, made up, not of a single crisis, but of a succession of
ordeals, is made necessary by the passions which characterize his
tragedies and is *vraisemblable* with their context. Each dramatist
adopts his own particular practice for the sake of arousing particular emotions in the audience: pity, fear, 'admiration'. But other
emotions are occasioned by drama — suspense, surprise — and to
them I must now turn.

CHAPTER IV
Plot: Drama in tragedy

Nous ne cherchons jamais les choses, mais la recherche des choses. Ainsi dans les comédies les scènes contentes, sans crainte, ne valent rien, ni les extrêmes misères sans espérance, ni les amours brutaux, ni les sévérités âpres.

Pascal, Pensées (Lafuma, 773)

Protasis est, in qua proponitur et narratur summa rei sine declaratione exitus: ita enim argutior est, animum semper auditoris suspensum habens ad expectationem. Si enim praedicitur exitus, frigiduscula fit...Epistasis, in qua turbae aut excitantur aut intendentur...catastrophe, conversio negotii exagitati in tranquillitatem non expectatam.

Scaliger, Poetices Libri Septem (1561)

...The poet must be more the poet of his stories or plots than of his verses, inasmuch as he is a poet by virtue of the imitative element in his work, and it is actions that he imitates.

The most important [element] is the combination of the incidents of the story. Tragedy is essentially an imitation not of persons but of action and life, of happiness and misery. All human happiness or misery takes the form of action; the end for which we live is a certain kind of activity, not a quality.

...One may string together a series of characteristic speeches of the utmost finish as regards diction and thought, and yet fail to produce the true tragic effect; but one will have much better success with a tragedy which, however inferior in these respects, has a plot, a combination of incidents in it... Beginners succeed earlier with the diction and characters than with the construction of a story...

In spite of Aristotle's insistence, in the sixth and ninth chapters of the *Poetics*, on the construction of plot being the dramatist's prime function, it is rarely studied seriously: from a reading of most criticism — of Corneille or Racine (or Shakespeare) — one would imagine that what really matters is poetry (in the narrow sense) or psychology or ideas. So there is abundant discussion of verse techniques or imagery, of the characters' Stoical self-assurance or hopeless despair, of Cornelian heroism or Racinian fatalism. Such things have their importance, and are often treated with erudition and sensitivity, but they are surely secondary to what is the most fundamental reality of drama — that it is something to be experienced theatrically, something to be enacted, whether on the actual stage or on that of the reader's imagination. It is then first necessary to analyse how certain effects are produced in the theatre and what responses they evoke (and how they evoke them) in the audience.[1]

Aristotle clearly believed this to be true: from his own real experience of plays he endeavours, in the *Poetics*, to understand how they work and what are the principles of their creation. For him, the dramatist was first and foremost the maker of his plots, and plot consists in the ordering of the material — of the 'traditional story' in tragedy — and in the creation of incidents so as to produce a particular dramatic and emotional effect, the creation of 'incidents arousing pity and fear'. The emphasis usually placed on pity is revealing, too, of critical prejudices. Both pity and fear arise, in one sense, out of the spectator's fellow-feeling with the character 'of middling virtue' like himself, with whom it is easy to sympathize, but one may lose sight of the fact that terror may be occasioned not only for the character on the stage but for the spectator in his seat by the sense of their both being suddenly overwhelmed by some unexpected or wishfully ignored, but terrifying, event. Such an event is often brought about by one character's plotting against another or by the dramatist's plotting, as it were, against his characters, laying traps for them, involving them in terrible incidents over which they have little or no control. Sometimes the spectator is given a privileged view or hint of that plotting as it occurs or before it strikes its victim; sometimes it is hidden from him and produces as much surprise (usually unpleasant) for him as for the victim.

From Saint-Évremond and La Bruyère onwards, discussion of the tragedies of Corneille and Racine, and comparisons between the dramatists, often concerned the extent to which surprises are sprung on the spectator as well as on the dramatic character. Modern theories have tended to assume that such surprise and the concomitant suspense and dramatic interest are incompatible with the tragic. Occasionally, some analysis may be made of these qualities in Corneille but it is quickly abandoned in favour of a study of the ideology they supposedly illustrate; their very existence is usually denied in Racine, or at least in what is deemed his best work. Already, before the end of the seventeenth century, Longepierre was criticizing 'les petites surprises ... qui ne servent dans la tragédie qu'à amortir et à éteindre le pathétique qui en est l'âme'.[2] For him, as for La Mesnardière almost sixty years earlier, pathos was paramount, though the sixth chapter of his treatise shows that this did not exclude 'l'impatient désir de voir à quoi aboutira la faute qu'il [the spectator] a vu commettre', nor the surprise which is 'l'un des plus beaux effets que produise le théâtre'. La Mesnardière's *Poétique* was published in 1639, before the flourishing of the new form of tragedy based on the dramatic concepts of tragicomedy and heralded by *Le Cid* and *Horace*. He was no doubt thinking of a continuation of the line represented by Rotrou's *Hercule mourant* and Mairet's *Sophonisbe*.[3] Longepierre, on the other hand, clearly refers contemptuously, as Boileau, writing in the same period, does not,[4] to what so much tragedy had developed into in the hands of Quinault, Thomas Corneille, and other authors of 'tragédies galantes' — romanticized travesties of tragic or potentially tragic themes in which 'les petites surprises', catching the audience off its guard and thereby creating a purely dramatic excitement, was perhaps the most sought after and noteworthy feature. Did Longepierre, as the would-be disciple of Racine, supposedly the master of pathos and therefore rarely if ever producing surprise, include Corneille by implication among playwrights like his younger brother? Certainly, in his view of things Racine stood in fairly sharp distinction from the Corneille who admitted to creating surprising incidents and was presumably in good critical company with the D'Aubignac who, although ironically enough in some respects the heir to the theories of La Mesnardière, was critical of the Greeks because their prologues so often eliminated 'tous les agréments d'une pièce, qui consistent presque toujours en la surprise et la nouveauté' (*Pratique*, p. 163). And Saint-Évremond, in his essay *De la tragédie ancienne et moderne* (1672) thought that Corneille had substituted for pity and fear the emotions of surprise and curiosity. Then Sainte-Beuve, who was so influential over such a long period, was to affirm: 'Ce qui distingue Racine ... c'est la suite logique, la liaison ininterrompue des idées et des sentiments; ... jamais il n'y a lieu d'être surpris de ces changements brusques, de ces *volte-face* dont Corneille a fait souvent abus dans le jeu de ses caractères et dans la marche de ses drames.'[5] It is in such statements, every part of which is open to question, that lies the germ of the contention of eminent present-day critics that whereas Corneille deliberately sought to produce dramatic effects, Racine equally deliberately avoided springing surprises on the audience, because these would detract from experience of the properly tragic emotions occasioned by his poetry or from concentration on his study of the passions.[6] While Corneille followed in the tragicomic tradition from which the new tragedy had sprung, Racine converted that vehicle to properly tragic ends.

Although persuasively argued, these views have not gone unchallenged. They beg important questions. Need one simply say with Lanson[7] that Racine merely took over Corneille's tragic mechanism and refined it? Or that he rejected it because it was primarily, perhaps solely, fitted for the arousal of suspense and surprise? Did Corneille use it only for that purpose? If he did, what then sets him apart from his contemporaries whose use of the mechanism trivialized tragic subjects?

We have seen that both Corneille and Racine often depart radically from the situations and stories which they find in their acknowledged sources, the sources in reality of their catastrophes. The reasons, as we know, are based partly on the demands of *vraisemblance* and *bienséance*, but more fully on what Corneille called 'necessity', the exigencies of dramatic form, and on the needs of the 'subject' as they conceived it. Working in a contrary direction, they made use of many devices, drawn from an intimate knowledge and sometimes a secondary interpretation of history, in order to invest their dramatic situations and dramatized stories with an air of historical truth. What this amounts to is that this procedure creates new situations and plots while preserving some essential links with the source material and still bringing about the traditional catastrophe. That catastrophe is usually suggested if not clearly indicated by the title of the play, and to that extent the spectator going into the theatre knew what to expect, as D'Aubignac clearly saw (*Pratique*, p. 138): 'La catastrophe est connue... par le titre même qui enferme le dernier événement.' The situation with which the spectator was presented at the outset, however, in terms of the exposition, was new and unfamiliar. Professor May has argued that Corneille went progressively further out of his way to seek out stories which were unknown to most of his audience, in order to create the maximum of opportunity for surprise; Racine, on the other hand, chose the most familiar of themes in order to eliminate surprise. Corneille played upon the ignorance of his audience for the sake of achieving the greatest dramatic effect; Racine relied upon its knowledge to produce the opposite result: all because one was, as he admitted, interested in producing dramatic effect, while the other never mentioned it except by implication, and there quite abundantly, in his private annotations on the Greek tragic poets. And is it mere coincidence that he twice summarized that part of the sixth chapter of the *Poetics* where Aristotle stresses the importance of plot rather than character (Picard vol. ii, 931, i.e. comments on Heinsius's edition, pp. 249–50; and *Principes*, pp. 14–15)? The thesis that Racine had no regard for plot construction is undermined, in the first place, because one man's public admission of a fact does not mean that another's silence is tantamount to denial. Since Racine was not only the rival but quite obviously the disciple of Corneille, his silence is understandable: he would never admit a real debt to a contemporary, and in his practice — the actual construction of his plays — he is clearly indebted to him. We neglect at our peril the effect of the polemical situation in the history of seventeenth-century drama: the silences of a playwright are as eloquent as his statements if read with that in mind. Moreover, for Corneille the researches of Professor Stegmann among others show that what might have been regarded as unfamiliar territory in all probability was not; and for Racine Professor Knight and others have demonstrated that his infidelities in his treatment of history were such as to make his subjects to all intents and purposes new. This does not, however, involve us in a complete reversal of Professor May's view, but rather in saying that the outline of most if not all of the stories dramatized by both Corneille and Racine probably formed part of a common stock of knowledge in their day, but that their treatment of them was such as to renew them through the creation of unfamiliar situations and, necessarily, unfamiliar developments.

Such novelty of itself immediately puts the spectator in a state of

ignorance, uncertainty and doubt: it is a source of curiosity and
suspense, if only, to begin with, on an intellectual level. This suspense may take a variety of forms and can be expressed in four
types of question which may come to the spectator's mind, depending on the form the play takes:

1. What will happen next? I do not know.
2. I know what will happen next, but will it?
3. I can see what is going to happen and can see it inevitably
   happening, but how will the characters react when it does,
   and surprises them?
4. I know what is going to happen, but my sympathy with the
   victim causes me to close my eyes to it as he does. Will it not
   be terrible when it happens?

I do not propose to ask and answer these questions in this form,
but they are put in order to illustrate the complexity of this question of suspense. So I begin by asking: is it true, as Professor May
suggests, that Corneille deliberately keeps his audience in the dark
for as long as possible in order to exploit suspense and curiosity to
the full, and that Racine, on the contrary, by careful and, above
all, clear indications early in the play, suggests or even reveals the
impending course of events? (If he did, why did he go to so much
trouble in order to create new situations? Their existence cannot be
entirely explained by considerations of *vraisemblance*.) For Corneille,
the answer, derived both from his dramatic practice and from his
commentaries on it, is quite clearly in the affirmative. For Racine,
the evidence is, at first sight, less clear.

We are concerned here with a problem which exercised the
logical and rational minds of the seventeenth century a good deal —
that of dramatic preparation — and which took a paradoxical
form. On the one hand, dramatic excitement in the form of surprises was demanded; on the other, a logical cause-and-effect
dramatic structure. While such terms as 'coup de théâtre' are used
in comment and analysis, few critics ask the question as to who
experiences them as surprises: is it the audience or is it the characters or is it both? Even some of Aristotle's formulations are not
clear: 'Such incidents have the very greatest effect on the mind
[whose?] when they occur unexpectedly [to whom?] and at the
same time in consequence of one another...'. The discussion in the
ninth chapter of the *Poetics* ends with the example of the statue of
Mitys, whose purport is not helpful to us in this context although
in another, as we shall see, it is. In the sixteenth chapter Aristotle
writes: 'The best of all discoveries... is that arising from the incidents themselves, when the great surprise [to whom?] comes about
through a probable incident...'. This is obviously related to the

whole problem of the probable and logical connexions between the incidents, which are regarded as fundamental to the structure of tragedy, and yet, in the twenty-fifth chapter, we read: 'There is a probability of things happening also against probability.' The text of the *Poetics* itself, then, suggests the paradox and, in the ninth and seventeenth chapters, links it to the invention and ordering of the episodes. In terms of invention they may be entirely new and, therefore, conducive to suspense, curiosity, and surprise; in terms of ordering, they will appear in a logical and probable sequence, which may seem to preclude those feelings.

Professor May has subjected Corneille's tragedies to a detailed analysis from the point of view of the procedures adopted in order to maintain the interest of the audience by the introduction of some new element of uncertainty and surprise at crucial points in development, notably at the beginnings and ends of the acts. Other critics[8] have made use of similar methods for tragic drama more generally, and have noted the numerous and varied devices resorted to by playwrights, especially in the middle years of the seventeenth century, for the sake of producing these highly dramatic effects: dilemmas which admit of no rational choice between equally daunting alternatives, *quiproquos* and misunderstandings arising out of ambiguities or mistaken or hidden identities, unexpected threats to life and limb, concealment of the truth from characters and spectators, often related to the creation of doubt and uncertainty, and so forth. The danger inherent in this type of analysis is that it may simply note each of these features in a variety of plays and neglect to relate it to its context, that is, to the play as a whole. D'Aubignac's advice (*Pratique*, pp. 228–9) to the dramatist is equally applicable to the critic: he must 'envisager son sujet d'un trait d'œil et l'avoir présent tout entier à l'imagination, car celui qui connaît un tout en sait bien les parties, mais celui qui ne le connaît qu'à mesure qu'il le divise, se met en état de le diviser très mal et fort inégalement.' It is noteworthy that, when he discusses his own work, Corneille avoids this pitfall, as he evidently did in the creation of the plays themselves. In the matter of preparation of the incidents, he has in mind not simply the exposition or the first act in itself, but its effect on subsequent developments. So, he says (first *Discours: Writings*, p. 21), 'Une exposition doit contenir les semences de tout ce qui doit arriver ... en sorte qu'il n'entre aucun acteur dans les actes suivants qui ne soit connu par ce premier, ou du moins appelé par quelqu'un qui y aura été introduit.' It is not, however, simply a matter of avoiding the fortuitous intervention of a previously unmentioned character, like the part, condemned by Corneille, of the Corinthian shepherd in *Oedipus*

Rex: the first act should also contain 'le fondement de toutes les actions'. But Corneille immediately suggests that such preparation does not exclude the possibility of surprise: he cites one of his own plays, *Cinna*, to prove his point. The perfect exposition will ensure 'l'entière intelligence du sujet', and this is particularly important in plays, like the majority of those of the seventeenth century, in which two actions become interrelated. In *Cinna*, the dramatic effect of surprise occasioned by Auguste's summons to Cinna and Maxime to attend a discussion would be lost if the conspiracy had not already been revealed to the audience quite independently of the emperor's uncertainty as to whether to abdicate. The essential situation is revealed by Émilie, an invented character, who through the moral pressure she exerts upon Cinna causes the conspiracy and the possible abdication, once they are brought together by a dramatically contrived coincidence, to be inextricable one from the other. The information emerges in the soliloquy of the first scene, and it is presented neither as straightforward narrative, which would be boring, nor as lamentation, which would, as in sixteenth century tragedy, simply engender pathos, but as the source of an inner conflict occasioned by Émilie's desire for vengeance and her fears for the safety of Cinna if he yields to her persuasions to procure it for her. Émilie, like Sabine in *Horace*, also a new and unknown character, thus presents herself at the outset in terms of her involvement in the situation which she naturally begins to sketch in (naturally, because she is torn by her very involvement), but she also presents herself in such a way as to attract the interest of the audience in her and in the situation, and its sympathy. All these factors are important, as is Cinna's assurance that all is prepared for the assassination and his incidental revelation that Maxime is party to it (I.iii), in preparing for subsequent developments and, psychologically and emotionally, for the 'coup de théâtre' of Évandre's words (l. 280):

> Seigneur, César vous mande, et Maxime avec vous.

That is a surprise no less for Cinna and Émilie than for the audience. But it leads to another, which lowers the tension, when Auguste betrays no knowledge of the conspiracy (II.i), but heightens it again when conflict arises between Maxime and Cinna (II.ii). That conflict in its turn is partly responsible for the 'coup de théâtre' of Maxime's revelation of the conspiracy to Auguste (II.i), the tension then being increased by the emperor's whispered order to Polyclète (ll. 1099–100)[9] and uncertainty about his decision as to whether to punish or pardon (IV. ii and iii) and as to the import of Fulvie's revelations to Émilie (IV.iv), particularly in regard to the entirely plausible report of the disappearance of Maxime. That, too, leads immediately to another surprise, in his unheralded return (IV.v) and his proposal to substitute himself, pardoned, for Cinna who, he believes, will die a traitor's death, in her affections. Uncertainty and suspense are intensified at the end of each of the first four acts,[10] and are not finally resolved until the very last scene, when Auguste's forgiveness comes as if divinely inspired — though it has been prepared by his self-doubting (IV.ii) and ambiguously expressed trust (l. 1258) — and suddenly brings about the 'conversion' of the conspirators, all of whom have, however, at some point expressed their perhaps grudging respect and admiration for him.

The succession of surprises and the mounting tension arise unexpectedly each in its context, but all are in some way prepared, and all have their origins in the doubts and conflicts revealed in the first few scenes of the play, and are related one to another. After the completion of the exposition, no new factor is introduced. Yet it is not possible to foresee developments with certainty. The initial doubts and conflicts, which lie at the emotional heart of the play, are the source of all the subsequent suspense, tension, and surprise. Corneille may occasionally, as in the emperor's whispered order to Polyclète, make use of somewhat crude melodramatic devices in order to heighten tension and arouse curiosity, but their use is not gratuitous, as it so often is in the work of lesser dramatists, because it plays on the fear of the audience for characters presented in a sympathetic light. The establishment of that sympathy from the outset — and from this point of view the form of the first soliloquy is crucial for our response to Émilie, this hitherto entirely unknown character — is an important part of the exposition, for without it the full emotional effect of the surprises would not be felt, and the curiosity of the audience, resulting from the uncertainties and consequent surprises, would remain purely intellectual. That Corneille, in his earlier tragedies at least, arrived at something else, is quite clear from what he writes of the two visits paid by Rodrigue to Chimène in *Le Cid* (*Examen: Writings*, p. 104). He defends these episodes against allegations that they offended against *la bienséance*, because, he says, 'il s'élevait un certain frémissement dans l'assemblée, qui marquait une curiosité merveilleuse, et un redoublement d'attention pour ce qu'ils avaient à se dire dans un état si pitoyable.'

Two observations are called for at this point. The first is that although, in *Cinna*, Corneille does make use of ambiguity, suspense, and surprise in such a way as to arouse sympathetic feeling as well as intellectual curiosity, in other plays, particularly in the later politico-matrimonial tragedies, this may not be so. In the preface to *Othon*, for example, he can write of the play only in these

terms (*Writings*, p. 169): '...je puis dire qu'on n'a point encore vu
de pièce où il se propose tant de mariages pour n'en conclure
aucun. Ce sont intrigues de cabinet qui se détruisent les unes les
autres.' In the same way he had, both in the original preface and
in the *Examen* (*Writings*, pp. 140, 189), been half-apologetic for the
complexity of the plot and for the hasty dénouement, which is justified, however, by the audience's purely intellectual satisfaction at
discovering the identity of the real Héraclius.

The second observation is that the characters who are caught
unawares by the dramatic surprises are not directly responsible for
them and are not able (any more than is the spectator) actually to
foresee them. Auguste's summons to Cinna and Maxime is not
foreseeable at that precise point, and its motive is unknown,
though characters and audience may guess at it — wrongly. The
other surprises in the play are no more predictable than this one
and no more under the control of their victims. Yet they do arise
out of earlier episodes and are all consistent with the data of the
exposition: in that sense they are prepared.

So are the ones we find in *Horace*, although in that tragedy the
'coups de théâtre' are occasioned, not by the activities of the characters themselves (in that respect *Cinna* represents an important
advance on Corneille's first truly tragic masterpiece), but by events
enacted and decisions taken without reference to them. Auguste, it
may be argued, has, by his past conduct, brought the conspiracy
upon himself, but Horace and Curiace have not occasioned the war
between their cities. That initial situation is a clue to the nature of
its development: the battle will, after all, not take place (I.iii), but,
second surprise, the Horace brothers are chosen (they do not
choose themselves) to represent Rome (II.i). This is followed at
once by the announcement of the parallel choice of the Curiace
brothers to champion Alba (II.ii). Again, a truce is called, not
because the champions refuse to fight, but because an outside
force, the army of either side, rebels against so inhuman a choice —
another surprise (III.ii). But (III.v) consultation of the gods results in the combat being after all engaged. This is followed by the
false report of the defeat of the Horace brothers (III.vi) which
shatters their father. The suspenseful second scene of Act IV is
contrived in such a way, thanks to the angry emotions of their
father, as to make Valère's announcement of the truth of Horace's
victory redouble the old man's relief and delight but at the same
time to dismay his daughter, particularly by its unexpected sud-
denness. The 'coup de théâtre' of Horace's rounding on his sister
and killing her in Act IV scene v, which Corneille in his *Examen* —
a tactical concession to his critics? — considers to be ill-prepared,

is of a different nature from the others: in spite of its being technically unprepared it does arise from within the two characters' own passions as we know them to be: it is not occasioned by some external force. The earlier 'coups de théâtre' are none the less prepared in that each one of them springs in some way from the whim of destiny in the context of war, a context which is firmly established from the outset.

Corneille had already achieved similar effects in *Le Cid*, where the succession of doubts and uncertainties is generated from within the initial situation, even though, as in *Horace*, the characters are not in a moral sense responsible for the outcome of each and every crisis (the duel, the battle). So uncertainty over the duel leads, when its outcome is known (II.vii), to uncertainty as to the king's decision whether to punish Rodrigue (II.viii). Rodrigue's fear of arrest causes him to conceal himself and then suddenly to irrupt (III.i, iv): this dramatic effect is enhanced by his questioning of Chimène's feelings for him (III.iv) and of her intentions. That questioning takes place also in the mind of the audience, which is left in doubt by the ambiguous nature of the lovers' parting. Then (IV.v) Don Fernand, after hesitation and suspense, allows the duel to take place: again, what will be the outcome, and what will its effect be on Chimène? When Don Sanche comes in bearing the sword (V.v) the audience, no less than Chimène, believes that Rodrigue is dead, only to be disabused in the last scene. Each of these surprising turns of events arises out of what is already in the situation in some form or out of an earlier development, but surprises most of them remain.

The technique used by Corneille in *Horace* and *Le Cid*, that of 'coups de théâtre' springing from within the general dramatic situation but not always from decisions taken by the characters, is one which is found in perhaps most of his tragedies, and in particular those which, like *Horace, Pompée, Sophonisbe*, and *Sertorius*, concern war. Exactly the same thing can of course be said about *La Thébaïde* (in many ways closely modelled on *Horace*) and *Alexandre le Grand*, and also, to a less marked extent, about *Bajazet* and *Mithridate*. Careful preparation for the final battle in *Le Cid* is made as early as Act II scene vi, which also, when followed by Act III scene vi, connects it with the king's pardon. The battle, and the actual result of the duel, are external events in themselves. Whether or not the characters are in some way directly responsible for the surprises, these frequently bring about *péripéties*, changes of direction in the development of the plot, which I shall consider at a later stage. It is, however, important to observe meanwhile that both psychologically and externally occasioned 'coups de théâtre' may be due to chance or coincidence: the conjunction of the ripening conspiracy and of Auguste's consideration of abdication, the superior military prowess of Horace, his chance meeting with Camille as he returns to the city elated with his victory, Rodrigue's strength and skill as a duellist ... That Corneille was himself aware of the dramatic emotion to be derived from the playwright's contriving the chance event or coincidence is clear from his remarks on *Cinna* in the first *Discours* (*Writings*, pp. 21–2):

La conspiration de Cinna et la consultation d'Auguste avec lui et Maxime n'ont aucune liaison entre elles, et ne font que concurrer d'abord, bien que le résultat de l'une produise de beaux effets pour l'autre, et soit cause que Maxime en fait découvrir le secret à cet empereur. Il a été besoin d'en donner l'idée dès le premier acte, où Auguste mande Cinna et Maxime. On n'en sait pas la cause; mais enfin il les mande, et cela suffit pour faire une surprise très agréable, de le voir délibérer s'il quittera l'empire ou non, avec deux hommes qui ont conspiré contre lui. Cette surprise aurait perdu la moitié de ses grâces s'il ne les eût point mandés dès le premier acte, ou si on n'y eût point connu Maxime pour un des chefs de ce grand dessein.

In this passage, where Corneille so meticulously analyses his own remarkable and fully conscious craftsmanship, the relationship between preparation and surprise is evident: the one does not preclude the other. For D'Aubignac, at any rate, preparation is necessary in order to avoid offending the audience's sense of *vraisemblance* (*Pratique*, p. 135):

... bien que le spectateur veuille être surpris, il veut néanmoins l'être avec vraisemblance; et bien que l'événement ne soit pas moins vraisemblable en soi, encore qu'il n'en fût rien dit, que si l'on en avait parlé, le spectateur ̈veut néanmoins qu'on en ait auparavant jeté les fondements ...

This of course, which in part echoes Aristotle's remark about the probability of the improbable happening, raises the question as to whether the spectator feels the need for *vraisemblance* during the excitement of the performance or only afterwards, when it is completed. D'Aubignac, however, does not discuss the point, but he does, in the same chapter of *La Pratique du théâtre*,[11] discourse at some length on the problem of reconciling the needs of preparation with those of surprise: he distinguishes between 'les incidents préparés' and 'les incidents prévus'. Forestalling possible objections to his doctrine of preparation, he writes:

Car, dira-t-on, s'il faut que les incidents soient préparés longtemps auparavant qu'ils arrivent, sans doute ils seront prévenus, et pourtant ils ne seront plus surprenants, en quoi consiste toute leur grâce [Corneille, in the passage on *Cinna* just quoted, seems to echo this expression], et ainsi le spectateur n'en aura aucun plaisir et le poète aucune gloire. A cela je réponds qu'il y a bien de la différence entre prévenir un incident et le préparer; car l'incident est prévenu lorsqu'il est prévu, mais il ne doit pas être prévu encore qu'il soit préparé . . . La préparation d'un incident n'est pas de dire ou de faire des choses qui le puissent découvrir, mais bien qui puissent raisonnablement y donner lieu sans pourtant le découvrir . . . Ces discours et ces autres petites considérations qu'on emploie pour préparer un incident le renferment si secrètement et le cachent si bien qu'on n'en puisse rien prévoir.

The need for preparation is justified by D'Aubignac's reference to Aristotle (*Poetics*, chapter X), supported by Scaliger, and by Donatus who had explicitly referred to the importance of holding the audience in a state of expectation. Later in the seventeenth century, Rapin was to be equally explicit:

Il est important de remarquer que préparer un incident, ce n'est pas tout à fait dire les choses qui le puissent découvrir: mais c'est dire seulement ce qui peut donner lieu à l'auditeur de le deviner . . . Le plaisir des spectateurs est d'attendre toujours quelque chose de surprenant . . . Et rien ne doit tant régner au théâtre que la suspension . . .[12]

Rapin, however, is clearly thinking here only in terms of the pleasure to be derived from intellectual curiosity and the excitement of anticipating the coming surprise. It is important to notice that Corneille understands the limitations of such a view. In his second *Discours* (*Writings*, p. 46), when discussing the possibility of inventing episodes for plays on historical subjects, he admits that they are acceptable provided that they are *vraisemblables* in their context. He then goes on: '. . . ceux qui aiment à les mettre sur la scène peuvent les inventer sans crainte de censure. Ils peuvent produire quelque agréable suspension dans l'esprit de l'auditeur, mais il ne faut pas qu'ils se promettent de lui tirer beaucoup de larmes.' Rapin appears to contradict himself on this matter because he also writes (chapter XXI, p. 107) that 'ce ne sont pas les intrigues admirables, les événements suprenants et merveilleux, les incidents extraordinaires qui font la beauté d'une tragédie'. But it seems as though in this passage he has specifically in mind the type of play, so popular in mid-century, whose sole interest was an exciting plot. On the one hand he follows D'Aubignac and Corneille, on the other La Mesnardière.

The connexion, established by Corneille, and perhaps implicit in this apparent contradiction between the invented episode and suspense, is important: one, at least, of the functions of invention is to create curiosity and suspense within a foreknown framework. As soon as a dramatist creates a new initial situation as a basis for an original version of an old story (as Corneille does in *Nicomède*), or introduces a new character into such a story (as he does in *Horace* or *Cinna* or as Racine does in *Bérénice* or *Phèdre*), or radically alters a well-known character (as Racine does in the case of Oreste and Hermione),[13] attention is assured and curiosity provoked. But if the playwright is to go beyond that point, he must also contrive his incidents so that they will produce pathos and present his characters in such a way as to arouse emotional interest and sympathy for them. The opening scenes of *Horace* and *Cinna*, of *Andromaque* and *Bérénice*, are not merely expository in a technical sense: they are masterpieces of the techniques of presenting new creations for whom emotional interest and sympathetic curiosity are at once aroused. However, as Corneille implies, and as these examples show, suspense and pathos are not mutually exclusive — quite the contrary in fact. As early as 1630, Chapelain[14] had insisted on the importance of suspense and surprise, and nowhere does he suggest that they exclude pathos. Neither does La Mesnardière nor the author of the *Discours de la tragédie* which prefaces Scudéry's *L'Amour tyrannique* (also of 1639), nor will La Bruyère, much later.[15]

Yet the desire to prove that Racine has a virtual monopoly of tragic sense, a desire which goes back through Voltaire to Longepierre and has become a kind of academic orthodoxy, has led to views which take the mutual exclusion of suspense and pathos almost for granted. So we find Professor May insisting that Racine 'ne se contente pas de préparer, il prévient délibérément son spectateur', in order precisely to eliminate suspense and surprise and to concentrate on expression of the passions. The roots of that idea, but without the exclusion, may be found in D'Aubignac's view of the purpose of tragedy, and it is contrasted with Corneille's definition of the function of the exposition (and Donatus's)[16] as being to allow the audience to understand the subsequent action and to lay its foundations, if only ambiguously, but not to foresee its precise development nor the way in which it will move towards an already familiar catastrophe. However, any discussion of the plots of Racine's plays and of the possibility of their being constructed so as to produce dramatic effects must surely be set against the background of the important changes he made to the source material for the ostensible subject of the plays. The first step in this direction is of course to consider the situation, as we have already done, but to relate it to the plot.

Let us recall some of the more outstanding changes made by Racine. In *Andromaque* the heroine is given power to resist Pyrrhus's amorous advances; Hermione and Pyrrhus are unmarried; the motive for the assassination is changed; the life of Astyanax is prolonged, and there is no Molossus. New motives and intrigues are created by the invention of Junie in *Britannicus*, even though this is perhaps of his secular plays the one in which Racine adhered most closely to history. Although in the first preface Racine attempts a justification from history for the introduction of Junie, it is far-fetched, and his reference in the same passage to the invented Sabine and Émilie of Corneille's plays is much more to the point. The Junie of the tragedy is moreover presented 'plus retenue qu'elle n'était', but Racine does not believe 'qu'il nous fût défendu de rectifier les mœurs d'un personnage': this argument would support the changes made to the characters in *Andromaque*. In the same way, Antiochus is historically justified, though in fact an original creation, and the morals of Bérénice and Titus 'rectified' and their relationship endowed with a new purity. The prolongation of the life of Monime makes possible the crucial love rivalries in *Mithridate*. *Iphigénie* provides us with one of the most important and controversial of Racine's inventions, that of Ériphile, which enables him to produce not only a new dénouement but a new love intrigue based on a traditional type of triangular jealousy plot. Equally novel and controversial is the creation of Aricie: the result is a love-lorn Hippolyte and, again, a jealousy plot whose effect on Phèdre's motives is profound. In the expositions of these plays, Racine carefully presents all these new and altered elements and lays the foundation for an understanding of what is to follow. At the beginning, however, the spectator, unfamiliar or familiar with the traditional story, cannot fail to ask what part the new or changed characters will play and how the dramatist will manipulate the story. Does Racine, then, seek in some way to dissipate the uncertainty and destroy the natural curiosity and suspense? Or does he sharpen our sympathy for his characters by exploiting these other feelings?

As with Corneille, the touchstone is perhaps the 'coups de théâtre'. In Act II scene iv of *Andromaque* Pyrrhus suddenly announces to Oreste, without prior warning, that he intends after all to marry Hermione and to deliver Astyanax to the Greeks. It is not until the following scene that Pyrrhus, in conversation with Phœnix, provides an explanation: 'la raison d'état' prevails over love. But Phœnix sees only too well that the victory is fragile. The announcement itself is certainly unprepared as far as the development of the plot is concerned up to that point, and it is equally certainly a 'coup de théâtre' for the unsuspecting Oreste. Is it one for the audience? It ought not to be — even for Oreste: Pylade has given him (and us) a warning about Pyrrhus's propensities:

> Ainsi n'attendez pas que l'on puisse aujourd'hui
> Vous répondre d'un cœur si peu maître de lui:

Il peut, Seigneur, il peut dans ce désordre extrême
Épouser ce qu'il hait, et perdre ce qu'il aime.   (ll. 119–22)
That was in the very first scene of the play. The issue is in doubt,
to say no more: the audience is put into a state of uncertainty. The
warning is almost immediately followed by the entry of Pyrrhus
(I.ii), who at once rejects Oreste's ambassadorial overtures, a decisive and resolute rejection followed in its turn (I.iv) by his threats
to Andromaque through which he hopes to persuade her to marry
him. Now the issue does seem clear: Pyrrhus appears to have made
up his mind. It is in this context and on that understanding that
Oreste's interview with Hermione (II.ii) takes place. So, in the
next scene but one, Pyrrhus's announcement:

D'une éternelle paix Hermione est la gage:

Je l'épouse,   (ll. 618–9)

falls as a bolt from the blue on Oreste — on the audience, too,
because Pyrrhus's earlier decision seemed so forceful as to cancel
Pylade's warning. Of course Oreste wants to believe what Pyrrhus
says to him in Act I scene ii, and so accepts it as final, but at the
same time the spectator also, confronted with a novel situation and
a hitherto unknown set of relationships between new, passion-
driven characters, has temporarily forgotten or set aside Pylade's
statement in the light of the subsequent development. Although the
incident of Act II scene iv is, in a technical sense, prepared by that
statement, it is not, by the time it occurs, 'prévu' or 'prévenu': it is,
in its dramatic effect, a real 'coup de théâtre'.

The situation is again reversed, as Phœnix foresaw (II.v), and
Pyrrhus returns to Andromaque. He is not, however, in any position to force a decision upon her, and so he offers horrifying
alternatives:

Je vous le dis: il faut ou périr ou régner...

Songez-y; je vous laisse, et je viendrai vous prendre

Pour vous mener au temple où ce fils doit m'attendre;

Et là vous me verrez, soumis ou furieux,

Vous couronner, madame, ou le perdre à vos yeux.

(ll. 968, 973–6)

The threat sharpens Andromaque's uncertainties and hesitations:
they are new to the traditional story, in which she had no alternative to submission to her captor. This being so, the spectator cannot foresee what her decision will be. The famous last scene of Act
III not only maintains the uncertainty over that decision, but over
the outcome of the whole play, since all depends on her choice. At
the same time, the sympathetic presentation of Andromaque is
such as to endow this scene with the deepest pathos, as Racine so
poetically has his heroine recall in her imagination the horrors of

the sack of Troy and the conjugal love that bound her to Hector. But the poetic evocation is not gratuitous: it is a mental action on the part of Andromaque, and we are never allowed to forget that it is set in a highly tense and dramatic context which calls for immediate action in the form of her decision. But the act ends, as do so many of Corneille's as well as Racine's, on a note of uncertainty and expectancy.

The opening of Act IV seems to bring resolution and relief, but (ll. 1072–3) the calm is shattered, for the audience no less than for Céphise, by Andromaque's real decision, her 'innocent stratagème' which still leaves the fate of Astyanax, at any rate, and the possible marriage of Pyrrhus to Hermione after all in the balance. So the uncertainties are prolonged through Act IV. It is only at the very end of scene iii (ll. 1249–52) that Oreste finally agrees to accept Hermione's commission to assassinate Pyrrhus. She, however, is no more predictable than is the king: almost immediately after Oreste's departure to fulfil his mission, she is prepared to withdraw her command (ll. 1273–4), the advent of Pyrrhus being enough to make her hesitate again. He, however, has now really come to his decision, and Andromaque, he believes, is ready to obey him: so he is in no mood for polite delay or compromise or even apology. The trap is sprung, and Hermione determined to punish an unfaithful and disdainful suitor. She leaves him with the ominous threat:

Va, cours; mais crains encor d'y trouver Hermione.   (l. 1386)

This time there is no turning back, though Hermione still deludes herself, at the end of her soliloquy (V.i) into believing that there is yet time to do so. Her powerlessness at that point is exactly paralleled later by that of Thésée, whose prayer cannot be recalled (*Phèdre*, IV.iii; V.iv–vi). Hermione's illusion, like Thésée's, is quickly broken, however, by the immediate arrival of Cléone, with her incomplete account of the marriage ceremony and of the assassination. (The incomplete and two-stage account is closely parallel — but swifter, because it comes nearer the end of the play — to the split narrative of the fight in *Horace* (III.vi and IV.ii), which creates similar tensions and emotional upheavals.) It remains, then, for Oreste to complete it and to suffer his second bolt from the blue in the form of Hermione's angry rejection of him. Act V scene iii provides another, and final, 'coup de théâtre' — certainly for Oreste and perhaps also for the audience. He has indeed understood her feelings for him as early as Act II scene ii:

Je vous entends. Tel est mon partage funeste:

Le cœur est pour Pyrrhus, et les vœux pour Oreste. (ll. 537–8)

He has also seen Hermione return willingly to Pyrrhus when the opportunity was offered (III.ii). Yet he has allowed himself (IV.iii)

to believe in her hatred and repudiation of Pyrrhus. Again the
effect of earlier warnings has been obliterated, and again the incident has been prepared, but it is not foreseen by Oreste even if the
audience may not have entirely forgotten the warnings.

What this analysis reveals is Racine's consummate stagecraft:
the audience is put in a state of uncertainty, first by the novelty of
the situation, then by the indecisiveness of the characters which, in
its manifestation in the development of the plot, keeps the spectator in a state of excitement and suspense as the 'coups de théâtre'
succeed one another. The effect of their sequence, with the ambiguities of the warnings and 'preparations', is to create dramatic
tension, but, because of the sympathetic treatment of the characters which involves the spectator emotionally in their fate, that tension is partly occasioned, thanks to the pathos generated, by pity
and fear. In some respects, all this is very close to what we have
seen in the plays of Corneille, but with one notable difference: all
the 'péripéties' and 'coups de théâtre' are brought about, not by
some external event beyond the control of the characters, but by decisions taken and reversed by the characters themselves. Corneille's characters have their doubts and hesitations, but before arriving at a decision (Rodrigue, Horace, Cinna). So do Racine's. But
Corneille's do not usually have second thoughts whereas Racine's
do (Pyrrhus, Hermione, Agamemnon, Thésée). It is this distinction which the nature of the plot and the 'coups de théâtre'
express. One should add here that it has been pointed out[17] that,
although, as is well known, *Andromaque* shows many situational and
episodic similarities to earlier seventeenth-century plays (Rotrou,
Hercule mourant, 1636; Desfontaines, *La Perside*, 1644; Corneille,
Pertharite, 1652; Thomas Corneille, *Commode*, 1658, and *Camma*,
1661), it is not characterized, as they are, by real passion being
masked by false passion, by secret rivalries or misunderstandings,
by unknown family relationships, by death coming from unexpected sources, by some element of reciprocated love. Motives are
clear to the audience even if feints are adopted, for the sake of
saving pride or keeping up appearances. They are, however, tragic-
ally self-defeating.

Racine's tragedies are not always fully characterized by the technique of internally derived 'coups de théâtre'. In both *Mithridate*
and *Phèdre* he has a presumably dead character return alive and
unexpectedly to the scene of the dramatic action. Similar effects of
surprise are produced by other chance events — Roxane's discovery of Bajazet's letter to Atalide, Agamemnon's messenger failing
to deliver his message to Clytemnestre and Iphigénie, the use of
the sword in *Phèdre*. Chance does play its part in Racine's

tragedies, as it does in Corneille's. Does that lower their tragic stature? Surely not: is Oedipus not a victim of chance in his encounter with his unknown father and in his marriage with his unrecognized mother? Chance, fortune, destiny, whatever we call it, plays an important part in human life: it is for the dramatist to present it in a tragic light. Oreste's arrival at the court of Pyrrhus at a particularly crucial moment is a chance coincidence, but Racine makes of it the starting-point of an inescapably tragic sequence of events, no less tragic for Oreste than for the other characters. The arrival of Iphigénie and her mother at Aulis is also the result of chance and again has fateful consequences.

Raymond Picard drew attention[18] to one or two other highly dramatic features of *Iphigénie*, including the dénouement. Is it occasioned by a 'coup de théâtre'? Professor May strenuously denies it.[19] He hopes to prove that by the end of Act IV the spectator can no longer be in doubt as to the identity of Ériphile and of the victim in the sacrifice. For this he relies on a number of key passages. First, the oracle (ll. 57–62); but this is by no means clear — it refers to the victim as being simply, 'une fille du sang d'Hélène'. This obscurity — or ambiguity — is specifically the subject of a comment from Clytemnestre in another of the key passages (ll. 1266 ff.). First, she throws doubt on the meaning of the oracle:

Un oracle dit-il tout ce qu'il semble dire?

Then she refers to Theseus's clandestine union with Helen, from which was born a princess. The key figure here is Calchas (l. 1283) who, therefore, presumably knows her identity. Even Ériphile, who claims to be in Aulis to consult Calchas — what irony, to discover that he who, she hoped, would release her from ignorance, is to be the agent of her sacrifice! — clearly states that she does not know who she is:

J'ignore qui je suis; et, pour comble d'horreur,
Un oracle effrayant m'attache à mon erreur,
Et, quand je veux chercher le sang qui m'a fait naître,
Me dit que sans périr je ne puis me connaître.   (ll. 427–30)

Certainly the catastrophe is prepared by this passage, but it contains only a supposition and does not, in Ériphile's mind any more than in anyone else's, indicate that Iphigénie will not be the victim in the sacrifice or that the occasion for the revelation of Ériphile's parentage will be that same forthcoming sacrifice. It is true, however, that implicit faith is placed in Calchas's reliability and divinatory powers (ll. 455–9). Even Ériphile's further statement about the oracle does not, on the other hand, couple the sacrifice with the problem of her own identity:

Tu verras que les dieux n'ont dicté cet oracle
Que pour croître à la fois sa [= Iphigénie's] gloire et mon
                tourment.   (ll. 1110–11)

All these passages certainly provide preparation — in D'Aubignac's sense — for the dénouement which, therefore, arises, as
Racine says in his preface, 'du fond même de la pièce'. But in none
of them can the dénouement be clearly seen: all are ambiguous, as
Clytemnestre says that the oracle itself is. It is of course true that
the very invention of Ériphile throws the audience as well as the
characters into a state of doubt and uncertainty: because of her
mysterious origins, she is herself an ambiguous figure, and remains
so until very near the end of the play. Until the last act, no one
really suspects that the catastrophe will follow any pattern other
than that of one of the several versions of the traditional story. For
once, Racine's dénouement is not one of those which might have
been expected from the title of the play: there is indeed a human
sacrifice, and the victim is 'cette autre Iphigénie', so there is also a
substitution. Both sacrifice and substitution are found in the
ancient versions of the story. However, the fact that Iphigénie,
daughter of Agamemnon and Clytemnestre, is neither slain nor
swept away to Tauris, does mean in fact that the dénouement is
not the one we might have expected. The part played by Ériphile
herself must be seen not only in relation to the dénouement but
also within the triangular love plot which is a complete invention
on Racine's part, not only as far as Ériphile is concerned, but also
in Achille's feelings for Iphigénie, though the latter are suggested
in Rotrou's play,[20] where Achille, vainly seeking to prevent
Agamemnon from sacrificing his own child, sees her for the first
time and falls in love with her. Racine has his Achille already in
love with her, and that is his motive for trying to save her.
Ériphile, then, is situated within that part of the play: it is a prominent feature of the first four acts, and indeed, in Act II, contains a
further highly dramatic component — the misunderstanding over
the object of Achille's love. Ériphile's involvement in this traditional triangular love-and-jealousy situation is such that the audience
does not focus its attention on the implications, *for the sacrifice*, of
the mystery of her identity. Far from allowing us to foresee the
elaborately prepared dénouement, Racine seems deliberately to
create ambiguity, even to distract our attention from the forthcoming catastrophe in so far as it concerns Ériphile, and thereby to
enhance the dramatic effect, the surprise, of the final disclosure.
On this evidence, Racine would not quarrel with Corneille's remarks on surprising the audience in his *Examen* of *Héraclius* (*Writings*, p. 138).

Two further points may be briefly made about *Iphigénie*. The first is that the doubt about the identity of Ériphile is not exactly a *pseudo*, as Mornet called them, of the kind most commonly found in seventeenth-century French drama. She is not mistaken for someone else, nor is the revelation the means of disentangling some sentimental imbroglio: it does not change Achille's feelings towards her or for Iphigénie. Racine is making use, then, of a variation on an accepted dramatic device, which is scarcely less common in the tragedy of mid-century than in comedy: mistaken identity becomes unknown identity, and while the recognition dramatically resolves the problem of the play it does not show that any of the characters have been courting the wrong person. Although, then, Ériphile is involved in a conventional triangular love-plot, it is not in that that her identity matters, but in what remains, in spite of Racine's creation of that plot, the new dramatization of the traditional story. None of this lessens the effect of suspense, uncertainty, and dramatic surprise which the introduction of Ériphile entails and was presumably calculated to engender, but it does raise its significance above the trivial use of similar devices in the hands of lesser dramatists.

The second point concerns the problem of ambiguity. The meaning of the oracle concerning the sacrifice is taken to be clear by all the characters except Clytemnestre, and it is in a vain attempt to avoid the realization of its prediction, as he understands it, that Agamemnon is engaged. Oracles and prophetic dreams are of course commonplace in tragic drama, and they are a means of dramatizing a futile struggle against the inevitable, which is one of the ingredients of tragedy. It carries with it the discovery that what is predicted is in very truth inevitable. Racine compounds that discovery by another — that of the actual meaning of the oracle which, as Clytemnestre suspects, admits of more than one interpretation. The discovery of the meaning, while it is prepared, comes nevertheless as a surprise, but .it is curious that Ulysse, in his account of Calchas's disclosure (ll. 1731 ff.) does not mention Agamemnon's reaction to it and that the king does not reappear in order to make it clear to the audience — curious, that is, if one forgets that the play does concern the sacrifice or safety of his daughter. Yet it is he who leads the struggle against the supposed prediction. This is another ambiguity.

Now the oracle obviously has the function of creating a sense of foreboding, but in *Iphigénie* it also provides the motive for Agamemnon's search for ways of avoiding its consequences. Racine endows it with a kind of dynamic power, as he does also Athalie's dream. It is the dream which drives her to the temple, to commit sacrilege by entering where she should not go, and which brings her ultimately to recognition of Joas. Her terror — so clear in her horrifyingly poetic account of the dream (ll. 487 ff.) — and her nascent recognition of the irresistible power of the God of the Jews drive her to take desperate measures which will result in the realization of the prophecy. Racine had noted the dream of Clytemnestra in the *Electra* of Sophocles. On line 417 he wrote (Picard, vol. ii, p. 848): 'Songe de Clytemnestre. — Ce songe de Clytemnestre vient bien au sujet, pour envoyer Chrysothémis au tombeau d'Agamemnon, où elle trouve les cheveux d'Oreste, qui y a été aussi...'. The dream, like the oracle, is given a dynamic and dramatic function, and it results eventually in the great surprise.

It is interesting to note that Corneille makes use of oracles and dreams by way of dramatic preparation and to provide some sense of foreboding and some surprise, but that he does not invest them with this dynamic power, except in the case of *Œdipe*, where interpretation of the oracle, particularly through suppression of the Sophoclean interview between the hero and Teiresias and the introduction of Thésée and Dircé, each in turn suspected of Œdipe's guilt, provides the major motive for the actions of the characters and is not clarified — for characters and spectators — until the end of the play. This play contains more than one interesting parallel with *Iphigénie*: apart from the mystery of the oracle, there is that also in one sense of Œdipe's identity, and Corneille has introduced the love-plot, with an invented Dircé. Like Agamemnon, Corneille's characters believe that they have understood the meaning of the oracles and dreams, and their interpretation is based on the events in which they find themselves involved. More characteristic of Corneille than *Œdipe*, as far as the function of the oracle is concerned, is *Horace*. Immediately after Camille's account of the oracle and her dream in Act I scene ii, which she takes to be prophetic of death and separation from Curiace, he appears before her totally unexpectedly, with news of the truce. She interprets the oracle in its most obvious and literal sense, and the dream, as Julie advises her (l. 223), 'au contraire sens'. Thus, as Corneille says in his *Examen* (*Writings*, pp. 112–3):

L'oracle qui est proposé au premier acte trouve son vrai sens à la conclusion du cinquième. Il semble clair d'abord, et porte l'imagination à un sens contraire; et je les aimerais mieux de cette sorte sur nos théâtres, que ceux qu'on fait entièrement obscurs, parce que la surprise de leur véritable effet est plus belle.

Although the ultimate result is, then, the same as it is in *Iphigénie*, the effect of the misinterpretation on the behaviour of the characters is quite different, because the oracle and dream do not drive Camille to action (and she is dead before they are correctly understood) as the oracle drives Agamemnon. For Camille, the most important consequence is that she is thrown into contrary emotions, despair and hope, a powerful source of pathos.

That is also, in part, how Pauline's dream in *Polyeucte* is used. In the same passage in the *Examen* of *Horace*, Corneille distinguishes between the form of the dream in that play where 'il ne fait qu'exprimer une ébauche tout à fait informe de ce qui doit arriver de funeste' (and indeed it is couched in the vaguest and most confused terms, though none the less terrifying for that), and the form of that in *Polyeucte*, 'où il marque toutes les particularités de l'événement'. Polyeucte's allusion to the dream in his first conversation with Néarque (I.i) is followed by Pauline's detailed account of it (I.iii) to Stratonice. It has, however, by then obviously begun to work on the feelings of both husband and wife and has created emotional tensions. Pauline's account contains two elements, the return of Sévère and the death of Polyeucte. In a manner exactly contrary to what we have seen in *Horace*, the real meaning of the dream is immediately shown to be the fatal one — in part — because in Act I scene iv Félix announces to his daughter the return of Sévère 'from the dead', covered with glory, as Albin's narrative makes clear, thus vindicating in detail parts of the dream:

> Il semblait triomphant, et tel que sur son char
> Victorieux dans Rome entre notre César. (ll. 227–8)

> De nouveau l'on combat, et nous sommes surpris.
> Ce malheur toutefois sert à croître sa gloire,
> Lui seul établit l'ordre et gagne la victoire... (ll. 308–10)

The second prophetic part of the dream is not fulfilled until the end, and then in a way which does not correspond literally to its own terms. But since the first part is clearly and immediately justified, the question hangs over the rest of the play, and over Pauline in particular: will the second part also come to pass? Because of Pauline's deep emotional involvement, and of her uncertainty over her own feelings, although the dream contributes to the creation of suspense as to the outcome, the suspense also generates pathos.

Pathos is aroused also around Agamemnon, too, in *Iphigénie*, in whom we watch a man struggling in vain against what he takes to be inevitable, but is so, ironically, in a different sense from the one he understands. His conduct in the face of the situation as he sees it is highly ambiguous and it gives rise to ambiguities whose real purport is misunderstood. At the same time, his deceitful behaviour — adopted with the best of intentions, that of saving his daughter's life — catches him out, but by pure chance, or the will of the ever-present gods. A kind of poetic justice arises from the deceiver being thus deceived. At one time or another he lies to everyone, only to be tripped up by forces beyond his control. He oscillates between acquiescence in and resistance to the demands of the oracle. When the father in him predominates over the king, he fears his allies, represented by Ulysse and Calchas, and takes action behind their backs. When the king prevails, he lies to Clytemnestre, to Iphigénie and Achille. This ambiguity in his behaviour springs from his inability to choose between the alternatives, both equally attractive and at the same time repugnant, which are those of his passions — 'la gloire' and paternal affection — and dominated by another passion — fear. At the beginning of the play, the father prevails: at the end of the first scene, Arcas is sent to head off Clytemnestre and Iphigénie from arriving at Aulis for the projected marriage with Achille, and Arcas is told to lie. But scarcely has one messenger departed when another arrives (I.iv) to announce that not only Clytemnestre and Iphigénie, but Ériphile as well, are about to reach Aulis, evidently not having received the first message. So, lying to wife and daughter, Agamemnon is forced, since Ulysse has heard the news, to let the king prevail. His subterfuge persists throughout Act II and until (III.iv) Arcas discloses the truth to his family and has to face them (IV.iv and vi) and after many hesitations orders Clytemnestre and Iphigénie (IV.x) to flee in secret. But again his design is frustrated (V.i) and the sacrifice will indeed take place — though eventually with another victim. This sequence of episodes — and each is a surprise — is Racine's creation, as is the Ériphile sequence. It is generated by Agamemnon's response to the situation in which he is placed and is itself characterized by ambiguity. The oracle, apparently so clear in its demand, imposes on Agamemnon a choice which he finds it impossible to make. Dramatically, he is placed between Ulysse and Achille, and he practises subterfuge, now with one, now with the other. This very human drama, however, concerns more than a resolution of a problem of relationships — and this goes also for the love plot — because it arises out of a divine imperative and is set within the context of the Trojan war, of events, that is to say, which transcend mere personal desires and passions. The dramatic oscillations and surprises of the play spring from ambiguous responses to the initial demand of the gods, which is itself ambiguous. It is there, and not in certainty as to the outcome or in clarity of dramatic preparation, that I think we find the dominating feature of the structure of the play.

In *Phèdre* also, crucial ambiguities exist, particularly in the heroine's search for Hippolyte's motives in rebuffing her, and Thésée's quest for the reason for his wife's and son's strange behaviour on his return. Like Ériphile, Thésée pursues the truth, an answer to his bewildered questions:

Que vois-je? quelle horreur dans ces lieux répandue

Fait fuir devant mes yeux ma famille éperdue? (ll. 953–4)

Although Racine adheres closely in this play to many of the essential aspects of the traditional story, his alterations to it are nevertheless such as to put the audience, as well as Thésée, in a state of uncertainty. Hippolyte — against the whole tradition — is in love, and in love with an invented Aricie. These changes provoke questions: how will Racine use them? How will they impinge upon the known situation and the development of the plot? In such uncertainty, the return of Thésée from from the underworld becomes a 'coup de théâtre', though it is prepared by the doubts expressed as to his whereabouts (ll. 17–18, 618–22, 729–30). The return is of course in the ancient sources, but Racine had used it before — in a perhaps more obvious 'coup de théâtre' — in *Mithridate*, as of course had Corneille, in *Pertharite*. Phèdre has allowed herself, in the first two acts, to be persuaded that he is dead and she has ignored Hippolyte's warning by saying:

On ne voit point deux fois le rivage des morts,

Seigneur... (ll. 623–4)

The audience may indeed note the warnings and remember, as Phèdre cannot, that in the traditional stories Thésée does return. But to the uncertainty occasioned by the change to Hippolyte and the invention of Aricie is added uncertainty as to when Thésée will reappear — if he does, because the radical nature of the other changes suggests that something equally radical might be altered here, too. Assuming that the death of Phèdre is part of the sacrosanct catastrophe (though it is really the death of Hippolytus in Euripides) will Thésée return before or after it? To put it another way, will Racine follow Euripides or Seneca? Such questions may indicate intellectual curiosity on the part of the spectator, but, as with the instances we have seen in Corneille, that curiosity is not without a sense of pathos, because the characters are presented in such a way and placed in such situations as to involve us emotionally in the action. We should not assume that curiosity, suspense, and surprise are at variance with the generation of such emotion.

The kind of ambiguities we have noted in *Iphigénie* are to be found also in two of Racine's perhaps more obviously dramatic plays, on which brief comment seems called for. In *Bajazet*, two notable features show that the dramatist was far from neglecting surprises, and indeed one of them, at any rate, allows us to see that he sometimes went out of his way to create them. These two features both concern Orcan, who is the immediate agent of the catastrophe. In the first place, he is not even mentioned until the end of Act III (l. 1101). His name comes as a 'coup de théâtre' to Roxane, for it has sinister associations. His arrival at the Byzantine capital is not only a complete surprise to audience and to characters; it not only performs something of the function of dreams and oracles in filling us with a sense of foreboding; but it also occasions a decisive turn in the development of the plot, because Roxane will now try to avert danger to herself either by allying Bajazet to herself or by obeying the Sultan's order to have him killed: the decision has become urgent. In the second place, as Cuthbert Girdlestone suggested,[21] Orcan is the unseen agent whose killing of Roxane comes at an entirely unexpected moment (V.x): the two preceding scenes turn our attention away from her to Bajazet. Then (V.xi) Osmin describes the death of Roxane in some detail: our attention is fixed on her, when, at the very end of the speech and almost incidentally (ll. 1692–4), the death of Bajazet is announced. Osmin is surprised (l. 1696) that Atalide does not know of it already. Orcan's primary and ostensible function was to carry a message from the Sultan, and so, as Professor Scherer[22] points out, his function as a hired assassin 'n'était ni prévue ni prévenue'. One should add that Racine's handling of the swift succession of scenes (V.ix–xi) is a masterly piece of stagecraft calculated to produce the maximum of surprise and dramatic effect. Suspense and tension are built up by the rapid entries and exits and by the disappearance of both Bajazet (end of V.iv) and Roxane (end of V.vii). Then the audience's attention follows Atalide's and Acomat's curiosity about the fate of Bajazet (V.viii, ix), only to find itself suddenly arrested (V.x) by Zaïre's announcement of the death of Roxane. Confirmation and more elaborate details then come (V.xi) from Osmin whose almost casual remark about Bajazet suddenly reverses the direction taken by our attention (and by that of the listening characters) in the previous scene. In something of the same way, Clytemnestre's allusion to the uncertainty of oracles which is, as we have seen, one of the elements of preparation for the dénouement of *Iphigénie*, is swiftly lost sight of: what matters to the queen at that moment is not the identity of Ériphile but the unworthiness of Helen to be a cause of war with Troy and, indirectly, of the death of the pure and innocent Iphigénie (see ll. 1265 ff.).

The other play which calls for brief comment is *Mithridate*, often assumed to be the most 'Cornelian' of Racine's tragedies, almost certainly, as we have seen, owing something of its situation to *Nicomède*,[23] and set apart by Professor May[24] as falling short of its author's ideal of tragedy as deduced from his prefaces. The reason for this alleged falling short is precisely that the play is found to be full of surprises and *péripéties*, including the quite unprepared return of Mithridate which is contrasted with the preparation for Thésée's but is similar to it in dramatic function, that of creating 'le nœud'. All the surprises are part of a newly-created dramatic action which bears as little resemblance to the historical accounts as does the situation itself. The supposed death of Mithridate occasions the declarations by his sons of their love for Monime, a rivalry quickly complicated, when the king unexpectedly returns, by their political differences. Their presence at Nymphea, against their father's orders, arouses his suspicions which he seeks to have confirmed or dissipated by his quest for the truth as to their motives and their attitude to him. Like Agamemnon after him he resorts to subterfuge which gives rise to tense and suspenseful episodes, when the audience is put in a state of anxious curiosity as to whether Xipharès and Monime will be trapped. All this is made possible by Racine's inventions and added to the rising of the troops and the arrival of the Romans in such a way as to intensify suspense and tension. Details, like the ambiguous use of the word 'bandeau' (diadem and noose) — a parallel will come in 'autel', in *Iphigénie* (place of sacrifice and of marriage) — enhance these feelings. Although *Mithridate* may well be the most 'Cornelian' of its author's plays, the dramatic technique he uses in it is not in fact different in kind from that of his other tragedies: it may be, for him, the extreme limit of his use of this technique, but the other plays, as we have seen, are marked by a similarly 'Cornelian' character, if not always quite so obviously. Indeed, the dénouement of *Iphigénie* may·be considered more dramatic and surprising, and perhaps less obviously prepared, than that of *Attila*, for example, where Corneille carefully relates the fatal haemorrhage (V.iv) to what has already been established as a characteristic fit of anger. Even *Bérénice* is not without its uncertainties and surprises. They begin early in the play because of the assumption, on the part of Bérénice and Antiochus, that Titus will in fact marry the Palestinian queen. That assumption is not shared by Phénice in lines (292–6) which foreshadow the dénouement, but whose effect is effaced by Bérénice's confident day-dreaming. In Act II, the uncertainty is maintained: Titus does not rejoice in freedom and mastery — he is melancholy in his need to send Bérénice away, and his resolution to do so appears shaken by her presence (scene ii). Will

he send her away? How will he bring himself to do it? When? How
will she react? The questions are not simply intellectual: they are
sympathetic. The characters are likeable; their situation is fraught
with pathos. It is the hesitation and silence of Titus which lead to
Bérénice's total misunderstanding of his feelings, to her distress
and dismay and anger. Then, for Antiochus, comes the double unexpected blow: he has just announced (a surprise) his imminent
departure, only to hear that Titus will not after all marry Bérénice —
Antiochus's hopes rise — and that he wishes Antiochus to break
the news — they falter only to rise again with Arsace's encouragement (III.ii). Can Antiochus face Bérénice? He tries to evade her
(ll. 841 ff.), but is dramatically prevented by her immediate entry.
Again he tries to evade carrying out Titus's request, but then is
forced to do so, and begins his own stumbling declaration. A 'coup
de théâtre', for him, results — Bérénice banishes him from her
sight. Now she is left in a state of agitated uncertainty (IV.i), for
she cannot believe (ll. 976–8) that Titus really will dismiss her
from his court. Racine maintains that uncertainty in the spectator's
mind by then deftly inserting Titus's long soliloquy (IV.iv) in
which he hesitates before the inevitable. He bows before it and at
once has to confront the queen. Will she, as she threatens, kill
herself? Can Titus save her? It is at that moment (IV.viii) that
Rutile announces that the people are clamouring to see him. There
is now no turning back. Yet uncertainties continue, especially for
Antiochus (ll. 1299–300) and the unseen Bérénice. Then come all
those threats of suicide, and the startling revelation to Titus that
he had a rival (ll. 1441–2). It is only at line 1482 that Bérénice's
false trail — and the drama — end: the suspense and uncertainty
are not dissolved until the truth is known to all. The quest for it is
entirely of Racine's making: it is fraught with uncertainties, dramatically exploited, always for the characters, sometimes for the spectator also.

On the evidence we have so far seen, it seems clear that Racine,
no less than Corneille, creates suspenseful and highly dramatic
plays and does spring surprises on his audience as well as on his
characters. One may, however, ask why it is that spectators were
and are willing to see such plays more than once, since on the
second occasion they obviously know what is going to happen and
how it will happen. I think it is unrealistic to suppose that it is only
plays with some deep psychological or moral interest which attract
spectators for a second time. It is true, of course, that Corneille
himself clearly saw the distinction between plays characterized by
'la nouveauté' and 'la surprise' and those possessed of 'quelque
chose de plus solide'.[25] The great box-office successes of the seven-

teenth century, like Thomas Corneille's *Timocrate* or *Camma*, in
which such interest is much less developed than in the tragedies of
Pierre Corneille or Racine, and which have always been noted for
their purely dramatic qualities, must, however, have drawn a fair
number of spectators twice or three times. In the *Examen* of *Horace*,
Corneille provides some explanation when he writes of 'l'attachement de l'auditeur à l'action présente' (*Writings*, p. 112). In the
Examen of *Cinna* (p. 116), he again says: 'L'auditeur aime à s'aban-
donner à l'action présente.' In the first instance, he is using the
statement to suggest that such attachment prevents over-
scrupulous criticism; in the second, to explain the difficulties of
handling complex plots. The psychological observation underlying
these remarks, as it underlies Corneille's ideas about *vraisemblance*
and the unities, is that the spectator goes to the theatre prepared to
be excited and enthralled by what happens in the play as long as it
lasts. He is willing to be surprised anew at what D'Aubignac calls
'des accidents imprévus' which produce 'une merveilleuse satisfaction, . . . une attente agréable, et un divertissement continuel' (*Pratique*, pp. 70–1). In another passage (p. 139), the same critic
writes:

Bien que la catastrophe ainsi que tous les autres événements . . . soient par-
faitement connus, ils ne laissent pas néanmoins de plaire et d'avoir toutes
les grâces quand il [the play] paraît sur le théâtre, parce qu'en ce moment
les spectateurs ne considèrent les choses qu'à mesure qu'elles passent. . . .
Ils renferment toute leur intelligence dans les prétextes et les couleurs qui
les font mettre en avant, sans aller plus loin; ils s'appliquent à ce qui se dit
de temps en temps et, étant toujours satisfaits des motifs qui les font dire,
ils ne préviennent point celles qui ne leur sont point manifestées, si bien
que, leur imagination se laissant tromper à l'art du poète, leur plaisir dure
toujours.

In the eighteenth century, another critic, Marmontel, analysed the
same phenomenon and used his experience of seeing *Rodogune* on
more than one occasion to illustrate it:

Il ne faut pas croire que l'art de rendre l'événement douteux et de laisser
le spectateur dans ce doute ne soit utile qu'une fois. L'illusion théâtrale
consiste à faire oublier ce qu'on sait, pour ne penser qu'à ce qu'on voit . . .
J'avais beau . . . savoir [le secret de la dernière révolution], je me le dis-
simulerais, pour me laisser jouir du plaisir d'être ému . . .[26]

Marmontel's is an acute observation and one which critics so often
fail to make: it is all too easy so to familiarize oneself through study
with great works of drama as to forget that their primary purpose
is to appeal, excite, and delight in the theatre, and that those pleasures are the ones the theatre-goer seeks when he pays for his seat.

The plays of Corneille and Racine surely gave — and give — them to him.

There are of course pleasures other than those of suspense and surprise. All great tragedy uses traditional stories in order to present a view of human life. By that I do not mean that it moralizes, though the view presented is, in the broadest sense, a moral one. It does not need to moralize if the dramatist gives it an ironical dimension. And here I should like to suggest that, in spite of the similarities of dramatic technique such as those we have analysed, in one aspect of plot-making Racine seems to differ quite markedly from Corneille. We have asked whether what are surprises for the characters are surprises for the audience as well, and have found that in the plays of both dramatists this is often so. This leads us to consider whether there are instances of the audience's being granted a privileged view of the situation denied, in part at least, to some of the characters. That kind of view is essentially ironical: it presents a double focus. The spectator sees the situation as the characters see it, but he also sees it from another angle and in so doing becomes aware that the characters' view is in some way distorted, untrue, that they are blind to what it really is. There is, of course, another kind of irony, which the characters themselves use in order to say things whose ostensible meaning may be harmless, but whose underlying sense — grasped only by those who are armed with a particular kind of knowledge or sensitivity — may be injurious, threatening, or simply sarcastic. Racine shares with Corneille the techniques of dialogue aimed at this sort of effect, but it is not my concern here.[27] While, as we have seen, both Corneille and Racine achieve suspense and surprise thanks in part to putting the audience into a state of ignorance and uncertainty by creating new situations and new plots, and while Corneille consistently puts the audience in the same degree of ignorance as his characters, Racine introduces features which add another dimension, that of irony in the audience's view of the events unfolded in the play. I use the word 'add' advisedly, because a tendency exists to think that dramatic effects like those we have been studying cannot co-exist with this ironical view.

In general, Racine's debt to Euripides, for whom he expressed admiration, has been fully recognized. In this particular matter, he seems primarily indebted to Sophocles.[28] Apart from the references, polemical in intent, in the first preface to *Britannicus* and the preface to *Bérénice*, the most interesting source of information about Racine's view of Sophocles is of course his marginalia. One of the most prominent features of the tragedies of this poet is that they so largely concern characters who are ignorant of some vital truth

about which they are ultimately enlightened. Racine was clearly
interested in this aspect of his plays.

The ignorance of the characters may be due either to their being
deceived by others or to their being blinded by circumstances, or
destiny, or their own passions or lack of judgement, or wilfulness.
The first category is the more straightforward both to achieve and
to understand. It is into this category that some of Racine's comments on *Ajax* fall (ll. 648 and 658, p. 845):[29] he notes that 'Ajax
trompe le Chœur et feint de vouloir vivre' and that his departure
from the stage is made possible by his deceiving the Chorus as to
his motive. Surprise will follow, of course, for the Chorus, when the
truth — that Ajax has gone out to kill himself — is discovered. The
most striking example of this kind of deliberate deceit is to be
found in *Electra*, a play which obviously — to judge from his co-
pious comments — fascinated Racine. He notes (l. 77, p. 846) how
Orestes and his companions deliberately conceal their arrival at
the palace. But Sophocles contrives the episode in such a way as to
show the arrival to the audience while concealing it from Electra,
Clytemnestra, and Aegisthus. It is against this background, with
this knowledge, that the audience listens to Clytemnestra's account
of her disturbing dream and sees a possible tragic interpretation of
it. (Racine makes use of a similar device in *Athalie*: when, in the
second act, the queen recounts it, we already know (I.ii) that Joad
is about to discomfit her by the revelation of Joas as the rightful
king.) Orestes and his tutor continue their deceit when the latter
gives his elaborate — and false — account of the death of his
young master: 'Le gouverneur d'Oreste vient faire un faux récit de
sa mort, pour surprendre Égisthe et Clytemnestre, et pour décou-
rir en même temps ce qui se passe.' (l. 660, p. 849; cf. l. 681) The
surprise is being reserved for Aegisthus and Clytemnestra. The
spectator knows the truth — Orestes is alive and, what is more,
already at the palace — and watches the contrasting reactions of
Electra and her mother, in emotions which will for both of them be
overturned when the truth is disclosed. But he has already, as he
listens to the narrative, an ironical view of the joy experienced by
Clytemnestra and the sorrow felt by her daughter. Racine also remarks on Orestes's pretence when he meets his sister (ll. 1098 ff.,
pp. 851–2), expressing admiration for the wonderful pathos of this
moving incident, 'où le poète s'est épuisé pour faire pitié'. 'Il n'y a
rien de plus beau sur le théâtre que de voir Electra pleurer son
frère mort en sa présence...'. And Orestes himself is overcome by
his sister's display of affection, to the point that (l. 1202) he reveals
himself for who he really is, the recognition being 'bien amenée de
parole en parole'. Racine's admiration for the technique of the

Greek poet is manifest, but that technique is not confined to the manner in which the recognition is gradually brought about: it concerns also — otherwise the effect on the audience would be quite different — the way in which the identity of Orestes is disclosed in advance to us but not to Electra. Finally, brother and sister prepare the deception of Clytemnestra and Aegisthus. Racine remarks on the resultant ambiguities of Electra's words — 'à double sens' — to the king (l. 1448, p. 854). The French poet makes use of similar devices: we have already noted the double meanings of 'le bandeau', in *Mithridate,* and of 'l'autel' in *Iphigénie.* In the latter play, we are made privy to the subterfuges of Agamemnon, so that his words to Iphigénie — 'Vous y serez, ma fille' — about the forthcoming ceremony assume their full pathos because she does not imagine what the audience knows, that it will be a sacrificial ceremony in which she is to be the victim. 'Électra,' writes Racine (l. 1439), 'veut tromper Égisthe en lui parlant plus doucement que de coutume.'

Deliberate deception is also practised in the *Trachiniae.* Racine devotes a whole sequence of notes (ll. 225 ff., pp. 867–9) to Lichas's concealment from Deianira of the truth of Heracles's love for Iole. First (l. 248) Lichas gives a false account of the situation. Racine comments: 'Faux récit de Lichas. Il y a déjà dans l'Électra un récit qui est faux tout entier... Je ne sais si ces narrations si longues sont assez dignes de la tragédie quand elles ne sont pas sincères.' This is one of Racine's very few criticisms of Sophocles's technique. It is noteworthy that he himself never engages a character in this kind of false narration on stage: the most remarkable example, perhaps, of his skill is to be found in Œnone's accusation of Hippolyte. At the end of Act III scene iii she assures Phèdre of her ability to save her honour; when, immediately afterwards, Thésée appears, she accompanies Phèdre in her hurried departure, to reappear with Thésée at the beginning of Act IV, her accusation already made and believed in the interval: that act opens with the king's expression of horrified astonishment. It is necessary only for Œnone in a short speech (ll. 101–22) to add the circumstantial evidence which inflames Thésée's anger and precipitates his fatal imprecations.

In the *Trachiniae,* then, Deianira is kept in ignorance of Iole's relationship with Heracles. This produces the irony of the meeting of the two women, on which Racine (l. 307) remarks: 'Déjanire s'adresse a Iolé, et la plaint [because of her captivity] beaucoup plus que les autres sans savoir que [c'est] sa rivale.' This is a situation which Racine himself exploits between Iphigénie and Ériphile when we first see them together (II.iii). But Deianira questions

Iole. Since the answers are likely to distress and anger the questioner, Lichas tries to spare her by interrupting and so (ll. 320–8) 'empêche Iolé d'instruire Déjanire de la vérité'. A similar interruption is to be found in *Phèdre*, when the heroine flees from her husband on his unexpected return (III.iv), thus making it possible for Œnone to fabricate her accusation against Hippolyte. The consequences are prolonged and more dramatically developed by Racine than by Sophocles, because Thésée's interrogation of Hippolyte (IV.ii) and of Aricie (V.iii) still fails to enlighten him. By these devices, Racine keeps Thésée in ignorance of what the audience knows, and so prolongs the suspense: will Thésée discover the truth? When will he do so and how? will it be too late? Racine notes that after the altercation with Teiresias in *Oedipus Rex*, the hero falls a prey to disquiet: 'Tirésias le laisse sans l'éclaircir' (ll. 430–7, p. 856). A further parallel in Phèdre is in knowledge of Hippolyte's love for Aricie being revealed to the audience but withheld from the heroine herself. The motive for the silence of the characters who know the truth is similar to that of Lichas: fear of the consequences. Racine's Agamemnon, as we have seen, behaves in the same manner, caught as he is between Clytemnestre and Ulysse. Racine, however, adapts the Sophoclean procedure of the false narrative in a highly dramatic way when he has his characters create false or deceptive situations in which traps are set for others: Roxane seeking the truth about Bajazet's relationship with Atalide, Mithridate seeking the truth about Xipharès's relationship with Monime. The audience, endowed with foreknowledge, watches the victim being invited to enter the trap: the spectacle is intensely dramatic because, whatever his forebodings as to the outcome, the spectator's fears are mingled with the hope he cannot help feeling for the victim's sake and with curiosity as to his or her response to the test and as to whether the reality of the situation will become apparent in time to prevent the disaster. Corneille, in his second *Discours*, notes that however hopeless the situation may appear for sympathetic characters, 'on espère toujours que quelque heureuse révolution les empêchera de succomber' (*Writings*, pp. 36–7). Curiosity goes hand in hand with a sympathy which is sharpened by the sense of uncertainty. A variation of the same kind of trap is to be seen in Athalie's temptation and interrogation of Joas, and yet another — for very different motives, and with the multiple ambiguities it involves — in Agamemnon's concealment of the truth from Iphigénie. With different motives yet again, Titus withholds the truth from Bérénice, who inadvertently falls into error as to his feelings for her and is thus moved to resist the truth when it is unfolded. The situation arouses curiosity over her reaction as we sympathetically watch Titus struggling with himself and eventually resolving to do what has become inevitable.

Agamemnon and Titus may be said to withhold the truth for what seems to them a laudable reason: the prevention of suffering. That could scarcely be said of Néron when he seeks to entrap Junie and Britannicus (II.iii–vi). In acquainting Junie with his plan — of what amounts to torture for her —, Néron also informs the audience of it. Tense, curious, and sympathetic, we watch Britannicus drawing the wrong conclusions from Junie's feigned indifference and incriminating himself before Néron, but we also watch Junie helplessly watching Britannicus and Britannicus misjudging her, and the emperor, unseen, watching them both. The pathos aroused for Britannicus is obvious: as for Junie, she is caught between the need to save him from Néron's jealous anger and her desire not to be misunderstood by, and inflict suffering on, the man she loves. There is nothing more truly theatrical than this drama of the watchers and the watched, but it is also full of the kind of pathos on which, as we shall see, Racine comments in some of his annotations on the Greek authors. Here it helps to create one of the most excruciatingly tense moments in Racine's drama[30] and, like the others I have mentioned, its effect depends on the kind of technique he comments on in Sophocles, by which the audience is made aware beforehand of the deceit about to be practised. Within the framework of the Britannicus story it is of course entirely invented (since there was originally no Junie). The new, additional motive — that of jealousy — for the assassination requires this proof. We know — or ought to know — that the young hero will die, and yet, since the whole episode is new, we do not know for certain that he will fail in this particular ordeal. But even if we did, our sympathy is such as to make us still hope for him — in suspense and growing tension. Néron's threats and warnings, and the awareness of his unseen presence, suffice to strike terror into our hearts.

Perhaps the most elaborate and truly Sophoclean example of a deception and revelation found in Racine's tragedies is the one in *Athalie.* Joad really stage-manages with skill and intelligence — and with implicit faith in the outcome — the disclosure of Joas as king and of Athalie's consequent discomfiture. As the great moment approaches, tension and suspense are heightened, exactly as they are in *Cinna*, when Joad (V.iii) whispers in Ismaël's ear. Then when Athalie and her escort enter the sacred place, he draws a curtain before the young king and his protectors. Full of pride and anger (V.v), Athalie storms at the high priest: she is certain of success. But then Joad draws back the curtain to reveal first Joas enthroned and then the armed Levites. It is a highly spectacular moment: the surprise and disarray of the queen are total. The episode is exactly parallel to that which we find in *Electra*, which Racine must surely have had in mind when he created it, for this is his comment on lines 1458–74 of the Greek play:

Égisthe commande qu'on ouvre les portes.
> (*Athalie*, l. 1699: Montez sur votre trône et ... Mais la porte
> s'ouvre.)

Ce commandement d'Égisthe marque un homme insolent qui ne craint plus rien et qui veut que tout lui obéisse;
> (*Athalie*, ll. 1705 ff.: ... Te voilà, séducteur,
> De ligues, de complots, pernicieux auteur,
> Qui dans le trouble seul as mis tes espérances,
> Éternel ennemi des suprêmes puissances!
> En l'appui de ton Dieu tu t'étais reposé:
> De ton espoir frivole es-tu désabusé?
> Il laisse en mon pouvoir et ton temple et ta vie ...
> Ce que tu m'as promis, songe à l'exécuter ...)

et en même temps cela prépare aux spectateurs le plaisir de la surprise d'Égisthe, qui, au lieu du corps d'Oreste, découvre le corps de sa femme.
> (*Athalie*, ll. 1731–2:
> Où suis-je? ô trahison! ô reine infortunée!
> D'armes et d'ennemis je suis environnée!)

The reversal for Athalie is complete, as it is for Aegisthus.

The ignorance of the dramatic character is not always the result of deceit on the part of some other character: it can stem also from blindness, particularly the blindness caused by passion — anger, hatred, jealousy, love. Oedipus, ignorant and blind to the truth, utters terrible imprecations, as Racine observes (l. 236, p. 855), against the murderer of Laius: 'Bel artifice du poète, qui fait qu'Œdipe s'engage lui-même dans d'effroyables imprécations' — an artifice which Racine uses himself when Thésée calls down the wrath of Neptune on the falsely suspected Hippolyte. Anger and ignorance, haste and pride, a failure to investigate and judge before yielding to passion, these are what eventually bring about the search for truth — but too late. Meanwhile, the spectator already knows it and watches the searcher in his rush to his doom. That foreknowledge results from a contrivance in the plot, the ordering of the dramatic material, the motives for Phèdre's and Hippolyte's flight from Thésée being disclosed before his return.

When Teiresias tells him the truth, Oedipus — how ironically, as it turns out! — accuses (l. 371, p. 855) the blind prophet of ignorance and calumny. The purport of Teiresias's remarks is lost on Oedipus, but not on the spectator, who knows that in accusing

the blind man he is incriminating himself: morally, it is Oedipus
who is blind. Like Oedipus, Racine's Bérénice is so caught up in
her own passion that she cannot (will not?) see the truth of which
the spectator is made aware at the outset of the play: she clutches
at every piece of falsely interpreted evidence in order to maintain
her illusion, her blindness to what she does not wish to see, and so
she radically misjudges Titus. In the last part of *Andromaque*, the
sequence of scenes is so ordered as to give the audience awareness
of developments which are hidden from the characters. In Act IV
scene v, Hermione's last hesitation is removed by the unrepentant
pride of Pyrrhus. She angrily sends him to the temple for his marriage to Andromaque: he disregards her threats and Pylade's warning (l. 1392), and has thoughts only for his bride. He does not
know that Oreste has already made his pact with Hermione
(IV.iii), nor that when she goes out uttering threats she presumably intends to countermand her stay of execution (ll. 1273–4).
But the audience does know. In Act V, when Oreste rushes back
from the assassination (scene iii) to claim his reward, he does not
know, as the audience does (scenes i and ii), that Hermione has
already begun to relent (as Thésée does, after his angry dismissal
of Hippolyte (ll. 1162, 1456)). Of course, Bérénice, Pyrrhus, and
Oreste ought to know what is in store for them but, like Oedipus,
they are blinded by their passions. By his careful ordering of the
plot, as by his invented situations and characters, Racine arouses
dramatic suspense and curiosity and at the same time provides the
spectator with an ironical double view of events. That view is,
however, not always complete, as Professor Scherer has pointed
out.[31] At the end of Act II of *Britannicus*, 'on sait ce que veut en
réalité le personnage [Néron], mais on ne sait s'il parviendra à
réaliser ses projets.' Even within the ironical framework, a measure
of ignorance may be maintained.

But there is more to irony than that. The ironical vision which is
vouchsafed to us by our privilege of omniscience or near-
omniscience is compounded by pathos, because we are almost always watching victims powerlessly caught in a web from which
there is no escape. Racine notes the pathos of Nestor's speech
(*Iliad*, Bk. VII, l. 248, p. 719) in which he laments his inability,
through old age (cf. l. 316), to accept Hector's challenge to single
combat (for an almost exact Cornelian parallel see Don Diègue's
lament over his powerless old age — *Le Cid*, ll. 697 ff.); of Electra's
inability (ll. 986–9, p. 850) to recover her honour by avenging
Agamemnon's death, and of Jocasta's surrender to her inability to
save her sons (Euripides, *Phoenissae*, ll. 1428 ff., p. 879); his own
Jocaste expresses the same resignation at the end of her fruitless

attempt to reconcile her sons (*La Thébaide*, IV.iii). It is precisely
that kind of pathos which, because the spectator is aware of the
tightening web while the victim still imagines that he may escape,
Racine creates for those who suffer from deception or self-
deception. Eugène Vinaver[32] asks whether Racine simply makes
use of pathos as 'un des ressorts de l'intrigue' and subordinates it
to the arousal of curiosity, or whether, on the other hand — and
unlike his contemporaries — he tries to create it independently of
'tout souci de composition dramatique'. The question seems to me
to put the terms the wrong way round. What my admittedly incomplete analysis appears to suggest is that Racine, far from being
uninterested in dramatic technique, makes masterly use of it precisely in order to stimulate pathos and pity and fear. Without the
plots there would be pathos only through the poetry of lamentation, which would take us back to sixteenth-century tragedy.
Racine's is certainly not that. As Professor Knight has said, 'c'est
dans une action toujours en progression que Racine semble voir la
source, la seule source pour lui admissible, de l'émotion
tragique'.[33] D'Aubignac made it clear that, by 1657 at the latest,
any other conception of tragedy was unacceptable (*Pratique*, p. 90):

... depuis l'ouverture du théâtre jusqu'à la clôture de la catastrophe ..., il
faut que les principaux personnages soient toujours agissants, et que le
théâtre porte continuellement et sans interruption l'image de quelques desseins, attentes, passions, troubles, inquiétudes et autres pareilles agitations ...

That Racine was a highly-skilled craftsman of the theatre there can
be no doubt from what we have seen in this chapter. That he did
not despise or disregard dramatic effects usually associated with
Corneille is equally clear. At the same time, of course, pity and fear
would not be aroused were the characters not, in Aristotle's terms
and as Racine never tires of telling us, 'men like ourselves'. The
fact that we can identify ourselves with them means that we are
drawn into their plight and, while possessing our transcendent
view of it, are able to experience with them their blind hope and
ultimate despair. We watch them in their blindness and know that
it leads to ruin, but we are also with them as ruin overtakes them
and surprises them.

Corneille found it difficult to come to terms with the need to
create characters of middling virtue and to accept that pity and
fear were the only specifically tragic emotions. Yet he refers several
times to pathos and 'commiseration'. Most of his heroes do not
arrive at their heroic decisions without some struggle — 'les combats intérieurs' — and they arouse pathos because an apparently

moral imperative, which they usually call duty, makes an ultimately irresistible demand upon them: apparently moral, because duty, whatever form it takes, is a passionate desire for self-fulfilment. Patriotism for Horace is a passion, not a rationally apprehended duty, and when that passion is crossed, as it is when Camille taunts him in the very moment of its fulfilment and victory, he becomes a figure of pity and pathos, diminished by foul murder. Other characters, Camille herself, Sabine and Curiace, all arouse pity because they are torn by contrary and equally demanding emotions. But the occasion for the struggle is imposed by external circumstances and its successive phases are marked by developments in those circumstances, often, as we have observed, quite unforeseeable, and they renew pathos, as the hero's will is subjected to new tests and others look helplessly on. The hero's survival of the test — or simply his courage in facing it — is a source of admiration, both for him, emotionally and morally, and for the skill of the dramatist, intellectually. That admiration for the hero, connected as it is — see the etymology of the word — with awe and wonder, is linked to Corneille's conception (second *Discours: Writings*, pp. 38–9) of the sublime, while sympathy for the hero in his ordeal and for the helpless onlookers couples it with 'le touchant', a word which Racine several times uses in his prefaces. If Racine's characters are not motivated by what one might call the heroic passions, such tests are replaced by the ever-tightening web of contrary emotions experienced within a set of close relationships.

If, in Corneille's plays, the 'coups de théâtre' serve to astonish the spectator and the characters, and to put the characters' will to the test, are they compatible in Racine with the ironic view of the omniscient? Can they catch the spectator unawares? D'Aubignac and Marmontel — and Corneille — all suggest that, in spite of prescience, he can still experience surprise. Sophocles would have suggested it, too, if we are to believe modern commentators. G. Germain writes: 'On a beau connaître le dénouement, on ne peut éviter, à mesure qu'il approche, un sentiment d'attente porté à l'extrême, voisin de l'angoisse, celui que l'on éprouve devant un acrobate ou un dompteur dans une situation périlleuse: tombera-t-il? sera-t-il mangé?' Professor T. B. L. Webster points out that the spectators' prior knowledge 'is outweighed by their sympathy with Oedipus... Knowledge yields to feeling and the spectator identifies himself with Oedipus.' On *Electra*, he comments: 'The audience know the truth, but they have so identified themselves with the chief character that their sympathy leads them to disbelieve what they know.'[34] Is this not true of Racine's drama, too? In this re-

spect, as in others, it seems like a fulfilment of D'Aubignac's pre-
scription (*Pratique*, p. 138):

Il faut conduire de telle sorte les affaires du théâtre que les spectateurs
soient toujours persuadés intérieurement que ce personnage dont la fortune et la vie sont menacées, ne devrait point mourir, attendu que cette
adresse les entretient en des pressentiments de commisération qui devien-
nent très grands et très agréables au dernier point de son malheur; et plus
on trouve des motifs pour croire qu'il ne doit point mourir, plus on a de
douleur de savoir qu'il doit mourir...

We have seen, however, that what might be termed sympathetic
surprise and suspense are accompanied, in Racine's plays, by the
kind of 'coups de théâtre' found in Corneille's. These obviously
heighten dramatic tension and excitement, but they can also fulfil
an authentic tragic function. They surely remind us that, for all
our sense of being omniscient, we, like the tragic victims, can be
caught unawares by forces within or beyond us which threaten us.
Indeed, the very feeling (or illusion?) of being omniscient can lead
us into the same *hybris*, the same pride of certainty and blindness of
passion, which leads to *nemesis*, as the tragic characters we watch.
In retrospect, moreover, we see that, like the surprising fall of the
statue of Mitys, the 'coups de théâtre' form part of a probable and
necessary sequence of events. One of the difficulties in our way
when we attempt to appreciate this is that, possessing that retrospective view, intellectually and after studying the plays closely,
we lose sight of the theatrical reality which yields that view only
after and by virtue of the surprises dramatically experienced.

While it is commonplace to accord some importance to plot-
making as far as criticism of Corneille is concerned — and indeed
it is sometimes suggested that he sacrificed other aspects of tragedy
to dramatic effect — the reverse is frequently the case with Racine,
as though to admit that he is a superb theatrical craftsman would
be in some way to diminish his stature as a tragic writer. 'The
prejudice against plot', as Eric Bentley calls it, [35] is widespread
among modern critics and theorists of tragedy. Yet classical scholars tell us that we ought to take Aristotle at his word about the
primacy of plot-making.[36] For him, Sophocles's *Oedipus Rex*
appears to have been the archetype of tragedy. Why should one
seek to deny to either Corneille or Racine, who was evidently a
disciple of both Corneille[37] and Sophocles, a place among the masters of tragedy, simply because they wrote dramas that were, after
all, dramatic? And, if Racine was a 'modern' no less than
Corneille[38] — and in part thanks to him — should we be surprised? It does not lower his achievement in the art of tragedy or

his sensitivity to what was most essential in the Greeks he revered. If Racine took his cue at the beginning of his career as a dramatist, not simply from Corneille directly, as I shall attempt to show in part of the next chapter, but from criticism of Corneille, a statement in D'Aubignac's dissertation on *Sertorius* may have provided a hint of the hybrid kind of tragedy he might achieve — both Sophoclean and Cornelian — the kind of tragedy my analysis here appears to have revealed: 'Il faut que les spectateurs de bon sens sachent toujours clairement ce qui s'est passé, et qu'ils ne prévoient rien de l'avenir, afin que la certitude de ce qu'ils savent rende leur plaisir plus grand quand on leur découvre ce qu'ils n'avaient pas prévu.'[39]

CHAPTER V
Simplicity: Situation and dénouement

Une tragédie est, premièrement et de naissance, une œuvre
dramatique, c'est-à-dire le lieu d'une action qui, opposant le
héros tragique aux circonstances adverses, l'isole des autres
personnages, amis ou ennemis, enserre sa vie d'un nœud toujours plus roide, parfois relâche le nœud au cours de péripéties
où se détend notre angoisse, mais au dénouement le livre soli-
taire et nu à la menace que bravait son effort.

André Bonnard, *La Tragédie et l'homme*

Some of Racine's remarks on *la vraisemblance* suggest, as we have seen, that he associated it with simplicity — a small number of 'incidents' which he accommodated without difficulty within the limitations of the unities of time and place —, but we have also found that his practice as a maker of plots is not such as to exclude dramatic qualities, 'coups de théâtre', or surprises arising precisely out of the 'incidents' which constitute the plot. Yet Racine himself, during the first half of his career as a dramatist, claims consistently to achieve simplicity, and makes his claims in the context of attacks on Corneille whose plays, by allusion and implication, are characterized by complexity and by extraordinary episodes. And successive generations of spectators, readers, and critics have taken their lead from Racine's remarks and perpetuated and developed the simplicity-complexity distinction between his and his rival's tragedies. But what do they mean, and what did Racine mean, by simplicity?

Almost from the outset of his career as a dramatist, and certainly from the beginning of his critical writing, Racine shows a particular interest in simplicity. It could well be that, apart from his natural bent, he saw in it, if he erected it into an aesthetic principle, a means of distinguishing his work from that of his great elder and of superseding him. The opportunity certainly seemed to offer itself.

In the early months of 1663, the year in which Racine was working on *La Thébaïde*, his first play to reach the stage, the long-standing quarrel between Corneille and D'Aubignac reached its climax with the publication of the abbé's four *Dissertations*, devoted chiefly to an attack on Corneille's latest plays (*Œdipe*, *Sertorius*, and *Sophonisbe*), and of the many pamphlets they provoked. The mediocre success of *Sophonisbe*, first performed in January and published in April, served as a pretext for D'Aubignac's revenge on the playwright who, in the *Discours* and *Examens* of 1660, had criticized some of the views expressed in *La Pratique du théâtre*, published in 1657 though conceived and written — at least in part — about 1640. But Corneille had also adopted some of the abbé's ideas, while failing to mention the book and to name its author. Some passages in the *Discours* patently allude, ironically, to the *Pratique*, notably the end of the first and third *Discours*, where the word *pratique* (verb and noun) is pointedly used.[1] At the same time, the fact that D'Aubignac had paid Corneille the compliment in his work — whatever his criticisms may have been — of drawing most of the examples for his theories from the plays of the man who dominated the French tragic stage for so long obviously exacerbated the abbé's hostility: we know moreover from Martino's edition that he

intended to publish a second edition from which the complimentary remarks were to be removed.

This was the situation in which Racine saw and seized his
chance. He possessed and briefly annotated a copy of the *Pratique*.
He certainly read the *Discours* and *Examens* as well as Corneille's
plays, and he cannot have failed to read the *Dissertations*. For an ambitious young dramatist aiming no lower than toppling Corneille
at the first blow from his pre-eminence, they must have been
interesting indeed. However, by the time the *Dissertations* appeared,
Racine was already embarked on writing *La Thébaïde*, which bears
many resemblances to *Horace*, though painting the strife within the
family in more sombre colours, unrelieved by motives, whether real
or illusory, of idealism. *La Thébaïde* substitutes for the relationships
by marriage and betrothal those of blood-brotherhood, and while
each in his own way fans the flames of war, Créon is a more sinister figure than le Vieil Horace and, as king, fails — unlike Tulle —
to restore order at the end; the women attempt to play a reconciling and restraining part in both plays, but in Racine's both perish,
as do Ménécée and Hémon, who is betrothed to one of them. Not
only, however, is Racine's the darker tragedy; it is also the more
complex in its larger number of main characters and in its triangular love plot, as it turns out (Créon–Antigone–Hémon), being less
closely integrated into the central action than is the corresponding
one (Curiace–Camille–Valère) in *Horace*. Yet, on the other hand,
and from the point of view of simplicity of the action itself, Racine
improves on the *Antigone* of Rotrou which is another of his models;
and, over ten years later, when he comes to write his preface to *La
Thébaïde*, he draws attention to the fact in these terms: the subject,
he says,

... avait été autrefois traité par Rotrou, sous le nom d'*Antigone*; mais il
faisait mourir les deux frères dès le commencement de son troisième acte.
Le reste était, en quelque sorte, le commencement d'une autre tragédie,
où l'on entrait dans des intérêts tout nouveaux; et il avait réuni en une
seule pièce deux actions différentes ... Je compris que cette duplicité d'action avait pu nuire à sa pièce ...

From this point of view, the comparative structural simplicity of
La Thébaïde may be seen as a response to D'Aubignac's criticism of
Œdipe, the more so when one remembers that like Rotrou's *Antigone*, its subject is drawn from the cycle of Oedipus myths. Corneille's play is the subject of D'Aubignac's third *Dissertation*, in
the course of which he complains of the duality of action he finds
there: 'Encore est-il vrai que la confusion de ces deux fables est si
mal ordonnée que l'unité d'action ne s'y peut rencontrer, quoique

M. Corneille demeure d'accord de cette règle qu'il faut observer' (ii, pp. 42–3).[2] This criticism is an application of the general principle already (only two years before *Œdipe*) enunciated in the *Pratique* (p. 83): '...il est impossible que deux actions (j'entends principales) soient représentées raisonnablement par une seule pièce de théâtre.' Corneille, like Rotrou by Racine, is being found guilty by D'Aubignac of infringing the unity of action. Racine's response, in practical terms, is to produce a play drawn from the same general sources as Corneille's *Œdipe*, to base it on the first half of Rotrou's *Antigone*, whose important scenes he rewrites, in exactly the same order, and thus to note D'Aubignac's criticisms of Corneille and to avoid falling into the same pitfalls. Even the rather clumsy way in which he attempts to knit the Créon–Antigone–Hémon plot to the main plot is indicative of his desire to avoid Rotrou's error of treating successively in the same play the subject of the *Phoenissae* of Euripides and that of the *Antigone* of Sophocles. Likewise Corneille had attached to the subject of Sophocles's *Oedipus Rex* characters drawn probably from *Oedipus at Colonus* (Thésée) and from Seneca's *Oedipus* (Phorbas) and, more importantly, the Thésée–Dircé love plot which is to all intents and purposes his own invention. Yet Racine, feeling like Corneille the need for a strong love interest, retains in however truncated and modified a form the Antigone story. He has not yet found his way to the more radical simplification which would be an important factor in the unification of later plays. Had he already begun work on *La Thébaïde* before the appearance of the *Dissertations*? And was he trying to modify his original conception of the play to avoid D'Aubignac's criticism of *Œdipe* and *Sertorius* when he complained in a letter to Le Vasseur in November, 1663, that work on the tragedy was progressing slowly (Picard, vol. ii, p. 457)? It was a desire for unity of another kind which presumably led him, in the course of the following months, to reduce and modify the *stances* with which he had earlier been so pleased (Picard, vol. ii, pp. 458, 459–60).

In arguing in this way, however, we are in danger of confusing simplicity with unity. Professor Scherer[3] has pointed out that they are not the same thing. Unity was a fundamental principle of the so-called classical doctrine and, needless to say, of Aristotle's concept of the work of art as an organic whole, complete in itself, and therefore unlike the continuum of ragged ends that is real life. The principle was firmly established, and Corneille accepted it in his own terms of unity of peril in tragedy. Simplicity, on the other hand, was — for Racine, at least some of the time, and for D'Aubignac — a desirable aesthetic characteristic which might also contribute towards the achievement of unity. In *La Thébaïde*,

Racine appears to be attempting less to simplify what he found in Rotrou's *Antigone* (or in Corneille's *Horace*) than to keep the two stories while giving them greater unity. In order to achieve this, he retains the altercation between Antigone and Créon, but changes the motive: Antigone is not punished for the burial of Polynice but kills herself out of sorrow and despair at the loss of her brother, her mother, and her betrothed, and finally at Créon's heartless proposal of marriage to her, so that he remains the agent of her death. Although Créon's proposal is ill-prepared and the *dénouement* awkward, they reveal Racine's concern for structural unity and for the avoidance of D'Aubignac's 'deux fables principales'. At the same time, it may be said that the omission of the burial story represents a simplification, and one which permits a degree of unity neither attempted nor attained by Rotrou. I would suggest that the example of Corneille's treatment of the Oedipus theme and D'Aubignac's criticism provided both the spur to the ambitious young playwright and the direction in which he should move to achieve success.

The effects of his partial acceptance and partial rejection of the example of Corneille and of other contemporaries on the evolution of his dramatic practice, at least up to 1670, have been studied in detail.[4] An imitation and implicit criticism of both Rotrou and Corneille may be observed in *Andromaque*. Although the parallels are less close than Gustave Rudler suggested,[5] Racine's third play does bear certain analogies to at least the first half of Rotrou's *Hercule mourant* (1634). The point at which the resemblances cease is when Déjanire goes mad (III.iv), the counterpart of the madness of Oreste in the last scene of Racine's tragedy. The similarities are chiefly those of situation: Pyrrhus, in love with his captive, Andromaque, abandons his betrothed, Hermione, as Hercule, in love with Iole, rejects his wife, Déjanire. Astyanax constitutes an obstacle to Pyrrhus, as does Arcas to Hercule. Hermione commands Oreste to murder Pyrrhus: he acquiesces with reluctance. Philoctète, though not in the same relationship to either of the female characters, acquiesces with the same reluctance in the murder of Arcas. That the similarities cannot be taken much further than this does not diminish the possibility that Racine was again deriving from Rotrou a suggestion which he corrected by simplifying and thus achieving greater unity of structure.

Confirmation, rather than refutation, comes from the closer parallel, noted by Voltaire, between *Andromaque* and *Pertharite*. Grimoald, like Pyrrhus, falls in love with his captive, Rodelinde, wife of the supposedly dead Pertharite, and abandons his betrothed, Edüige. She, jealous, like Hermione with Oreste, incites Garibalde, her suitor, to murder Grimoald. While the motivation is different in almost all important respects, this situation and its development run parallel in the two plays — until in Act III scene iv Pertharite reappears and causes a new situation to arise and a new problem to be solved.[6] *Andromaque* represents an advance on *La Thébaïde* in the sense that, while in both plays we may see a simplification of models found in Rotrou and Corneille, the later one excludes any equivalent of the Antigone–Créon development which, in a modified form, Racine had apparently felt unable to sacrifice in writing his first tragedy.

He was very probably shown the way in which to move forward from *La Thébaïde* by the example of two plays by Thomas Corneille, *La Mort de l'empereur Commode* (1657) and, more particularly, by the quite recent *Camma, reine de Galatie* (1661). The first of these, in its situation, resembles *Andromaque*, with the emperor parallel to Pyrrhus, Marcie to Hermione, Helvie to Andromaque, Électus to Oreste; the threat to Andromaque's son is similar to that to Helvie's father. But the play has ambition as its central theme. *Camma*, although dominated by a revenge plot occasioned by the murder of the heroine's husband, is in some ways closer to Racine's tragedy: the heroine's name furnishes the title of each play; Sinorix, betrothed to Hésione, forsakes her in favour of Camma; Sostrate becomes the instrument of her revenge, the dominant theme of the play; Camma agrees to marry Sinorix, but, like Andromaque, with the intention of killing herself before the marriage is consummated; the marriage proves fatal to him as it does to Pyrrhus. However, both the female characters are bent on vengeance against Sinorix — one for murder, the other for unfaithfulness — and both choose Sostrate to effect it. Sinorix's survival beyond the end of Act III is due to the entirely fortuitous entry of Sostrate in scene iii at the moment when Camma, a dagger in her hand, is about to murder Sinorix. The dagger falling to the floor, Sinorix is unable to decide whether it had been drawn by Camma or by Sostrate who has restrained her. This touch of the technique of the detective play, though cleverly handled, accentuates the importance of mystery, deceit, dissimulation, and misunderstanding which outweighs that of the psychological sources of the action but accounts for its intensely dramatic and suspenseful quality and, doubtless, for its enormous popular success in its own time.[7] The closeness of the situational parallel with *Andromaque* is probably not entirely accidental,[8] but Racine did not need a chance event (the unexpected arrival of Sostrate) to maintain or renew dramatic interest any more than he needed the return of Pertharite from the dead. Thomas Corneille's device, however, is more internal to the play than is his brother's: Sostrate appears as early as Act I scene v, and no one supposes him dead. Racine interiorizes even further: once Oreste has arrived at the court of Pyrrhus, thus actually creating the initial situation, no further chance event arises in *Andromaque*. This in itself, as we shall see, provides one clue to the relative simplicity of Racine's drama, at least during the first half of his career.

If his concern from the outset for that particular quality is evident from his treatment of situations found in plays by Rotrou and the Corneille brothers, his technical skill in achieving it is acquired only gradually, but he could never have been accused, as Corneille was, of attempting to dramatize two equally important stories in the same play. While it is true that the treatment of the death of Antigone depends on the final blow of Créon's unexpected offer of marriage and monarchy it is closely subordinated to her devotion to Polynice, without the need for the discovery of the burial and all the claims to guilt which ensue. The barbarous enmity between the brothers and the suffering of their mother still provide the main plot and interest. It is not some new chance intervention which links to it the Antigone–Créon conflict of the end (dispatched, one observes, in the last three scenes): Antigone has lost all — brothers, mother, lover — and Créon's preposterous proposal, not out of keeping with the unsympathetic presentation of his character through his behaviour in fanning the flames of conflict, merely precipitates her suicide. Racine's references to the tragedy in his correspondence, as *Les Frères ennemis*, show where he placed the emphasis, and it is to this main theme that he refers in his preface, several years later, as 'le sujet le plus tragique de l'antiquité'. The love interest, as he also writes there, is necessarily of minor importance — an implied criticism, perhaps, of Corneille's introduction into the story of 'les incestes, les parricides et toutes les autres horreurs qui composent l'histoire d'Œdipe et de sa malheureuse famille', of the prominent invented love plot which, for D'Aubignac, had ousted the traditional Oedipus story from its position of primary importance.

Technically, Racine knits his two plots and interests together by means of a character who is responsible for the tragic end of both — Créon. Although psychologically and emotionally the least interesting of the principal characters, mechanically he occupies a central position: he is ambitious for the throne, and incites the brothers to fight, hoping himself to inherit the kingdom after their deaths; and he is in love with Antigone, and sees the death of his son as a way to marriage with her. In the first of these desires he is successful, in spite of the contrary endeavours of Jocaste, Antigone,

and Hémon; but although the brothers, their mother, and his
son — the obstacles in his way — all die, he fails in the second, as
Hémon, striving to be worthy of Antigone, fails in his endeavour to
separate Étéocle and Polynice. The success of Créon's first desire
leads directly and ironically to the failure of his second: his weakest
adversary, left alone, deprives him through her death of the fruits
of victory.

If the main plot of *La Thébaïde* may be thought of as a Cornelian
one and the secondary plot as deriving from the tradition of 'galan-
terie', the proportions and emphasis are reversed in *Alexandre le
Grand*, whatever Racine implies in his second preface. Heroism
there is in Porus and Axiane, but Alexandre's heroic stature is
made less prominent than it might be by virtue of his love (and the
nature of his love) for Cléofile. It is perhaps not surprising that
Saint-Évremond thought that Porus had become a greater figure
than his Macedonian conqueror, and that in spite of Alexandre's
act of magnanimity.[9] But the heroic element does feature in the
play, particularly in the Porus–Axiane–Taxile plot, in which the
nature of the rivalry is characterized by the superior courage and
will-power of one man over that of the other. As in *La Thébaïde*, it is
the least appealing character, Taxile, who serves as the means of
achieving mechanical unity, and this thanks to an invented family
relationship with Cléofile whom Racine supposes to be his sister.
He thus provides the link between two couples and is caught between them, belonging to both but wanted by neither. They can be
freed in their love only by his disappearance: his death is the condition of the dénouement. *Alexandre* reveals an advance on *La Thébaïde* because the two plots are worked out simultaneously, not successively. In *Andromaque*, Racine's adoption of a device found in
many of the novels and pastoral plays of the period, the chain of
unreciprocated loves, marks a further advance towards inner
coherence and unity and reveals a further simplification: since all is
made, for all the characters, to depend on the decision of the
heroine, the playwright is enabled to dispense with the equivalent of
Créon and Taxile, and, as already in *Alexandre*, the resolution of the
dramatic conflicts is one and simultaneous. Moreover, he has sim-
plified the relationships themselves: none of his characters is now
bound by both family relationships and passionate love. He has
moved away from the invention of family relationships (Cléofile
and Taxile) in order to ensure unity of structure, in which he had
perhaps followed Corneille (Dircé as the daughter of Jocaste and
the sister of Œdipe, as well as the betrothed of Thésée). The most
important achievement, in terms of dramatic structure, may be
that of a cohesion which still permits of more than one interest, not

only in the form of the sound Aristotelian principle of the necessary subordination of one to the other, but also in that of their indivisibility and simultaneity. Racine attains this, at this point in his career — and he will never lose it, even though after *Bérénice* some apparently more Cornelian complexities can be seen in his work —, thanks to a simplification of his actual material and of the relationships in which his characters are placed one to another, and to a progressive reduction in their actual number.

That he had in mind D'Aubignac's strictures on Corneille's recent plays, *Œdipe*, *Sertorius*, and *Sophonisbe*, there can be little doubt. The fact that those strictures were so negative must have constituted for Racine an invitation to improve upon Corneille. Part of the *Dissertation* on *Sertorius* implicitly contains a veritable programme for an aspiring playwright to fulfil. Corneille, says the abbé,

...a plusieurs fois péché contre les règles de la vraisemblance la plus sensible, et choqué les esprits les plus communs; il s'est relâché souvent en des sentiments peu raisonnables, introduit des passions nouvelles et peu théâtrales, et souffert des vers rudes, chargés d'obscurités et de façons de parler peu françaises... (p. 266).

These complaints are directed against a lack of *vraisemblance* and *bienséance*, arousal of passions other than pity and fear, and bad style and versification — obscurity (and complexity?) in particular. It is the first of these targets which concerns us here, because Racine connects it with a particular kind of simplicity. In the preface to *Bérénice*, for example, he writes: '... il n'y a que le vraisemblable qui touche dans la tragédie; et quelle vraisemblance y a-t-il qu'il arrive en un jour une multitude de choses qui pourraient à peine arriver en plusieurs semaines?' The question is embedded in a discussion of simplicity as a challenge to the inventive playwright, and it clearly suggests that, the unity of time being accepted as an invariable rule, verisimilitude will not be achieved by attempting to confine within its limits events which are complex in the sense that they are, specifically, numerous. Of course, that depends in part on what one understands by 'events', but some initial clarification comes from *La Pratique du théâtre*, where D'Aubignac had written that 'les petits sujets entre les mains d'un poète ingénieux et qui sait parler ne sauraient mal réussir (p. 89). Racine doubtless set out to prove that he was such a 'poète ingénieux'. That he saw 'les petits sujets' as a challenge is highly probable, for the same preface contains the remark: 'Il y avait longtemps que je voulais essayer si je pourrais faire une tragédie avec cette simplicité d'action qui a été si fort du goût des anciens...'. Plautus is superior to Terence because of the simplicity of his subjects, but, 'combien Ménandre était-il encore plus simple, puisque Térence est obligé de prendre deux comédies de ce poète pour en faire une des siennes!' We have seen Racine doing precisely the reverse in *La Thébaïde* and *Andromaque*: taking half the subjects found in Rotrou and Corneille and making a full tragedy out of that.

It is important to avoid confusion of Racine's use and concept of simplicity with those of Aristotle (*Poetics*, X and XI), who, as D'Aubignac reminds us (*Pratique*, p. 95), employed the word to describe a plot without discovery or peripety. One or both of these featured in all French tragedies of the seventeenth century and, as we shall see later, they have a particular importance in Racine's. His references, in the preface to *Bérénice*, to the *Ajax* and *Philoctetes* of Sophocles seem, in the light of his own and his contemporaries' practice, to be confused or confusing, perhaps even specious. Those plays are simple — in the Aristotelian sense — in a way in which no tragedy written within a tradition largely derived from tragicomedy could be. Nor can simplicity be taken to imply singleness of plot or action in a technical sense. In the third of his *Discours* (*Writings*, p. 62), Corneille points out that unity is achieved by the sub-plot or -plots being so integrated and subordinated to the main plot that, as *acheminements*, they contribute directly to the *effet* characterized by the catastrophe, a sound Aristotelian principle enunciated (*Poetics*, chapter VIII) in terms of 'probable and necessary connexion' of a kind that 'the transposal or withdrawal of any one of them will disjoin and dislocate the whole'. That makes for organic unity: it does not imply simplicity, still less singleness. For all his advocacy of 'les petits sujets', D'Aubignac still suggests (*Pratique*, p. 87) that, as in life, so in painting and drama, no action is completely isolated and independent: 'Il n'y a point d'action humaine toute simple.' That Racine perceived the implications of this seems evident in his having added to the parting of Titus and Bérénice (or to the model, cited in the preface, of Dido and Aeneas) the part played, as we have seen, by Antiochus.

That he did not come to the notion of simplicity suddenly at the time of writing *Bérénice* will be obvious from what we have seen in his first three plays, and from the fact that his earliest actual statement on the subject, couched in the form of a rhetorical question addressed to his critics, occurs in his very first critical work, the 1666 preface to *Alexandre le Grand*: '... de quoi se plaignent-ils, si toutes mes scènes sont bien remplies ... et si, avec peu d'incidents et peu de matière, j'ai été assez heureux pour faire une pièce qui les a peut-être attachés malgré eux, depuis le commencement jusqu'à la fin?' Four years later, and still before *Bérénice*, he feels it necessary again to write that 'pour contenter des juges si difficiles',

... au lieu d'une action simple, chargée de peu de matière, telle que doit être une action qui se passe en un seul jour et qui, s'avançant par degrés vers sa fin, n'est soutenue que par les intérêts, les sentiments et les passions des personnages, il faudrait remplir cette action de quantité d'incidents..., d'un grand nombre de jeux de théâtre..., d'une infinité de déclamations...

In that first preface to *Britannicus*, those remarks — significantly — serve to introduce the famous tirade against Corneille, and they are again closely connected with 'le vraisemblable', as opposed to 'l'extraordinaire'. The preface to *Bérénice* marks the culmination, and indeed the end, of Racine's statements about simplicity. In it, in the year following the first preface to *Britannicus*, he is still attacking his critics:

Il y en a qui pensent que cette simplicité est une marque de peu d'invention. Ils ne songent pas qu'au contraire toute l'invention consiste à faire quelque chose de rien, et que tout ce grand nombre d'incidents a toujours été le refuge des poètes qui ne sentaient dans leur génie ni assez d'abondance ni assez de force pour attacher durant cinq actes [cf. the first preface to *Alexandre*] leurs spectateurs par une action simple, soutenue de la violence des passions, de la beauté des sentiments et de l'élégance de l'expression.

These phrases are a clear echo of the passage from the *Dissertation* on *Sertorius* in which d'Aubignac criticizes so strongly Corneille's handling (and indeed choice) of his subject. They also recall the abbé's comments on 'les petits sujets'. It looks, therefore, as though Racine — doubtless following also his natural bent — is consistently seeking up to this point to take up the challenge offered by hostile criticism of Corneille by dramatizing simple actions characterized by 'peu d'incidents', 'peu de matière', the episodes arising, not fortuitously nor from some extraneous agency, but from the emotions and interests of the characters, whose expression in discourse constitutes the action. It would, however, be mistaken to think that, for either Racine or D'Aubignac, that action is not dramatic. In drama, says the abbé (*Pratique*, p. 282), 'parler, c'est agir': the drama lies precisely in the clash of interests and passions which occasion not only conflict between the characters but strife within them. Corneille is in entire agreement when he states in the second *Discours* (*Writings*, pp. 41, 42) that 'les combats intérieurs' constitute the most powerful factor in the generation of tragic emotion. In order that such conflict may arise, however, more than one interest and incident is necessary, as D'Aubignac maintains in the *Dissertation* on *Sertorius* (pp. 237–8), thus creating 'les nœuds', the difficulties to be resolved, the obstacles to be overcome.[10] A surfeit of these 'nœuds', however, confuses the spectator, wearies his memory, and deprives the dramatist of the opportunity to develop the passions and their expression in such a way as to arouse the audience's emotions. So (p. 235) D'Aubignac writes of *Sertorius*, in words which are echoed in the passages we have examined from Racine's prefaces: 'Je ne puis comprendre pourquoi Monsieur Corneille a pris un sujet d'une si grande étendue. Est-ce qu'il ne se sent plus capable de soutenir de petites choses par la grandeur des sentiments?' Repeatedly returning to this theme, the critic points out that in *Le Cid*, *Horace*, and *Cinna*, Corneille had contrived simple plots where the characters were 'toutes ces grandes âmes agitées de contradictions intérieures [almost Corneille's own words], poussées de tant de nobles emportements et de tant de généreuses irrésolutions' (p. 236). Even after the publication of the *Discours*, D'Aubignac recognizes the greatness of Corneille; in fact, the *Dissertations* consist in a lament over the poet's fall from grace in the plays from *Œdipe* onwards. There are, however, as we shall see, aspects of the earlier tragedies which, in spite of their relative simplicity in terms of subject-matter, foreshadow later developments.

That Racine's definition of the nature and function of the simple plot corresponds closely to D'Aubignac's is evident when, in his pernickety way, the abbé distinguishes in *Sertorius*, not two 'fables principales' as in *Œdipe*, but no less than five plots, using, to describe them, a word he had explained in the *Pratique* (p. 284); 'les fables polymythes, c'est-à-dire chargées d'un grand nombre d'incidents'. Without the technical expression, Racine uses the phrase, as we have seen, in the first prefaces to *Alexandre* and *Britannicus* and in the preface to *Bérénice*. But neither writer has so far attempted a definition of the word 'incidents'.

D'Aubignac, at any rate, returns in his second *Dissertation* (pp. 230–1) to his idea of 'les fables polymythes', but the explanation he now gives of it is no longer what it was:

Le plus grand défaut d'un poème [dramatique] est lorsqu'il a trop de sujet et qu'il est chargé d'un trop grand nombre de personnages engagés dans les affaires de la scène, et de plusieurs intrigues qui ne sont pas nécessairement attachées les unes aux autres, ce que les Grecs nomment *polymythie*, c'est-à-dire une multiplicité de fables ou d'histoires entassées les unes sur les autres...

According to this etymologically more exact explanation *la polymythie* consists in a large number, not of incidents as episodes (as in the earlier definition), but of characters involved in a large number of stories. What Racine meant by 'la matière', D'Aubignac clearly means by 'le sujet' (in this context at least), but is Racine thinking

of a small number of *incidents* (as his own words suggest) or of a
small number of characters and stories? A consideration of his
plays in the light of D'Aubignac's criticism of *Sertorius* may enlight-
en us.

Above all, the abbé attacks what he understands by the *polymythie* of Corneille's play and in doing so foreshadows and perhaps
influences Racine's definition of the ideal tragedy as we find it in
the preface to *Bérénice*. D'Aubignac's adverse comments again form
part of his lament over Corneille's failure to fulfil the promise of his
early tragedies:

Cette polymythie nous prive encore d'un plus grand plaisir, en ce qu'elle
ôte à M. Corneille le moyen de faire paraître les sentiments et les passions:
c'est son fort, c'est son beau [*sic*] ... Il lui faut tant de temps pour expliquer les intérêts et les desseins de ses personnages qu'il en reste fort peu
pour mettre au jour les mouvements de leur cœur. (p. 238.)

Again, Corneille has '[accablé] la beauté de son génie sous le faix
et la multitude des matières' (p. 239). Already, in the *Pratique*
(pp. 83, 88), D'Aubignac had affirmed that the combination of two
equally important stories or, worse, 'plusieurs actions toutes fort
illustres', resulted in their real beauties on the tragic stage being
stifled ('étouffées'). As in the *Dissertation*, he had suggested that
each such story would suffice to make a play. It is in this context
that we find the well-known statement: 'Le poète doit toujours prendre son action la plus simple qu'il lui est possible...', a precept
which Racine seems to have taken to heart when writing two of his
first three plays.

D'Aubignac's analysis of the five plots he claims to find in *Sertorius*, while it may on several counts be unacceptable in itself, throws
light on the contrasting technique adopted by Racine, who may
have learnt something from it. That analysis is based in every detail on the idea that the simpler the dramatic situation, the easier it
is to understand, the less explanation it requires, and the more
fully the conflict of the passions can be developed.

The abbé begins with Sertorius, whose overriding passion is
ambition. Although elderly, he wishes to marry the Spanish queen,
Viriate, in order to secure his authority in Spain and a base from
which to prosecute the civil war which divides the Roman Empire.
But Aristie, whom Sylla has, for political reasons, compelled Pompée to divorce, is a refugee in Spain: Sertorius is tempted to marry
her to strengthen the alliance against Pompée. According to
D'Aubignac, Perpenna, Sertorius's unscrupulously ambitious
lieutenant and his rival for the hand of Viriate, is unnecessary in
this story, as is Pompée, whose reconciliation with Aristie might,

however, provide a suitable dénouement.

Similarly, Perpenna is the centre of a plot. Ignorant of Serto-rius's feelings towards Viriate, he actually asks him (as Pyrrhus will ask Oreste) to act as his spokesman and convey to her his desire to marry her. This alone, says D'Aubignac, would make a good play: neither Aristie nor Pompée need appear and the historical outcome, the death of Sertorius, would be safeguarded — and suitably mo-tivated by Perpenna's jealousy on discovering the truth as well as by political and military rivalry.

In her turn, Aristie is ambitious and bent on revenge for her divorce (for which she wrongly blames Pompée). The obvious candidate for an alliance — and for marriage — is Sertorius; but he loves Viriate. The problem could be solved by his marrying the Spanish queen and promising to reconcile Aristie with Pompée, whom she still loves. Neither Pompée nor Perpenna need be dra-matically involved.

A fourth play, from which Pompée and Perpenna would be ex-cluded, could be constructed around Viriate. Ignorant of his love for her, she wishes to marry Sertorius, but thinks that Aristie stands in her way.

But for D'Aubignac, the real centre of the play is Pompée. (This is as perverse, one may think, as the suggestion that the Thésée–Dircé plot is the more important of the two in *Œdipe*.) Although compelled, in order to cement his alliance with Sylla, to divorce Aristie and to marry Émilie, Pompée still loves his first wife. Not realizing this, Aristie suspects him when he seeks reconciliation: the obstacle, legal and emotional, is Pompée's second, and, he claims, unconsummated marriage. Reconciliation could be brought about, however, as it is in Corneille's play, with the death of Émilie, Sylla's abdication and the end of the civil war. This story suffices in emotions, suspicions, and ultimate enlightenment: Viriate and Perpenna — and perhaps Sertorius (D'Aubignac does not say) — are superfluous.

It is striking that in each of these proposed plays, the number of characters is reduced to three, an illustration of how to avoid 'la polymythie' understood as a large number of characters and stor-ies, but also perhaps a pointer for Racine to what he will eventual-ly achieve in *Bérénice*, in the preface to which 'ce grand nombre d'incidents' echoes D'Aubignac's earlier definition of 'la polymy-thie'. The abbé's analysis seems excessively cavilling when one realizes that his first, third, and fourth plots involve the same three characters, but he has conceived of each of them as having a diffe-rent principal character and of his or her interests as the dramatic centre of the story. These interests being, as he alleges, distinct and separable, the result is five plots, all combined in Corneille's play. One suspects that, since they are not bound together either by family relationships or by passionate love, but by political considerations, D'Aubignac found that the plots were not 'nécessairement attachées les unes aux autres'. This provides another clue to the devices used by Racine to connect his subordinate to his main plots: our analysis of the structure of his first three tragedies suggests its validity.

That there are two plots or interests in *La Thébaïde* and *Alexandre* is quite clear. It has often been said that this is true also of *Andromaque*, one involving Andromaque herself, Hermione, and Pyrrhus (the man caught between the two women), the other Pyrrhus, Oreste, and Hermione (the woman caught between two men). We can, however, construct an analysis along the lines proposed in D'Aubignac's criticism of *Sertorius* and suggest that there are four, not two, stories, each dominated by the 'interests' of one of the main characters. Oreste wishes to marry Hermione, but is frustrated by her love for Pyrrhus. In her turn Hermione finds her love for Pyrrhus frustrated by his for Andromaque. The obstacle to Pyrrhus's marriage to Andromaque is his long-standing bethrothal to Hermione. Finally, Andromaque's desire to remain faithful to the memory of Hector and to safeguard Astyanax is threatened by Pyrrhus's passion and menaces. Following D'Aubignac to a logical conclusion, we can suggest that Andromaque should be eliminated from the first of these hypothetical plays, Oreste from the second and third, and both Oreste and Hermione from the fourth. This kind of dismemberment is, however, not possible because Racine's play is made up of an endless chain of unreciprocated passions: Oreste loves Hermione who loves Pyrrhus who loves Andromaque who still loves her dead husband. It may be true that each of the characters pursues his or her own interests, but those interests are made absolutely dependent one on the other. Everything turns on whether Andromaque will submit to Pyrrhus — and the dramatic movements of the play spring from her inability to resolve the dilemma — and on whether Pyrrhus can come to a firm decision. His wavering, reported by Pylade to Oreste at the outset (ll. 121–2), arises out of hers but also of course out of his reluctance to carry into effect his own threats. It is therefore clear that even if one may say that in *Andromaque* there are 'plusieurs intrigues', they are certainly 'nécessairement attachées les unes aux autres', so much so indeed that one cannot speak of plot and sub-plot: the play is not made up of 'une multiplicité de fables ou d'histoires [or even two] entassées les unes sur les autres'. In this sense, then, *Andromaque* may be regarded as a marked advance on and a de-

velopment from the two earlier tragedies and as being characterized by simplicity. It is significant that the creation of the single
focus is due to bonds between the characters which Racine invented: Créon's passion for Antigone, Cléofile's relationship to
Taxile, and the endless chain of passions in *Andromaque*. None of
these is to be found in his sources. Such devices will continue to be
used even when, after scoring the final victory over Corneille with
Bérénice, Racine has ceased to proclaim the ideal of simplicity. In
Mithridate, for example, the play often regarded as his most 'Cornelian' creation, the invention of the king's amorous rivalry with his
sons and of their passion for Monime draws even the political and
military theme into the same single dramatic interest.

'...Combien de personnages, combien d'intérêts, combien de
choses!' exclaims the good abbé of *Sertorius* (*Dissertation*, p. 234)
shortly after his remark that 'le plus grand défaut d'un poème
[dramatique] est lorsqu'il a trop de sujet et trop de personnages'.
For him, one way of avoiding the defect of too many separate plots
and interests is to reduce the number of active characters: this is
implicit in his analysis of the five plots. Reducing the number of
characters is precisely what Racine does, and he does it progressively as the principle of simplicity becomes more assertive — six
characters in *La Thébaïde*, five in *Alexandre*, four each in *Andromaque*
and *Britannicus*, three in *Bérénice*. This process does not of course of
itself make for simplification: it does so only in combination with
the creation of necessarily interdependent interests and emotions
through the bonds of family or passionate relationships. They may
make for structural unity and simplicity, but not for emotional simplicity. And they need to be made dramatic, which is why Racine
found it necessary to introduce into the simplest of all situations,
that of the parting of Titus and Bérénice, the third character of
Antiochus.

We have seen D'Aubignac complain that in *Sertorius* 'il...faut
tant de temps pour expliquer les intérêts et les desseins [des] personnages qu'il en reste fort peu pour mettre au jour les mouvements de leur cœur' and that Corneille has thereby deprived us of
the pleasure of his earlier skill of 'faire paraître les sentiments et les
passions'. Corneille's own preface, published before the *Dissertation*
and therefore not a reply to it, opens with these remarks which
could well have provoked D'Aubignac's complaint:

Ne cherchez point dans cette tragédie les agréments qui sont en possession
de faire réussir au théâtre les poèmes de cette nature: vous n'y trouverez ni
tendresses d'amour, ni emportements de passions, ni descriptions pom-
peuses, ni narrations pathétiques. Je puis dire toutefois qu'elle n'a point
déplu, et que la dignité des noms illustres, la grandeur de leurs intérêts, et

la nouveauté de quelques caractères [cf. D'Aubignac's criticism of the 'passions nouvelles'], ont suppléé au manque de ces grâces. Le sujet est simple . . .

If Corneille distinguishes between 'les intérêts' and 'les emportements de[s] passions', stating that he has substituted one for the other, so does D'Aubignac, lamenting the substitution. This is precisely what Racine avoids, strengthening the inner bonds of 'tendresses d'amour' and passion and thereby being enabled to create a simpler dramatic structure in which 'les intérêts et les desseins' of the characters are not multiplied. Yet Corneille had the impertinence, as D'Aubignac doubtless saw it, to claim: 'Le sujet est simple'. He must have intended the remark to be taken seriously — though it is not developed — because he was never ashamed of admitting to complexity: he had done so long before *Sertorius* (Preface to *Clitandre*, 1632; *Au Lecteur* of *Héraclius*, 1647) and again quite recently (*Examen* of *Héraclius*, 1660). But can we take the claim seriously? I think we can — and it is instructive — if we apply it to the development and dramatic aspects of the plot rather than to the situation.

In some respects, *Sertorius* is a remarkably static play. Its situational complexity is undeniable and requires explanation, even if not to the extent suggested by D'Aubignac. If Donneau de Visé, who undertook to reply to the *Dissertation*, is to be believed, Corneille was chiefly criticized by the first audiences for having written a play which was, dramatically, too simple. Addressing D'Aubignac, he wrote:

Vous devez être persuadé que personne n'entrera dans vos sentiments, puisque tout Paris n'a pu s'empêcher de dire qu'il n'y avait presque point de sujet dans le *Sertorius* et que cette tragédie n'a été admirée que pour les beaux vers et la force des raisonnements qui s'y trouvent . . . Il faut que vous n'ayez pas bien observé *Sertorius*. Comme il y a peu de sujet, tous les actes excepté le dernier n'ont que trois ou quatre scènes. Il y en a même qui n'en ont que deux, et dans l'acte second Sertorius et Viriate sont ensemble une scène qui a près de deux cents vers, ce qui ne pourrait être si cette pièce était si pleine de sujet, puisqu'en celles où il y en a beaucoup les actes ont d'ordinaire huit ou neuf scènes.[11]

What the Parisian spectators appear to have disliked was the absence of 'incidents', since their number does usually bear some relation, within the framework of French stage conventions, to the number of scenes. (I shall return to the matter of the intensified movement of the last act.)

It is true that much of the dialogue is devoted to political discussions or dialectics of love and marriage within a context of ambition and public policy (the abbé's 'la grandeur de leurs intérêts'): none of these discussions results in firm decisions, resolutions of problems, or effective action. In spite of D'Aubignac's denial, the characters are subject to 'irrésolutions' as were those of the earlier plays, but lack perhaps the kind of 'générosité' which he coupled with that word. They remain uncertain after argument and discussion: of this the abbé was severely critical, particularly in the case of Sertorius who, he rightly claims, is murdered before he has arrived at any decision — his death is thereby deprived of the heroic quality and emotional appeal associated with that of Curiace or Polyeucte, and foreshadows the almost accidental disappearance of Suréna. The indecisiveness cannot solve the dramatic problem which is characterized by dilemma and 'impasse'. Corneille maintains interest during the first four acts by showing the characters vainly seeking a solution: dramatically, the action is simple. But the problem must be solved in a dénouement, and requires an external intervention to provoke it, the arrival of a letter from Rome (V.ii), informing Aristie of Sylla's abdication and Émilie's death. Pompée is thereby released, the uncertainty between him and Aristie is dispelled, and they are free to remarry. In the following scene the murder of Sertorius is announced. This, Perpenna's brief moment of triumph, is singularly undramatic. The death of Sertorius, said Voltaire, 'devait faire un grand effet, [mais] n'en fait aucun'. It is, however, the second stage of the dénouement, but instead of making possible the marriage and alliance of Perpenna and Viriate, parallel to the new possible marriage of Pompée and Aristie, it creates a new problem, that of the punishment or pardon of Perpenna. This can only be solved by the return of Pompée (V.v, vi) which does not produce any excitement in itself because it has already been announced (V.ii). Meanwhile, however, Viriate is released from her imprisonment by Perpenna who, when Pompée arrives, finds his accusations against the 'traitors' turned on herself (V.vi): this is certainly a dramatic moment, occasioned by Pompée's 'générosité', which also inspires his act of reconciliation towards Viriate and his respect for the dead Sertorius. These things, following one upon another in quick succession, seem to be what Racine castigates, in more general terms, as 'ce grand nombre d'incidents'. They are necessary, because of the polymythic structure of the play, in order to effect the dénouement and to leave the audience, as Corneille put it, in no doubt as to the outcome. They make for a crowded final act after the somewhat inconclusive arguments, in the first four, about conflicting feelings, ambitions, and interests. Those first four acts are notably lacking in *péripéties* of the kind which give rise to dramatic complexity by altering the course of events, creating new situations, and throwing the characters into disarray. It is on that, perhaps, that Corneille's claim to simplicity rests. Most of the apparent *péripéties* form parts of the dénouement, where they create at least one new problem which, like the pre-existing ones, needs to be solved before the play can end.

Hesitations before apparently insoluble problems, emotional, moral, and political, lie at the root of the 'combats intérieurs'. From *Le Cid* to *Pertharite*, however, Corneille's heroic characters are able to come to decisions, sometimes with disastrous swiftness — Rodrigue, Horace, Auguste, Polyeucte, César, Cléopâtre (in *Rodogune*), Nicomède. Once set on their courses, they hold to them whatever may befall. The later plays, including *Sertorius*, feature characters who generally find it difficult or impossible to arrive at positive decisions: their failure culminates in the wasted death of the last of them, Suréna. In spite of their hesitations, Racine's characters, too, are able to be decisive, even if they are also capable of changing their minds. *Andromaque* may, in the development of its plot, be characterized by the vacillations of the heroine and of Pyrrhus, but eventually Andromaque does conceive of her 'innocent stratagème', and Pyrrhus finally rejects Hermione. Pyrrhus can act with dramatic unpredictability, as when (II.iv) he suddenly announces that he will after all marry Hermione, but he is also capable of reversing that decision. The *péripéties* and the dramatic intensity of the play do not depend on external factors, but arise out of the nature of the emotions generated within a single and fundamentally unchanging situation from which no escape is possible. The decisions and actions are undertaken in the belief that escape is possible and that the situation can be changed. Since, however, the nature of the situation is determined by fate-given, ineradicable passions, that belief is illusory. Within it, the internally-provoked *péripéties* provide changes of direction, but each direction followed leads only to the blank wall which violently returns the deluded victim to imprisonment and eventually crushes him.

At first sight *Mithridate* seems to resemble Corneille's earlier plays: its development appears to be linear and its movement affirmative. Like that of *Sertorius*, its last act is notable for its 'coups de théâtre': the poisoned cup sent by the king to Monime, the order countermanded, the news that Xipharès is still alive, Mithridate's final act of pardon. Although the incidents are occasioned by factors over which the characters have no control — the course of the battle against the Romans and the fatal wounding of the king — their actual nature depends on what we already know of the characters themselves and of the situation in which they are caught or believe they are caught. Mithridate's sending of the poison

arises out of his possessive passion for Monime, and his change of
mind out of the knowledge that he is dying and has seen a worthy
and courageous successor in Xipharès; pardon and reconciliation
spring from the same sources, and he sees Monime as a reward
which he gives to Xipharès for his loyalty (ll. 1667, 1671–3). To
this extent, therefore, the final 'coups de théâtre' originate, like
Pyrrhus's change of mind, from within. The same applies to the
earlier ones and to the *péripéties* in the course of the first four acts.
An exception may seem to exist in the return of Mithridate from
presumed death in Act I scene iv. This occurs too early in the play,
however, for any real development to have taken place, because the
first three scenes show the new situation, consequent upon the
news of the king's death, as already in existence, albeit without the
knowledge of the characters involved, except in so far as each one's
own emotions are concerned. Retrospectively (l. 428), Mithridate
is seen to have imagined his own 'death', the false rumour he has
sown being characteristic of the dissimulation which is his most
prominent habit. Like Pyrrhus's sudden change of mind (II.iv),
Mithridate's reappearance is explained only after it has taken
place. Certainly it is a highly dramatic moment, a 'coup de théâtre'
and a *péripétie* already, reversing the newly-existing situation between Xipharès, Monime, and Pharnace. But it is part, too, of the
extraordinarily masterly exposition itself in that it both allows the
spectator to become aware of the 'combats intérieurs' to which it
subjects the characters and which will be the cause of the subsequent dramatic tensions, and also provides the motive for Mithridate's relentless search for the truth on which he will embark immediately (ll. 475 ff., 519 ff.) and which becomes the central strand
of the action. It is out of that search, and out of his alternating
certainties and doubts that the *péripéties* arise (see the sequence II.
iii; II.iv–v; III.ii–iv; III.v; IV.ii; IV.iv–vi), with the irony
that his use of subterfuge recoils upon him in the discovery of a
truth he had not expected to find (III.v). Even the successful
landing of the Romans is not only dramatically prepared (l. 338)
but is due to the treason of Pharnace which takes the form of rebellion at the crucial moment (IV.vi), and that treason springs from
jealousy and ambition. In this highly dramatic play, all the *péripéties* arise out of the passions of the characters and their interaction.
The final act, with its 'coups de théâtre', provides excitement to
the end, but the dénouement is not complicated by the need to
conclude a many-stranded plot: it is of quite a different character
from that of *Sertorius* in spite of a superficial resemblance. The fact
is that Mithridate can and does make decisions, too hastily indeed,
for once made they cause him do doubt his former certainty: he is a

distinct foreshadowing of Thésée. The 'incidents' arise from within and are in fact the expression of the conflict of the passions: that conflict not being resolved until the dénouement, it is up to that point constantly self-renewing and demands a succession of decisions and their reversal.

Here again, a passage from *La Pratique du théâtre* (p. 72) points to the interdependence of 'incident' and passion. D'Aubignac distinguishes three kinds of dramatic subjects — those composed of incidents, of passions, and of a mixture of the two:

> Il est indubitable que les mixtes sont les plus excellents, car les incidents renouvellent leurs agréments par les passions qui les soutiennent, et les passions semblent renaître par les incidents inopinés de leur nature, bien qu'ils soient connus, de sorte qu'ils sont presque toujours merveilleux et qu'il faut un long temps pour leur faire perdre toutes leurs grâces.

It is of course true that Corneille's tragedies fit this formula almost as fully as do Racine's, because D'Aubignac does not distinguish between incidents which are external in origin and those which are internal. This distinction, however, arises not only in this form but also in that in Corneille's plays, up to 1652 at least, we witness less a rebirth of passion out of the incidents than an illustration of its persistence in its course, whereas in Racine's the reverse is true. In *Horace*, for example, the series of externally-occasioned incidents is calculated precisely to show that the hero's passion for duty and glory is immovable and that it survives every test, even the crucial one that brings about his downfall, his confrontation with his mourning sister. Racine's first two plays are clearly modelled on the same principle. In *La Thébaïde* we find the same uncertainties as to the outcome of the actual events of the combat. Indeed, where in *Horace* it is the women's powers of persuasion which fail to prevent the battle, in *La Thébaïde* the actual death of Ménécée and of Jocaste cannot prevail over the brothers or their uncle to make peace. The death of both combatants, not one, seems to be required to prove their passionate hatred, and that of Hémon serves to demonstrate the fidelity of Antigone's love. Similarly in *Alexandre le Grand*, a number of parallel incidents are found: the Macedonian's victory, his supposed death, his 'resurrection' and final defeat, all necessary in order to put to the test and to reveal Alexandre's heroism and magnanimity, Porus's persistent and heroic resistance, and Axiane's fidelity and unswerving patriotism. It is evident, however, that in *Alexandre* the number of such incidents is greatly reduced, a fact which, if we are to judge by the first preface, aroused the disapproval of the critics. With *Andromaque*, all this has gone, for ever in Racine's work. 'Ce grand nombre d'incidents' seems to refer in part to those externally-occasioned episodes which serve less as turning-points in the plot than as tests by which the hero's rectilinear resolution is demonstrated, and in part to those which are required, from a technical point of view, in order that a polymythic play may be brought to an end. Presumably, Racine did not consider the *péripéties* which originate internally and which abound in his plays — more perhaps than in Corneille's — to be 'incidents' in that sense. Was an anonymous eighteenth-century writer referring to Racine's early plays when he made the assertion, so surprising at first sight, that 'le grand Racine a mis trop d'épisodes dans ses pièces'?[12] The nature of the episodes, from *Andromaque* onwards, rather than their actual number, is what prevents Racine from falling under the kind of criticism levelled by D'Aubignac (*Dissertation*, pp. 236–7) at the 'actions' in *Sertorius*: 'Je ne crois pas qu'aucun des spectateurs, après la première repré-sentation, les ait pu garder en sa mémoire pour en suivre avec plaisir les événements et les démêler', a remark which strangely echoes the *Examens* of *Cinna* and *Heraclius* (*Writings*, pp. 116, 140).

The process of simplification, in the sense that we may now understand it, is consistent in Racine's tragedies from *La Thébaïde* to *Bérénice*, whose success marks its author's final victory over his rival. The younger dramatist may well have been encouraged by D'Aubignac's criticism of Corneille to follow this particular direc-tion (indeed the prefaces contain too many echoes of the *Pratique* and *Dissertations* for us to think otherwise), but other dramatists had already begun to move away from the 'romanesque' fashion, with its complex dramatic structure' and the abundant external in-cidents and complicated dialectics of passion necessary for the sol-ution of its apparently insoluble problems. Thomas Corneille's *Commode* (1657) is already in this respect simpler than his earlier plays; Gilbert's *Cresphonte* (also 1657), Quinault's *La Mort de Cyrus* (1658–9), Boyer's *Clotilde* (1659) and above all Gilbert's *Arie et Petus* (also 1659) all, among others, show the same tendency. It was perhaps against this trend that Pierre Corneille reacted when he returned to the theatre with *Œdipe*, and against him in his turn that Racine reacted by proclaiming and following the dogma of simplicity.

In the *Dissertations*, D'Aubignac is clearly disappointed in Cor-neille's new vein: he had obviously hoped for more tragedies in the style of the early masterpieces. Those plays, for all the externally-occasioned incidents they contained, were certainly simpler in situation and dénouement than the later ones, and Racine seems to develop their possibilities along his own lines and in that sense to follow D'Aubignac's lead. It is not without interest that his first play takes up a theme analagous to that of Corneille's first truly tragic masterpiece and that his treatment of it is in many respects strikingly similar to the older dramatist's. In one important point, however, they differ. Whereas in effect the dénouement of *Horace* takes place when the hero kills Camille, the remainder of the play being concerned with his trial, moral and judicial, that of *La Thébaïde* is delayed until the last two scenes. While, in a technical sense, the murder of Camille infringes the unity of action, it is clear that D'Aubignac's criticism of the play (*Pratique*, p. 68) is paralleled later by the echoes of adverse comments on the end of *Britannicus* which we find in the first preface. One consequence of the kind of simplicity which allows of a simple dénouement seems to be that the dénouement itself need not be delayed beyond the end of the fourth or the beginning of the fifth act.

That is a point in the plays which invites comparison between the techniques of the two dramatists: it is particularly telling in *Sertorius* and *Britannicus*. At the end of Act IV in either play both Perpenna and Néron have arrived at the moment of decision over the fate of their victims. Each has consulted his evil counsellor: both Aufide and Narcisse advise immediate murder. Perpenna speaks the last half-line (1544) in the scene:

> ...Allons en résoudre chez moi.

Néron, too, speaks the last line (1480):

> Viens, Narcisse, allons, voir ce que nous devons faire.

Despite the obvious similarity, there is a world of difference in dramatic quality between these two moments. In *Sertorius*, we simply do not know what Perpenna will do: if we feel anything, it is curiosity. In *Britannicus*, we feel morally certain that Néron will have the murder committed: we experience tension, expectancy, foreboding, terror, pity. Why? Because in Corneille's play, so much time has been given up to explanation and argument that Perpenna has not emerged as a passionate, desperate man who can make up his mind and act with despatch. Néron, on the other hand, involved in a relatively simple situation, has been shown, largely through the proving-grounds of *péripéties* provoked in part by himself, to be precisely such a man, or at least to have taken the first step towards becoming one. The passions have already been shown in all their imperiousness to be the source of action within a situation which, because it is essentially simple, does not require lengthy explanation. For the same reason, Racine's expositions are remarkably swift and already characterized by and conveyed in action: he can embark quickly on the series of *péripéties* which represent the turning-points of the same action as it develops. One must not of course lose sight of the fact that in ending Act IV of *Sertorius* on a note of uncertainty which has not been dispelled up to that point Corneille is deliberately keeping the audience in the dark and in suspense and therefore ready for the surprises still to come in Act V, surprises which do not necessarily make it more dramatic than the final act of plays by Racine. It is interesting also that what an admirer of Corneille[13] found particularly praiseworthy in *Sertorius* was the great deliberation (III.i), entirely political in nature, between the hero and Pompée, to which might have been added the earlier one (II.ii) between Sertorius and Viriate. The preponderance given to the political dimension of the play necessarily detracts from its dramatic intensity and from its emotional power, particularly since the characters have lost something of the human warmth and passion which had made the deliberations in *Horace* and *Cinna* so appealing. The complexities of the political theme have replaced the authentic passion of Camille for Curiace or of Cinna for Émilie, Horace's ardent desire for glory, and Auguste's for inner peace. Néron may act in the political sphere, but his passion is purely self-centred and self-inspired: to wrest himself from his mother's tutelage and Junie from Britannicus's love. The political interest in Racine's tragedy is therefore not complex, and so Néron's passion is established in the spectator's mind as pitiless and irresistible by the time the fourth act is ended.

Other plays of Racine reveal the same quality as *Britannicus* at the corresponding part of their development, and for similar reasons. Hermione finally rejects Pyrrhus, after his rejection of her, in Act IV scene v: in spite of all her earlier hesitations, it is beyond doubt that she has now sent him to his death: her frustrated love has evidently turned to hate. That certainty in our minds gives its tragic poignancy to the remorse with which her love, unquenchable in fact, begins to relent in the famous soliloquy which opens Act V. Bajazet's fatal letter is discovered at the end of Act IV: Roxane's passion, like that of Hermione, is such that the discovery of the truth seals his fate. Titus's solitary discovery, in Act IV scene iv, of the final inescapability of imperial duty makes separation from Bérénice and, above all, his announcement of it to her in person, inevitable. These points in the plays are intensely dramatic and tragic precisely because the passions have been established and their victims brought to a high pitch of tension through the complex crises and oscillations of a plot whose basic situation is necessarily simple and coherent.

D'Aubignac criticized in *Sertorius* the multiple plots, all equally important as he saw them, which made for an unsatisfactory dénouement. All the knots must be unravelled, as near simultaneously as possible: the audience, say both the abbé and Corneille, is impatient to see how things will turn out. Unlike eighteenth-century critics,[14] they do not distinguish between dénouement and catastrophe. (The title of one chapter of D'Aubignac's *Pratique* — Book II chapter IX — in fact equates the two: 'Du dénouement, ou de la catastrophe et issue du poème dramatique'.) The distinction is that the dénouement unties the last knot, removes the last obstacle and, in so doing, provokes the catastrophe. Hermione's final dismissal of Pyrrhus is the dénouement: it causes his death, the accusation of Oreste, Hermione's suicide, and her suitor's madness. All these are catastrophe. One can apply the same analysis to Néron's last line in Act IV, to Titus's 'ne tardons plus... Rompons le seul lien...' (ll. 1039–40), to Ériphile's last line in Act IV also ('...A Calchas je vais tout découvrir': l. 1491), to Roxane's 'Sortez!', to Phèdre's jealous refusal to exculpate Hippolyte (IV.v). In *Sertorius*, on the other hand, a series of partial dénouements succeed each other rapidly: their effect is not dramatically felt, precisely because there is no room for catastrophe — the fault of the multiple plots, each of which must be at least mechanically resolved. No wonder Corneille needed to claim for the last act the 'privilege' of 'precipitation'. In the last act to *Nicomède* even the 'privilege' seemed to him later to have been strained.[15]

It was doubtless with Corneille's practice in mind that Racine's critics suggested, if we are to believe the first preface, that *Britannicus* should end with the announcement of the hero's death. He persistently refused to bow to such criticism because the immediate consequences of that event on the other characters must be known and those consequences, in the terms of our analysis here, make up the catastrophe. The first preface to *Andromaque* had already described the tragic characters as 'ceux dont le malheur fait la catastrophe de la tragédie'. Of *Britannicus*, he writes: 'Ma tragédie n'est pas moins la disgrâce d'Agrippine que la mort de Britannicus.' The consequences of Néron's cruelty and of the death of his rival have to be seen to be felt by Agrippine, Junie, Narcisse, and the emperor himself.

In the original version of *Andromaque*, the eponymous heroine made a last entry in Act V, in conformity with a traditional convention, to which Corneille refers in the *Examen* of *Nicomède* (*Writings*, p. 153). Writing of what was apparently a lost first version of his play, and referring to Prusias and Flaminius, he has this to say:

D'abord j'avais fini la pièce sans les faire revenir, et m'étais contenté de faire témoigner par Nicomède à sa belle-mère grand déplaisir de ce que la fuite du Roi ne lui permettait pas de lui rendre ses obéissances... Mais le goût des spectateurs, que nous avons accoutumés à voir rassembler tous nos personnages à la conclusion de cette sorte de poèmes, fut cause de ce changement, où je me résolus pour leur donner plus de satisfaction...

In the definitive version of *Andromaque* Racine, by then presumably confident enough to disregard the convention, reverses Corneille's procedure and devotes the last act entirely to the 'disgrâce' of Hermione and Oreste, the catastrophic moment of discovery for them. So, originally, it was for Andromaque, who was presented by Oreste as a prize — in a new captivity and a new widowhood — to Hermione. 'Deux fois veuve', she says, 'et deux fois esclave de la Grèce',

> Je ne m'attendais pas que le ciel en colère
> Pût, sans perdre mon fils, accroître mon misère...[16]

Presumably here Racine was conscientiously completing his catastrophe, but it was theatrically inconvenient: while Andromaque was speaking — at considerable length — Hermione was on the stage awaiting news of the death of Pyrrhus. Since, however, Andromaque's own recognition scenes were really Act III scene viii, and Act IV scene i (her resolution on the 'innocent stratagème'), Racine would have realized that her 'malheur' was already sufficient; but he maintained the completion of the mechanical dénouement with the report, right at the end (ll. 1587 ff.) that, in place of the dead Pyrrhus, she, as his widow, had taken charge of the situation in the city. A parallel exists in the suppressed reappearance of Junie in Act V scene vi, of *Britannicus*, the next play. The removal of these scenes represents in itself a measure of simplification: it does not affect the dénouement which has, as we have seen, already taken place at the end of Act IV of both plays. But the original versions are instructive because they show Racine dutifully attempting fully to work out the catastrophe and not yet confident enough that the simplicity of his plays was such as to allow him to forgo this and to concentrate at the end on the 'disgrâce' of Hermione, Oreste, and Agrippine and to leave the 'escape' of Andromaque and Junie to reports made in the final scenes. The revisions reveal the process of simplification which will culminate in *Bérénice* still going on after that play was written: the problem continued to concern certain complexities in situation, remnants of 'la polymythie', the reduction of the number of 'fils', as Professor Scherer calls them.[17] Nevertheless, even in the original versions, Act V of these two tragedies has a quality markedly different from the final act of Corneille's plays, because of an unspoken distinction between dénouement and catastrophe. But Racine also puts into catastrophe, as we shall see, a characteristic essential to his dramaturgy and to his conception of tragedy which is entirely absent from the work of Corneille.

CHAPTER VI
Simplicity: Peripety and discovery

... L'une des plus grandes beautés que le poète puisse former dans la structure de sa fable, c'est de faire que l'aventure qui doit finir tragiquement aille bien avant dans la joie, avant d'être troublée par les accidents funestes qui composent sa catastrophe, et que pour relever avec adresse l'éclat de ces renversements, il faut qu'au moins une fois on voie dans le bonheur et dans le plaisir ces personnes que le malheur doit accabler dans la fin de la tragédie ...

... La péripétie, c'est-à-dire ... cet événement imprévu qui dément les apparences et qui par une révolution que l'on n'attendait point vient changer la face des choses ...

[Saint-Ussans], *Réponse à la critique de la 'Bérénice' de Racine*
(1671)

'In tragedy, the principal source of pleasure for the spectator is in the Peripeties and Discoveries.' This statement, made by Aristotle in the sixth chapter of the *Poetics*, is on the one hand a seemingly valuable clue to his concept of tragedy, and on the other — as regards French criticism and dramatic writing of the seventeenth century — the cause of a great deal of confusion and obscurity.

The difficulty arises over the use made of the word *péripétie* in French criticism. Professor Scherer is right to distinguish between its connotations in the singular and the plural.[1] The singular is reserved for the word employed in Aristotle's sense, to mean, as most English translators have it, 'a change of fortune', though, as we shall see, there are objections to this rendering. It is found in this sense in, for example, Heinsius[2] (*peripetia* in Latin) and in La Mesnardière[3] and Sarasin[4] (*péripétie* in French). (Except when they are actually translating or glossing the text of the *Poetics*, I do not find that either Corneille or Racine ever uses the word, singular or plural.) As found in Heinsius, La Mesnardière, and Sarasin, the use of the term in the singular suggests that any given tragedy will have only one such peripety, and it is true that many of the plays of the first half of the seventeenth century conform to this pattern. This is, perhaps, why we tend to describe tragedies like Tristan's *Mariane* (1637) and Du Ryer's *Alcionée* (1640) as 'simple'.

Now the word *péripéties*, in the plural, obviously means something quite different from peripety, in the singular. As such, it is not to be found in the critical vocabulary of the seventeenth century. It has been vested in more recent times with its now familiar meaning:[5] the oscillations of the plot, the changes of direction, the rebounding of the action consequent upon some surprising event or 'coup de théâtre'. It is, however, true that D'Aubignac, in his definition of catastrophe (*Pratique*, p. 136), appears by implication at least to recognize that the word *péripétie* might be used to denote this latter sense. Of the term *catastrophe*, he writes:

Je sais bien qu'on le prend communément pour un revers ou un boule-versement de quelques grandes affaires, et pour un désastre sanglant et signalé qui termine quelque notable dessein. Pour moi, je n'entends par ce mot qu'un renversement des premières dispositions du théâtre, la dernière péripétie, et un retour d'événements qui changent toutes les apparences des intrigues au contraire de ce qu'on en devait attendre.

The last *péripétie* is equated, then, with peripety, and that is a characteristic of the catastrophe. We may infer from this that seventeenth-century writers understood current dramatic practice to involve a number of *péripéties* before the catastrophe was reached. D'Aubignac did, however (*Pratique*, p. 95), refer especially to the

Aristotelian use of the term, and did so correctly. La Mesnardière (*Poétique*, pp. 55 ff.) obviously sees peripety in the Aristotelian sense, but he attempts to accommodate the idea to contemporary French practice. Yet he criticizes modern dramatists for multiplying the *péripéties*, on the grounds that not more than one such sudden change can occur with *vraisemblance* within the twenty-four hour limit of the unity of time. In his moralizing way, he also connects peripety with divine retribution and cites the example of Aegisthus and Clytemnestra among others.

Having drawn his distinction, Professor Scherer goes on to concentrate on the use of *péripéties*, in the plural sense of the word, in French drama. He shows how they may be largely reserved for the closing scenes of a play, as in *Mithridate* (to which we shall return); how their source must lie outside their victims; and how characters cannot be victims of *péripéties* of their own making, springing from some change of will or intention on their part because then there would be no surprise or dismay for them, and such surprise is an essential characteristic of these *péripéties*. However, what is not stated, though the point is vitally important, is that peripety (in the singular) does arise out of the victim's own intentions or actions and that it consists precisely in the unexpected consequences of those intentions and actions. (If we do not accept this, then all Aristotle's statements about 'undeserved misfortune' will be unconvincing.) Moreover, the *péripéties* may be, and often are no doubt, productive of surprise in the spectator as well as in the victim: peripety frequently is not. Again, the *péripéties*, which cause the action to start off in a new direction, often do result from a character simply changing his mind, and, in such cases, they are reversible. So, for Andromaque, there is a *péripétie* when Pyrrhus announces that he will marry Hermione after all — but this decision can be, and is in fact, overturned, thus producing another *péripétie*. The same may be said of announcements of pieces of information which are later discovered to be false, or of obstacles which turn out to be unreal, or of interpretations of oracles or dreams which are subsequently found to be defective; and here we touch upon a rich arsenal of *péripéties*, the so-called *quiproquos*. These false obstacles to the resolution of the problems of situation and plot were immensely popular in all the dramatic genres. Rosimond actually entitled a comedy *Les Qui pro quo* (1673). They might be a source of curiosity, as Corneille suggested in the preface to *La Veuve* (1634: *Writings*, p. 176), but were often a means of achieving suspense and drama, as in *Nicomède* (I. ii), where the hero allows Attale to go through a whole scene without disclosing his own identity (this also constitutes a testing of Attale's loyalties); or as in *Le Cid* (V. iii), when Chimène, seeing the sword carried in by Don Sanche, believes that Rodrigue is dead; or as in *Horace*, where (I. iii), Camille misunderstands Curiace's presence or (IV. ii) le Vieil Horace misunderstands Valère's. The *quiproquo* may take several forms: identity mistaken, words misconstrued, events wrongly or incompletely reported, motives misunderstood; they may be deliberately contrived by one character to test or trap another. The audience may or may not know the truth. As we have seen, on knowledge or ignorance depends the emotion to be generated. While it is often true that the *quiproquo* may be the source of a *péripétie*, sometimes to the point of being identified with it, its discovery by the character concerned is not always a surprise for the audience, whereas such surprise is one of the main purposes of a *péripétie*.

Mention has been made of an apparent omission in Professor Scherer's discussion of peripety and *péripéties*. There is another: he omits to link peripety explicitly with discovery. It is on discovery that Eugène Vinaver[6] bases an important part of his argument that Racine, sharing this characteristic with Sophocles, alone of his contemporaries achieves true tragedy. The French dramatist creates, however, a new type of discovery, psychological discovery, unknown to the Greek playwright. Concentrating on the nature of discovery and the poetic means of effecting it, Vinaver minimizes the importance of the dramatic processes which lead up to it. He maintains that Racine had little or no interest in those processes, that if the audience has its attention directed to them the tragedy fails to arouse pathos, and that without pathos there can be no authentic sense of the tragic (which is, of course, something at once deeper and greater than the pathetic). We have seen, however, that for Racine the plot and its surprises and *péripéties*, far from precluding pathos, can be powerful agents in its creation.

Here, however, it is with *peripety* (singular) as used by Aristotle that I am primarily concerned. Most translators render his Greek either by transliteration or by formulas such as 'a change of fortune'. Aristotle, however, makes it clear that the latter would be unacceptable to him. In chapter X of the *Poetics* he writes of the kind of action in which 'the change in the hero's fortune takes place without Peripety or Discovery'. This has to be borne in mind when reading such statements as these: 'The change in the hero's fortunes must not be from misery to happiness, but on the contrary from happiness to misery' (chapter XIII); and 'A Peripety is the change from one state of things within a play to its opposite, in the probable or necessary sequence of events' (chapter XI). Aristotle uses the expression 'change' or 'reversal of fortune' as a definition of tragedy, and the word *peripety* as a definition of part of the action. Humphry House suggested[7] that we should translate peripety by 'a reversal of intention': you intend something to happen, but what in fact happens is its opposite. 'A reversal of intention' seems to me to be a little too limited and, although the formula is clumsy, I should prefer to say something like 'a reversal or inversion of the anticipated consequence of an action or intention'. What this means for the character involved is that his action or intention recoils upon him in an unexpected way, and, for the spectator on the one hand the generation of sympathy and, on the other, the spectacle of the direction of the action being changed or the psychological situation being altered. Aristotle himself (*Rhetoric*, II. viii) writes of the pity aroused 'if some misfortune comes to pass from a quarter whence one might have reasonably expected something good'.

With these preliminaries and precautions in mind, let us see what Corneille has to say about the question of peripety and discovery. In the first *Discours*, he is critical of the whole concept because he thinks that peripety must be provoked by some event unattached to the action, unprepared by the exposition, and, therefore, illogical (neither probable nor necessary, to use the Aristotelian terminology). The example he gives is the arrival, in *Oedipus Rex*, of the messenger from Corinth who reveals the identity of the King of Thebes. This is the discovery which brings about peripety for Oedipus: in the light of his recently-acquired knowledge that he it was also who killed Laius, his good intention of cleansing the city of the plague has unexpected and disastrous consequences for him. It is significant that *Oedipus Rex*, which was, for Aristotle, the discovery-tragedy *par excellence* and perhaps the masterpiece of the Greek theatre, comes in for a good deal of criticism from Corneille: it is, of course, obvious that he had no real understanding of either *hamartia* or *catharsis* and that he was, like all his contemporaries, hopelessly tangled up with moral and psychological concepts quite foreign to Greek drama. His whole dramaturgy rests on the foundation which he calls 'les combats intérieurs' in characters seeking less self-knowledge than self-mastery.

At the same time, he was also obsessed with the idea that discovery (*agnition* as he calls it) was necessarily the discovery of identity — personal, physical identity — as in *Oedipus*. This is no doubt explicable, in part at any rate, by the fact that Aristotle, in the sixteenth chapter of his treatise, where he discusses the various means by which discovery can be effected, confines himself precisely to considering this kind of recognition. In view of the incompleteness of the *Poetics*, at least as Corneille and we ourselves know the work, it is not possible to say whether Aristotle would have studied — or indeed did study — other aspects of the question. At all events, Corneille's conception of tragedy makes him critical of this kind of discovery: 'Je sais que l'agnition est un grand ornement dans les tragédies: Aristote le dit; mais il est certain qu'elle a ses incommodités.' Thus he writes in the second *Discours* (*Writings*, p. 42), and there follows a criticism of plays by Ghirardelli and Stefonio, prefaced by a significant statement the gist of which is surprisingly akin to the ideas of D'Aubignac: 'Les Italiens... affectent [l'agnition] en la plupart de leurs poèmes, et perdent quelquefois, par l'attachement qu'ils y ont, beaucoup d'occasions de sentiments pathétiques qui auraient des beautés plus considérables.'[8] Corneille's position is clear.

Although he never demeans himself, as he would see it, by mentioning them or even openly alluding to them, the condemnation of the Italians may well embrace the French authors of so-called tragedies — particularly in the 1650s — whose works were liberally enriched (if that is the right word) and enlivened by cases of mistaken identity. True, the Quinaults and the Boyers — *Le Feint Alcibiade* or *Oropaste, ou Le Faux Tonaxare*, etc., etc. — could, had they wished, have found some sanction in the *Poetics*, and even for the establishment of true identity by means of material objects: Aristotle specifically talks of 'signs' and instances necklaces and other tokens. Rings, swords, letters, portraits litter the plays of Corneille's contemporaries and rivals. With these he will have virtually nothing to do. Even in *Héraclius*, where Martian's ignorance of his identity lasts almost throughout the play, it occasions above all 'les combats intérieurs'.

This, however, is really a point of minor importance in itself, though significant as a symptom of Corneille's attitude to his art. The *Discours* and *Examens* are, of course, primarily, though not exclusively, concerned with problems of technique, as were all such treatises in the seventeenth century. As we have seen, Corneille has technical reasons for rejecting discovery — hence his condemnation of Sophoclean practice, as exemplified in *Oedipus Rex*, on the grounds that in that play the discovery is brought about by the mere chance arrival of the messenger from Corinth. But, whereas Corneille is critical of Sophocles's procedure, Aristotle praises it — and this shows how much the Italian and French theorists had narrowed the doctrine of the *Poetics*: 'The best of all discoveries', we read there (chapter XVI), 'is that arising from the incidents themselves, when the great surprise comes about through a probable incident, like that in the *Oedipus* of Sophocles...'. Here a fact of some importance should be pointed out, on which I shall enlarge later, namely, that Corneille finds himself discussing the question of discovery in the context of the construction of the play and, perhaps unconsciously, linking and almost identifying it with the dénouement (which is a problem of technique), whereas Aristotle, clearly though not explicitly, distinguishes discovery from dénouement.

However, if Corneille discusses the matter of discovery almost entirely in technical vein, we should not jump to the conclusion that he had no deeper reasons for rejecting it. Even while, in the second *Discours*, he is actually considering the best technique for effecting the 'catastrophe', as he calls it, something of these reasons breaks through: 'Lorsqu'on agit à visage découvert, et qu'on sait à qui on en veut, le combat des passions contre la nature, ou du devoir contre l'amour, occupe la meilleure partie du poème' (*Writings*, p. 41). Clearly, as Vinaver has pointed out,[9] this idea of characters acting 'à visage découvert' automatically precludes any kind of material or physical discovery. It is, however, evident also that Corneille is thinking not only in terms of characters who know their own and one another's personal identity, but of characters who know their moral identity as well, or who, at the very least, reveal that identity through their actions. When, therefore, he talks of surprise, suspense, 'un certain mouvement de trépidation intérieure', curiosity, he does not connect these things with discovery, as Aristotle associates surprise with it in the sentence just quoted, but with the audience's uncertainty as to the outcome of the plot, and, more important perhaps — as the plays themselves show (take Rodrigue or Horace or Nicomède) — specifically with uncertainty as to how the hero will stand up to the moral test to which he is subjected.

Now, what of Racine's attitude to the question of discovery? In the first place, he has little to say about it in his prefaces or marginalia. He does translate two fragments (*Principes*, pp. 32–4) from the sixteenth chapter of the *Poetics* where, as I have mentioned, Aristotle is discussing the various means of effecting discovery. I do not think that much can be deduced from the fact that Racine chose those particular passages. However, if we look at Racine's practice as a dramatist, we may say that perhaps these two fragments can be seen as supporting his implicit objections to the use of material objects to bring about discoveries of personal identity and his desire to make discovery at once a source of surprise and a logical dramatic outcome of elements of the plot. The only other part of his annotations on the *Poetics* which directly concerns us is on chapter XIV (*Principes*, pp. 21–6) where Aristotle considers the various kinds of action suited to tragedy. Whereas the Stagirite

merely compares (using the comparative degree) the different possibilities, we find Racine picking out one of them and rather obviously reinforcing the terms of comparison to express his own preference: 'A better situation', translates Bywater, 'is for the deed to be
done in ignorance, and the relationship [i.e. of agent to victim]
discovered afterwards...'. In Racine's version, this becomes: 'Mais
le meilleur *de bien loin*, c'est lorsqu'un homme commet quelque action horrible *sans savoir ce qu'il fait*, et qu'après l'action il vient à
reconnaître *ce qu'il a fait...*' (My italics). The substitution of ignorance of the nature of the act for ignorance of the identity of its
victim seems to me not only, as Vinaver rightly suggests,[10] an expression of Racine's retention of discovery in tragedy, while giving
it a moral and psychological quality, but also a rejection, no less
clear than Corneille's, of material discovery. The statement, taken
as a whole, is also a declaration of preference, stronger even than
Aristotle's, for tragedy in which an act is perpetrated in ignorance
of its real consequences and in the hope of other, illusory, consequences, and for tragedy in which ignorance is ultimately dispelled,
thus effecting or being effected by peripety. This seems to give
point to what looks very much like a chance remark or afterthought
in that section of the preface to *Bérénice* in which Racine defends
simplicity of plot by referring to Sophocles's *Ajax* and *Philoctetes*.
'*L'Œdipe* même', he then goes on, '*quoique tout plein de reconnaissances*,
est moins chargé de matière que la plus simple tragédie de nos
jours' (My italics). The concessive clause seems to have a double
significance: first, that simplicity being in this context considered a
virtue, discovery (even more than one) does not detract from it;
second, that discovery of the particular kind found in *Oedipus Rex* is
however not approved of.

If the annotations on Vettori's edition of the *Poetics* date from
some time after 1662,[11] and the preface to *Bérénice* from the end of
1670, it seems probable that Racine was, in the years following the
publication of the *Discours*, facing a problem raised by Corneille,
and not very satisfactorily solved. Corneille had expressly condemned *Oedipus Rex* and *Ajax*, and he had rejected the kind of action
defined above: the act perpetrated in error – enlightenment – peripety. He seems to have done this in order to give to some of his own
plays a semblance of orthodoxy, as his analysis of *Le Cid, Cinna,
Rodogune, Héraclius*, and *Nicomède* shows (*Writings*, pp. 12, 21, 39–
41). Racine had presumably found his rival's explanations specious
or at any rate unconvincing, but since he, too, could not accept
material discovery, he was trying to find some way of preserving
the substance of Aristotelian and Sophoclean discovery while
changing its form. This seems all the more probable because the

annotations on the *Poetics* were not intended for publication, and so
were not polemical: Racine does seem to be grappling with a problem discussed by Corneille, whose doubts he apparently shared,
but whose solution, which amounts to avoiding discovery
altogether, he could not adopt. Moreover, both playwrights, in
their theatrical practice, generally avoid material discovery. The
only exceptions are in *Héraclius*, where, however, the concealment
of true identity is part of a ruse and not in any way comparable to
that in *Oedipus*; in *Iphigénie*, the discovery of Ériphile's identity
occasions peripety only for herself and not for the central character
in whom we take most interest, Iphigénie's escape from it being in
that respect comparable to the self-discovery of Auguste in *Cinna*;
and in *Athalie*, where concealment is ruse again, but with the
complicity of the audience.

I have said that Corneille discusses the question of discovery
(and, by implication, of peripety) in the context of the dénouement. This in turn is related to the completeness of the action
which, he says, paraphrasing D'Aubignac — as usual without
naming him —, must be 'complète et achevée, c'est-à-dire que
dans l'événement qui la termine, le spectateur doit être si bien instruit des sentiments de tous ceux qui y ont eu quelque part, qu'il
sorte l'esprit en repos, et ne soit plus en doute de rien...'.[12] He
writes also of delaying the dénouement until the last possible moment in order to keep the audience on tenterhooks right to the end,
and of the audience's impatience to reach the end in order to see
what actually does happen. For this reason, too, he demands for
the last act what he calls a 'privilege', that of compressing a great
deal of action into it so that the knot may be swiftly unravelled
(third *Discours: Writings*, p. 73). It is clear from the *Examen* that
Corneille saw the establishment of the identity of Héraclius, not as
being connected with discovery and peripety, but as being connected with the dénouement. Again echoing D'Aubignac, he says:
'Comme il est nécessaire que l'action soit complète, il ne faut aussi
rien ajouter au-delà, parce que quand l'effet est arrivé, l'auditeur
ne souhaite plus rien et s'ennuie de tout le reste... Son attente
languit durant tout le reste, qui ne lui apprend rien de nouveau.'
Remembering D'Aubignac's strictures on *Le Cid* — 'la pièce n'est
pas finie' (*Pratique*, p. 140) — and his demand that the dénouement
being completed the play be ended, Corneille criticizes in retrospect two of his own early plays, *Mélite* and *La Veuve*, stating that
when he wrote them he was unaware of these rules.[13]

Now in that diatribe against Corneille which is what the first
preface to *Britannicus* really is, we find Racine taking the opposite
point of view. Discussing the last act of his tragedy, he writes:

Tout cela est inutile, disent mes censeurs: la pièce est finie au récit de la mort de Britannicus, et l'on ne devrait point écouter le reste. On l'écoute pourtant, et même avec autant d'attention qu'aucune fin de tragédie. Pour moi, j'ai toujours compris que la tragédie étant l'imitation d'une action complète, où plusieurs personnes concourent, cette action n'est point finie que l'on ne sache en quelle situation elle laisse ces mêmes personnes. C'est ainsi que Sophocle en use presque partout...

This, we may say, is merely an oblique objection, made by Racine the orthodox follower of D'Aubignac, to *Le Cid*. But does not Racine's phrase about 'en quelle situation elle laisse ces mêmes personnes' simply correspond to Corneille's 'le spectateur doit être bien instruit des sentiments des personnages'? Similarly, is not Corneille's 'l'action doit être complète et achevée' merely echoed by Racine's 'la tragédie [est] l'imitation d'une action complète'? It is all the more tempting to think so since Corneille's deprecation of Sophocles's *Ajax* — on the grounds that the action is completed with the death of the hero and that the subsequent dispute between Menelaus and Teucer is superfluous — finds an apparent and typical response by contradiction in Racine's self-defence behind the shield of Sophocles, and since, in his very next preface (*Bérénice*), he holds up not only *Ajax* but *Philoctetes* and *Oedipus* as supreme examples of tragic art. (Corneille echoes his own criticism of *Ajax* in his remarks on *Horace*, when in the *Examen* he acknowledges an infringement of the unity of action.)

What seems to be a fundamental difference between Corneille and Racine is suggested by some remarks of D'Aubignac (*Pratique*, p. 139): if the spectators 'ont raison de demander qu'est devenu quelque personnage intéressé dans les grandes intrigues du théâtre, ou s'ils ont juste titre de savoir quels sont les sentiments de quelqu'un des principaux acteurs après le dernier événement qui fait cette catastrophe, la pièce n'est pas finie...'. If we return to our questions and take the second first, it becomes clear that the word 'action' gives rise to ambiguity. In Corneille's sentence it really corresponds to plot. We ought in fact never to speak of 'unity of action' when we are concerned with dramatic technique: we should say 'unification of the plot'. In Racine's sentence, we have an Aristotelian statement not about plot — 'l'imitation d'une action complète' — but about the action.[14] One dramatist uses the word 'action' in terms of plot-construction, the other to refer to the nature of the tragic action itself. Plot is the means of representing a kind of action which is inherently tragic. If, then, we examine the first question, we see that likewise Corneille is talking in terms of the dénouement, i.e. part of the plot. The phrase about 'les sentiments des personnages' indicates what is their relationship to one another. Racine's expression 'en quelle situation elle laisse ces mêmes personnes' is, however, connected with the tragic action, not with the unravelling of the plot. The action is not complete unless we know what effect it has had upon the characters, particularly in a moral and psychological sense. Corneille seems to be looking on the *Poetics* — the ultimate source for both aspects of the problem — as a handbook of technical advice. Racine, ostensibly doing likewise, implicitly realizes that Aristotle provides him with enlightenment as to the nature of tragic action. Indeed, it may well be that Racine's failure to comment on the problems of dramatic construction springs from the fact that he recognized — though he would never admit it — that in the *Discours* Corneille had satisfactorily solved most of them. All this theoretical discussion on the part of the two playwrights is, however, only the reflection and justification of their own dramatic output, of which I shall now examine a few examples, beginning with two plays from whose prefaces I have quoted.

We have seen that the dénouement of *Britannicus* takes place at the end of Act IV and that, in Racine's terms, the remainder of the tragedy is taken up with 'la disgrâce d'Agrippine'. Two successive *péripéties* lead up to this 'disgrâce': Néron's agreement (IV.ii) to his mother's request for more honourable dealings with Britannicus and Junie brings about a change in her feelings and constitutes a *péripétie* for Agrippine: the decision is not hers, and it causes the action to move in a new direction, as is clear from Narcisse's report of her boastings in the final scene of this act. In turn, that report effects a *péripétie* for Néron: the actions of another character as narrated by Narcisse, cause him to reverse his earlier decision. Between them these *péripéties* are the immediate cause of the peripety which overtakes Agrippine: she has all along misjudged her son and she has committed the error of engineering his accession to the imperial throne, actions which recoil upon her. (Racine's portrayal of her excessive love for Néron could well have been suggested by Caussin, who insists on the cruel peripety she experiences when he plots against her.[15]) Her boastings after her apparently successful interview with Néron and the hopes she expresses to Junie in Act V scene iii are overturned by the discovery that her son has, after all, murdered his step-brother. Burrhus's account of the death of Britannicus would, presumably, in a play by Corneille, have brought the proceedings to a close: there is, in fact, a fine couplet on which, with a change of tense, to bring down the curtain:

> Et j'allais, accablé de cet assassinat,
> Pleurer Britannicus, César et tout l'état.   (ll. 1645–6)

Racine, however, must show the effects of the murder on Agrippine — it is her *hamartia* that has brought it about, and she must be enlightened; and so she is:

> . . . Arrêtez, Néron: j'ai deux mots à vous dire.
> Britannicus est mort: je *reconnais* les coups;
> Je *connais* l'assassin . . .   (ll. 1648–50)

Then, in the light of this discovery of her error, she makes her prophecy about the future of the emperor. When he leaves the stage, again the play could end; and again this will not do. There follows Albine's account of the escape of Junie and of the death of Narcisse. We are surely not much interested in the latter for his own sake but the point is that Albine's speech serves the same kind of purpose as the *récit de Théramène*: the fact that Néron has abandoned his most trusted adviser to the mercies of the Roman mob drives home the truth to Agrippine in showing her to what depths of depravity and disloyalty he can sink. It drives home to her the discovery of her error, the discovery not only of the true character of Néron, but of herself as well, and makes her more fully aware of the peripety which has overtaken her. The discovery is as much part of her 'disgrâce' as is the peripety. Indeed, they are inseparable, and they could not have been effected without the *péripéties*.

The same applies to *Bérénice*. Professor Scherer would have us believe that this is the ideal of the play without *péripéties*,[16] because nothing occurs to change the characters' feelings for one another. The term can, however, be used to indicate the creation of doubts about the nature or sincerity of those feelings. So, when Antiochus is asked to convey the fatal message to Bérénice, he begins to hope that she might now turn to him for love (III.ii). Having clearly suspected Titus of jealousy (II.v), she again, after Antiochus's announcement, doubts the sincerity of the emperor's love (V.iii; IV.i). These changes alter the direction of the action. Bérénice's original quest for the reason for Titus's apparent coldness after the period of mourning for her father is ended becomes a quest for his true feelings. Like Oedipus, or Neoptolomus in *Philoctetes*, she begins her search for the truth. Act IV is filled with alternations of hope and despair, and its end could be (and would be, no doubt, in the hands of Corneille) the end of the play: a separation of the lovers. But Act V, like the closing scenes of *Britannicus*, is necessary, for Bérénice has to discover that she has misjudged Titus (he does still love her), and his situation (he cannot marry her), and her own (she cannot find escape in suicide). Even within Act V — again as in *Britannicus* — there are *péripéties* (all those threats of self-immolation) which are required for this discovery and the accompanying peripety to be effected. Like the *Philoctetes* of Sophocles, mentioned by Racine in his preface, *Bérénice* is a play essentially of discovery. In more general terms than those I have so far used, the queen makes her error of judgement (*hamartia*) in coming to Rome in the first place in the hope or with the intention of being (or remaining) united with Titus. The major *péripétie* is the death of Vespasian, which is the prime cause for the discovery of her error, and the peripety consists in the inversion of the anticipated consequences of her initial act.

If, as seems evident, Sophocles's *Oedipus Rex* was for Aristotle the archetype of tragedy, Racine would appear to share his admiration — at least, if we may judge from his practice as a dramatist and not merely by his prefaces. This is not simply a matter of preference for tragedy with peripety and discovery — discovery of a different *kind* from that which features in Sophocles (or in any of the ancients), but discovery all the same. I have just mentioned that Bérénice searches for the truth about Titus. This is what Oedipus also does. His *initial* error of judgement, his *hamartia*, does not seem to me to consist in killing Laius in a fit of hot temper (though this is what is generally assumed), but in failing to pursue his quest for the truth about his parentage and leaping to an over-hasty conclusion which drives him out of Corinth to avoid killing Polybus, his supposed father, and marrying Periboea (or Merope), his supposed mother.[17] It is only when he is actually on his journey that he encounters Laius, his real father, and kills him. This incident, seen in this perspective, is part of the complication (the 'nœud', as the seventeenth-century critics call it). It is really a *péripétie*: it will eventually lead to Oedipus's search for the truth about the identity of Laius's murderer, but — and this is the peripety — that search will reveal the horrifying and quite unsuspected and unexpected truth about Oedipus's relationship to Laius and Jocasta. In the same way, on insufficient evidence, Bérénice concludes that Titus is either jealous of Antiochus or has ceased to love her altogether; but her nagging doubts drive her to seek out the truth: she finds it but — and here is the peripety again — she finds something much more heart-rending than a cessation of Titus's love: he loves her still, but she must be separated from him.

The same analysis may be applied to Thésée in *Phèdre*. On his return to Troezen, finding his family avoiding him or tongue-tied, he begins his search for the truth which is being concealed from him. On false circumstantial (and insufficient) evidence, like Oedipus he leaps to the wrong conclusion and acts in the haste of anger; but doubts nag him and he pursues his quest, finding the truth too late. It is that final discovery and the complete overturning of his assumptions which occupy the last act of the play. Athalie, in her quest for the truth about the presence and identity of Éliacin, makes the error of entering the temple. She questions the boy in a scene derived from Creusa's similar interrogation (with similar intent) in Euripides's *Ion*.[18] Frustrated in her attempt, Athalie invades the sacred place, where the truth is finally discovered to her. As Racine puts it in his preface, the play 'a pour sujet Joas reconnu'.

None of these examples, however, suggests any direct parallel in Corneille, except that, as we can see from *Britannicus* and *Bérénice*, he would probably have ended all these tragedies before the final discovery. And here, the last act of *Mithridate* seems to me to be particularly revealing. As we have seen, that play may, from the situational and thematic viewpoint, be compared with *Nicomède*. From the point of view of construction, however, Racine's tragedy may be more profitably compared with *Rodogune*. The closing scenes of both plays are fast-moving, rich in *péripéties*, highly dramatic; both arouse our curiosity; both make use of a material object — the poisoned cup — as one means of heightening the tension; in both, at least one *péripétie* is fortuitously or intentionally reversed. But there is nothing to be disclosed to Cléopâtre: indeed, far from searching for the truth, as Mithridate seeks out the real nature of Monime's feelings for Xipharès,[19] she attempts to conceal it to the end, falsely accusing Rodogune of the murder of Séleucus; and Antiochus's endeavours to discover which of the two women is guilty remain singularly ineffectual. There is, then, in *Rodogune*, no discovery: the final act simply unties the knot, satisfies the audience's curiosity, and shows how the characters stand in relation to each other. We may say that there is peripety, in the sense that Cléopâtre's machinations recoil upon her, but even she does not recognize that this is so; still less does she discover the root causes in herself of the disaster which overtakes her. She persists to the end in self-glorification and dire threatenings, and we find in her none of that realization of being humbled which results from a true conjunction of peripety and discovery.

As for Mithridate, he undergoes a twofold peripety and discovery, as does Oedipus: first, that of knowing not only that Monime is in love with Xipharès, but also that he, the trusted son, is disloyal to him to the extent of pledging his love to her; then, that of knowing military disaster — Mithridate's avowed intention to invade Italy recoils upon him (and he is made to realize it) in the form of a Roman invasion of his own territory. Racine, in his preface, makes this clear: '. . . ce projet fut le prétexte dont Pharnace se servit pour faire révolter toute l'armée, et . . . les soldats, effrayés de l'entreprise de son père, la regardèrent comme le désespoir d'un prince qui ne cherchait qu'à périr avec éclat. Ainsi elle fut en partie cause de sa mort, qui est l'action de ma tragédie.' In Corneille's hands, to judge from the parallel of *Rodogune*, the play would have ended, in Act V scene iv, on the words of Monime:

> Ah! que, de tant d'horreurs justement étonnée,
> Je plains de ce grand roi la triste destinée!   (ll. 1639–40)

But no — the dying king must reappear, like Phèdre. He still boasts, like Cléopâtre, about his glorious past, and still gives martial advice for the future; but, in strong contrast to Cléopâtre, there is intense pathos in the juxtaposition of those lines which also reveal that, although his pride will not allow him to admit it openly, he has discovered the truth of his defeat — and this is peripety for him — both in love and in war. There is, in the last act of *Mithridate*, much more than the dénouement, in the proper and technical sense, which we find in *Rodogune*; more, too, than the exciting *péripéties* (one in each of the first four scenes) which are partly responsible for producing the final discovery.

With its simplicity of situation and plot and, in the technical sense, its early dénouement, *Suréna*, which has often been described as Corneille's attempt at Racinian tragedy, might seem to fall into the same category as *Bérénice* or *Phèdre*. Corneille's last play has been compared to *Bajazet*, in particular for its dénouement and the certainty of the catastrophe from the beginning of Act IV.[20] It has no peripety and no discovery because throughout the last two acts every character is morally certain as to where every other character stands. It is true that in this tragedy one can see a quest for the truth, Pacorus's relentless interrogation in turn of Suréna, Eurydice and Palmis (Act II) being followed by Orode's questioning — prompted by the Machiavellian informer, Sillace — of Suréna and Palmis (Act III). Although the truth has not been admitted, the suspicions of the king and his son appear to be well enough founded by the beginning of Act IV for guards to be placed to prevent the escape or revolt of Suréna and Eurydice. Pacorus's quest for the truth about Eurydice's willingness to marry him turns into an attempt to extract from her a declaration of her love for him (IV.iii). Her steadfast refusal to make it is coupled with her unwillingness to persuade Suréna to marry Mandane as the only way of saving his life. Her *hamartia* is a real error of judgement: she resists all the evidence that Orode is in reality capable of destroying the greatest support of his own power (IV.ii) and is still resisting at the very moment when Suréna is being assassinated (V. iv). She does finally yield to the pleadings of Palmis (though she is not enlightened as to her own error) but it is already too late. The laconic announcement made by Ormène (ll. 1713–17) does not bring discovery either, nor any sense of peripety. Neither is Suréna enlightened, nor are we shown him undergoing peripety, but he

has committed two errors of judgement, first in being too frank
with Orode (see the latter's response, 1. 937), and then in defying
Pacorus's command that he marry Mandane (ll. 1309 ff.). He
knows all along what his real fault is in the eyes of his superiors: it
lies in his excess of 'vertu' and 'gloire' (V.ii). No amount of com-
pliance will save him in the end, a truth to which he returns immediately before his fatal exit (V.iii). He, too, like Eurydice, maintains that Orode would not risk destroying him. So Suréna does
not need enlightenment as to his 'defect', and he never achieves it
as to his mistaken judgement. Corneille's play, then, although
structurally akin to several of Racine's, does not include either of
the complicating elements of peripety and discovery. In this, it
bears a marked resemblance to Sophocles's *Oedipus at Colonus*, also
a play written in its author's old age. What we witness in *Suréna* is
an interrogatory quest for the truth, as in *Bérénice* or *Mithridate* or
Phèdre, but, once that is accomplished, we watch the net tightening
on the victims whose obstinate fidelity one to another prevents
them from ever seeing their errors of judgement. The actual catastrophe is withheld until the last short scene of the play and its
effects are not explored.

What, then, of *Horace*, whose critics alleged that the dénouement
came in Act IV scene v, before the death of Camille and at the
point of the hero's victorious return from the battlefield? Corneille
dwells at some length in the *Examen* on the way in which the murder constitutes an infringement of the unity of action because
Horace's new peril is not a necessary consequence of the first. Yet,
in spite of criticism, he maintained the end of Act IV and the
whole of Act V in successive editions of the play. It is certain that
the murder brings about peripety for Horace: he carries it out and
the deed recoils upon him in his falling from victory into disgrace,
or from happiness into misery, as Aristotle would say. Is the remainder of the play then devoted to a discovery of the kind we see
in *Andromaque*, *Britannicus*, or *Phèdre*? At first one might think so.
Horace confesses his crime and is willing to die. But it is as a crime
in the legal sense that he sees the murder. He first calls it 'un
châtiment' (l. 1321), 'un acte de justice' (l. 1323), 'un supplice'
(l. 1324), a punishment of Camille's 'crime' (ll. 1331, 1427) which
is in its turn a crime. He recognizes that both his father (V.i) and
his king (V.ii) have the authority to punish it, but neither immediately after the murder (IV.vi) nor at any later stage does he
accept moral guilt. In his encounter with Sabine (IV.vii), he takes
no blame upon himself, but blames her for not sharing in his glory,
exhorting her to follow his example:

> C'est à toi d'élever tes sentiments aux miens,
> Non à moi de descendre à la honte des tiens...
> Embrasse ma vertu pour vaincre ta faiblesse,
> Participe à ma gloire au lieu de la souiller...
> Fais-toi de mon exemple une immuable loi.

Horace is still concerned only with his 'gloire'. His wish to die at once springs from a desire to leave the world at the height of his glory and unsullied by the shame of the crime he has committed (ll. 1429 ff.). The king's power to punish is that of the law (l. 1537) to punish crime (l. 1541), and that is how Tulle himself sees it:

> Un premier mouvement qui produit un tel *crime*
> Ne saurait lui servir d'excuse *légitime*:
> Les moins sévères *lois* en ce point sont d'accord;
> Et si nous les suivons, il est digne de mort.   (ll. 1735–8)

The king recognizes the 'enormity' of Horace's crime (l. 1733) and makes no attempt to excuse it. If Horace is to be saved, he is not to be pardoned. No word of pardon is pronounced. Paradoxically, Tulle, referring only fleetingly (l. 1734) to the real guilt of Horace, treats this outlaw as the bastion of his own law-giving authority (l. 1745): which means that even he scarcely comes to any real discovery of the nature of Horace's guilt. Instead of being pardoned, Horace is put above the law:

> Ta vertu met ta gloire au-dessus de ton crime.   (l. 1760)

Meanwhile, Horace has said, referring to Valère:

> Il demande ma mort, je la veux comme lui.
> Un seul point entre nous met la différence,
> Que mon honneur par là cherche son assurance,
> Et qu'à ce même but nous voulons arriver,
> Lui pour flétrir ma gloire, et moi pour la sauver.   (ll. 1550–4)

He is, furthermore, quite explicit in not recognizing the depth of his guilt:

> Et si ce que j'ai fait vaut quelque récompense,
> Permettez, ô grand roi, que de ce bras vainqueur
> Je m'immole à ma gloire, et non pas à ma sœur.   (ll. 1592–4)

These are his last words, to be followed by Sabine's plea and that of le Vieil Horace (V.iii) and the king's final judgement in which, as we have seen, crime is redeemed by glory. The peripety is itself in part at least reversed, but of profound Racinian discovery of the truth there is not a trace.

In discussing the completeness of the tragic action (first *Discours: Writings*, p. 26), Corneille writes a revealing passage which is of great significance in this context. Of the last act of a tragedy he says:

... il faut, s'il se peut, lui réserver toute la catastrophe, et même la reculer vers la fin, autant qu'il est possible. Plus on la diffère, plus les esprits demeurent suspendus, et l'impatience qu'ils ont de savoir de quel côté elle tournera est cause qu'ils la recoivent avec plus de plaisir: ce qui n'arrive pas quand elle commence avec cet acte. L'auditeur qui la sait trop tôt n'a plus de curiosité, et son attention languit durant tout le reste, qui ne lui apprend rien de nouveau. Le contraire s'est vu dans *la Mariane* dont la mort, bien qu'arrivée dans l'intervalle qui sépare le quatrième acte du cinquième, n'a pas empêché que les déplaisirs d'Hérode, qui occupent tout ce dernier, n'aient plu extraordinairement; mais je ne conseille à personne de s'assurer sur cet exemple.

In the first place, the final sentence of this passage, in which Corneille treats Tristan's play of 1637 (the year of *Le Cid*) as a solitary example, must surely have struck Racine as a challenge, to which the construction of *Andromaque* and, more particularly, of *Britannicus*, seems to be a reply, proving that the author of those plays could indeed 's'assurer sur cet exemple'. The first preface to *Britannicus*, with its defence of Act V as being necessary, precisely, to complete the action, its reference to the practice of Sophocles and its indication of what needs to be accomplished after the death of the hero ('les imprécations d'Agrippine', 'la retraite de Junie', 'la punition de Narcisse', 'le désespoir de Néron'), points to a function of the last act unsuspected by Corneille. Racine argues that the spectators' attention does not languish — on the contrary — because they wish to know 'en quelle situation' the action leaves the characters. It is still a form of curiosity, but its object is knowledge less of what, in the simple sense, happens to the characters at the end, than of what effect events have upon them. A direct parallel may also be seen between Corneille's analysis of *Mariane* and Racine's (in his second preface) of his own play. In Tristan's tragedy, the heroine's death occurs before the beginning of Act V, bringing misery to all, including the husband who has condemned her and the mother who had not dared to defend her. The murder of Britannicus is reported in Act V scene iv: four further scenes follow. 'Les déplaisirs d'Hérode' find an echo in 'la disgrâce d'Agrippine'.

Since he thinks of the audience's curiosity as being principally aroused by the uncertainty of the purely dramatic outcome, Corneille not surprisingly demands for the last act the 'privilege' of 'precipitation' (third *Discours: Writings*, pp. 72–4). He cites *Le Cid* and *Pompée* as examples, and points out that they 'forcent la vraisemblance commune en quelque chose'. That very formula, however, suggests that 'la vraisemblance extraordinaire' is not infringed. Because the characters of Corneille's tragedies are extraordinary and are involved in extraordinary situations and actions, the 'precipitation' of Act V falls within the internal 'vraisemblance' of the plays. The final 'conversion' scenes of *Polyeucte, Théodore, Cinna*, and *Nicomède* clearly come into this category, but they also introduce a kind of instant recognition based on admiration for the suffering but triumphant hero. This, however, is discovery of exceptional moral qualities in others, not, as in Racine, of moral defects in oneself. Cornelian discovery, in this sense, is an immediate response to the spectacle of heroic action, while in Racine's plays the characters recognize the nature and moral consequences of their own usually unheroic acts after they have been committed. Cornelian discovery, where it exists, is part of the dénouement: it is in the course of the subsequent catastrophe that Racine's characters make their discovery. La Mesnardière (*Poétique*, pp. 230–1) pleaded for a single peripety placed not too near the end of the play, on the moral grounds of allowing the guilty and remorseful the opportunity to repent. Corneille's contrary insistence (prefatory *Epître* of *La Suite du Menteur: Writings*, p. 183) on the aesthetic rather than the moralizing function of drama could have been one factor in determining his conception of the last act of his plays.

Racine presumably grasped more clearly than did Corneille the import of Aristotle's separation of peripety and discovery on the one hand from dénouement on the other. He discusses the first two when dealing with the nature of the tragic action (chapters X and XI), and analyses dénouement (*lusis*) along with what English translators call 'complication' and the French 'le nœud' (*desis*), that is, in the context of plot construction (chapter XVIII). And it is to the plot that the *péripéties* belong.

Corneille appears to use the *péripéties* not merely for the sake of exciting curiosity and creating suspense — though that is what he himself says many times — but in order to arouse a certain kind of emotion (wonder, admiration) through the spectacle of heroic characters being subjected to exacting moral ordeals. Such characters, knowing themselves and their purposes, cannot usually be involved in Aristotelian discovery. This, incidentally, links up — Nicomède immediately comes to mind — with Aristotle's ideas on the perfect character knowing immediately, without deliberation, what course of action to adopt. If one considers Auguste, in *Cinna*, as a character involved in discovery — discovery that the pursuit of mere power entails the destruction of one's moral self — there is no peripety, because he still has the means to punish (that is, to maintain his power by 'orthodox' methods); but he refuses to do so, and thus saves himself from peripety. Self-discovery seems to preclude peripety and, on the evidence of Cléopâtre, in *Rodogune*, the converse is also true. Such moral discovery as we see in Auguste is an interiorization of Aristotle's third type of tragedy (*Poetics*, chapter XIV), in which the character makes the discovery of his relationship to his would-be victim 'in time to draw back'.

Racine similarly uses *péripéties*, but in order to arouse different emotions — curiosity, like Corneille, as we have seen, and even admiration occasionally (for Monime and Xipharès, for Hippolyte, for Joad), but primarily pathos. There is no greater source of pathos — Aristotle says so (*Poetics*, chapter XI) — than the spectacle of suffering, and, in dramatic terms, there is no more pathetic spectacle than that of a character involved in peripety and the discovery of a tragic destiny. Pity and fear are aroused by the sight of such characters rushing headlong, in ignorance, and as a result of some error of judgement, to that discovery and that peripety. The *péripéties* in Racine excite pathos precisely because the audience can see that each one of them is a step on that road to discovery, to that change, as Aristotle says, 'from ignorance to knowledge'. The true tragic emotions are complex: we stand outside the characters, knowing more than they do, but in fear and pity for them, and ultimately surprised with them, so to speak, because they are like ourselves, because we recognize ourselves in them, because, thanks to the manner of their presentation, we are allowed to become them and to be involved in their action. If this were not so, there could be no catharsis.

No doubt Racine, in order to make room, as it were, for discovery and peripety, simplifies his plot to some extent, and generally confines the *péripéties* to the first four acts. 'The action', writes Aristotle in chapter X of the *Poetics*, 'I call simple, when the change in the hero's fortunes takes place without Peripety or Discovery; and complex, when it involves one or other, or both.'[22] He is thinking, of course, of the complexity of the emotions aroused in the audience. And on that basis, the dramatic art of Racine is, paradoxically, less simple than that of Corneille, for it involves a combination of peripety and discovery.

CHAPTER VII

Rhetoric: Trial and judgement

...The drama abounds in trial scenes, good and bad...The
stage and the law court are two versions of the same thing —
our human refusal to obey the precept, Judge not that ye be
not judged. Plays are, in a symbolic sense, 'trial scenes'; and it
is inevitable that they run to thousands of trial scenes in a
literal sense. This being so, it is also inevitable that the language of the law court should creep in too. Judge, prosecutor,
advocate, plaintiff, defendant — what play could not be written with these five characters and a witness or two? Court-
room language is a dramatic language in that it finds itself
under the dual compulsion of the theatre: to keep things moving and to be at each moment esthetically impressive.

Eric Bentley, The Life of the Drama

The thought of the personages is shown in everything to be effected by their language — in every effort to prove or disprove, to arouse emotion (pity, fear, anger and the like), or to maximise or minimise things.

(Aristotle, *Poetics*, chapter XIX)

Cette partie a besoin de la rhétorique pour peindre les passions et les troubles de l'esprit, pour en consulter, délibérer, exagérer ou exténuer...

(Corneille, first *Discours*)

We have seen how, in the last act of Corneille's play, Horace is arraigned before his king, accused and defended by his family and exposed to the judgement of the audience. The procedure is apparently judicial with prosecution and defence by counsel, a statement from the accused and a verdict from the judge. It has been remarked[1] that such 'trials', whether or not they end in an actual judgement, as in *Horace*, are frequent in the plays of Corneille, but absent from those of Racine. The implications of the observation are far-reaching. Even in *Suréna*, for example, where no actual trial as such is to be found, procedures are adopted which have a judicial character: both Pacorus and Orode behave like investigating magistrates seeking evidence on which to base an accusation of Suréna and Eurydice, who attempt to conceal that evidence, not by denying its existence, but by proffering evidence of a different kind in order to justify themselves. Indeed, the construction and, in more senses than one, the plot of several other of Corneille's plays is based upon judicial or quasi-judicial proceedings: *Le Cid*, *Polyeucte*, *Nicomède*. With the exception of *Esther*, it has been said, 'Racine s'est interdit de mettre en scène un procès explicite... L'emportement de l'amour, de la haine ou de l'ambition est... trop prompt pour qu'un procès puisse s'y instaurer. La tragédie racinienne est le lieu du crime. La tragédie cornélienne est celui du châtiment.'[2] Attractive though this formula may be, it is not substantiated by an analysis of plot and dénouement as we have pursued it. True no doubt it is that some of Corneille's plays bring punishment to the guilty — or occasionally pardon, as in *Cinna* —, but the implication that in Racine's the force of passion is manifested in a precipitation (to use Corneille's word) which precludes everything, including judgement and punishment, except the crime itself, is clearly not tenable in the light of our findings so far. It is, after all, Corneille who claims the right to hasten the action in the closing act of his plays, and Racine who seeks to defend himself against criticisms of alleged slowness and superfluity. It is hazardous to attempt to isolate the particular characteristics of a work of art without reference to their relationship to the whole or indeed without reference to the nature of that whole and its total effect, and in drama in particular to neglect structure and plot, the 'soul of tragedy', as Aristotle puts it.

The fact is that Corneille's plots take the form of a series of trials (whether actually judicial or not) and tests through which the characters are held up to the judgement of the audience if not of a stage judge: tests of resolution and moral strength, arousing interest and curiosity not simply as to whether the characters will succeed in them but also as to the manner in which the tests are set and in which the characters face them. Usually their behaviour is such as to arouse our admiration and wonder, sometimes such as to arouse our sympathy as well. The outcome even of the political trials is not of a kind to bring the accused characters to moral self-discovery. Horace discovers nothing about himself and makes no admission of *hamartia* in the course of his trial: in a legal sense he admits to having committed the crime of murder, but he does not in a moral sense recognize this as being the result of a blindness occasioned by the passion of his 'gloire' carried to an inexcusable extreme. Indeed, as we have seen, he explicitly justifies himself by continuing to put the blame on Camille. We watch him survive successive trials of his will and courage (his 'vertu') and see his heroism grow, thanks to his iron self-control, until he reaches the final test, the encounter with his sister, in which he fails precisely because his self-control fails. But this he does not acknowledge as such a failure but seeks to justify it by appealing to the abstract principle of patriotism.

If it can be said that Racine's characters are also put to the test, it must be admitted that the test is of a very different kind. In the first place, it is not imposed upon them by external circumstance or by authority, but by themselves in their relations with others. It is Oreste's own wish to undertake the embassy to Pyrrhus, not in the hope of being successful but in that of meeting and marrying Hermione. He sets himself an impossible test by thus creating a situation which ends in his failure both as ambassador and as lover. Pyrrhus likewise puts himself to the test by retaining Hermione at his court while rejecting her love for the sake of his passion for Andromaque. Titus undergoes a test of his own making by first bringing Bérénice to Rome and keeping her there after the death of his father. On the other hand Néron puts Junie and Britannicus to the test in the spying scene: Roxane does the same to Bajazet and Atalide, Mithridate to Xipharès and Monime, Thésée to Hippolyte and Aricie. Little of this, however, can be said to have a judicial character, though it is all calculated to arouse curiosity and sympathy. What we have seen in our analysis of the endings of Racine's plays is not a verdict passed on the characters, but their discovery, as a result of the tests occasioned by the *péripéties*, and failure in them, of the enormity of their own *hamartia* and indeed of themselves. If judgement there be, it is they themselves who make it. To that extent it is true that Racine does not include trial scenes or procedures as such in his tragedies and that one may see in their absence a valid distinction between his plays and those of his rival.

But this distinction raises its own problems. Our earlier analysis of *Cinna* showed us Auguste narrowly avoiding peripety by making a crucial discovery about the nature of power and about himself in relation to it. He withdraws from the brink of further punishment and proscription because he comes to the realization that the true glory resides not in mastery over the universe but in mastery over himself. This is, therefore, the discovery as much of an abstract principle or ideal as a recognition of his own inner proneness to a blinding passion. In terms of plot and structure, it is not Auguste who has been on trial, but Cinna, Maxime, and Émilie: the emperor is in fact their judge after the discovery of the conspiracy. The development, concealment, and discovery of the conspiracy constitute the beginning, middle, and end of the plot. The conclusion is, however, totally unexpected in the sense that Auguste judges himself by his very act of pardon. That act discovers his true greatness to those who had, willingly or reluctantly, plotted against him, and it is that discovery, of his moral authority, rather than of themselves, which brings about their immediate and spontaneous conversion to a new loyalty to him. This is not at all the same kind of discovery as Hermione or Oreste, Agrippine, Bérénice, Mithridate, Thésée, or Phèdre are finally brought to accept: nowhere in Racine, except perhaps in Athalie, do we find recognition of the moral authority of another character, and even in his last play, the queen may say:

> Dieu des Juifs, tu l'emportes! . . .
> David, David triomphe; Achab seul est détruit.
> Impitoyable Dieu, toi seul as tout conduit,   (ll. 1768, 1773–4)

but there is no question of her conversion. *Cinna* is not the only play by Corneille whose ending is characterized by conversions. The recognition of celestial glory in the martyrdom of Polyeucte converts Pauline and even Félix to loyalty to his God. The immovable devotion and courage of Nicomède bring the other characters to recognition of his ideal of true kingly authority. Recognition of this kind arises out of awe and admiration for unswerving service to something transcendent and for its vindication in the tests, judicial or other, to which it has been subjected. Like Auguste's own final act of clemency,[3] which springs from his discovery, such recognition of the ideal in others is instantaneous and dramatic, but not technically unprepared. The inner self-discovery in Racine's plays differs radically from this not only, as we have seen, in nature, but also in manner. Born of failure to achieve a passionately desired end, it represents only the last stage in a recognition which, in spite of resistance, is shown to be inevitable. Self-deception gradually yields to enlightenment: through crises which subject him to contrary emotions, Oreste ultimately discovers what he has known all along, that Hermione cannot reward his service, even if it includes regicide at her behest, with the love with which she desires Pyrrhus. Bérénice's doubts, suspicions, and accusations eventually give way to recognition that, as she has really known from the beginning, Titus does love her devotedly and faithfully. He, in his turn, makes the full and final discovery, in his great soliloquy in Act IV, that he has to find it in him to accept the truth which he already knows and to acquaint Bérénice with it and its inevitable consequences. This process, which extends over the whole play and is expressed in the first place in the nature of the plot, through which the characters are inescapably caught in an ever-tightening web, seems to be Racine's equivalent of the recognition which he so much admired in Sophocles's *Electra*, 'bien amenée de parole en parole' (l. 1202: Picard, vol. ii, p. 852).

Such distinctions between Corneille and Racine are evident from the handling and nature of plot and its relation to the function of the dénouement. Let us now consider in detail two examples from Corneille in which judicial procedures are actually used, one from *Horace*, the other from *Cinna*.

The last act of *Horace* is devoted almost entirely to the trial of the hero. As we have seen, it follows a quasi-judicial procedure and it can easily be shown that the forms of the speeches are cast in the mould of the accepted legal rhetoric of Corneille's day.[4] That is perfectly natural, not only in view of the great importance attached to the study of rhetoric, with abundant examples and exercises, in seventeenth-century education in France, but also because of the poet's own training and practice in the legal profession. The object of the rhetorical plea was to persuade, and all the devices available to the lawyer or orator are to be found in the speeches of Corneille's characters, which are set out in a clear order and make use of argument and counter-argument, compliments (even flattery) and irony, examples and maxims, appeals to precedent, and of course rhetorical questions, interjections, and other figures of rhetoric. Yet although that is true in general terms, examination of the pleas and of the judgement reveals other characteristics, unexpected perhaps within this framework.

Let us look first at Valère's case for the prosecution. In relation to the sources on which Corneille drew (Dionysius of Halicarnassus in this instance) he has become the single representative of the Roman nobles who, in the historical accounts, pressed for the punishment of Horace. The individualization, with a proper name, of an anonymous group is in itself indicative of the fundamental shift from history to drama, as is the reduction, within the play, of the three brothers to one character on either side. More important is the fact that Valère is placed in a clear emotional relationship to Camille and Curiace and through Camille to Le Vieil Horace. The consequences of these inventions will be obvious in the nature of part of the arguments he advances. He begins, however, in the conventional way, by paying compliments to the judge, the King (ll. 1481–91). His main contention, introduced in the last lines of the exordium, is that Horace's example, if it goes unpunished, will be a perpetual danger to the Roman people. Although this may have the appearance of an argument from principle, it is in fact simply a forecast of the probable consequences of present leniency. The argument is reinforced by the suggestion that, because so many Romans are bound to Albans by ties of family, marriage, and friendship, many of them might behave as Camille has done and so provoke similar punishment. Valère is turning back to front the usual procedure in legal rhetoric: he is projecting a future possibility on to a present case, not arguing that case on the grounds of precedent.

Behind the seemingly disinterested and patriotic plea, there lies moreover an element of egocentricity. In the situation as Corneille has imagined it, Valère is necessarily pleading for the punishment of the murderer of the woman he loves and whom, once his rival has been killed, he might have hoped to marry. The background to Valère's speech here is his behaviour in Act IV scene ii, where he has recounted the death of Curiace (and his brothers) with barely concealed satisfaction and without regard for the feelings of Camille as she listens, and where his account is followed by ingratiating remarks calculated presumably to dispose her father in his favour as the rival and possible successor of the Alban warrior. The underlying motive for Valère's plea accounts for the touch of evident irony with which he presents it. Of Horace, he says:

> Faisant triompher Rome, il se l'est asservie;
> Il a sur nous un droit et de mort et de vie;
> Et nos jours criminels ne pourront plus durer
> Qu'autant qu'à sa clémence il plaira l'endurer. (ll. 1507–10)

Even the arguments of political expediency, in the absence of those of precedent, are thus rendered suspect by an unexpressed wish for personal revenge which cannot fail to colour our interpretation of

Valère's final argument that the sacrifice of thanksgiving arranged by Tulle for the morrow will be rendered sacrilegious if performed by guilty hands. Again, no actual principles or precedents are invoked but rather, in the peroration, a demand that the possible consequences weigh in the balance against clemency:

> En ce lieu Rome a vu le premier parricide;
> La suite en est à craindre, et la haine des cieux:
> Sauvez-vous de sa main, et redoutez les Dieux.   (ll. 1532–4)

Valère here comes dangerously near to committing sacrilege himself, because the King is not only the ruler of his people but the representative and interpreter of their gods. Not only does Horace recognize the peculiar status of the King (ll. 1538–40), but so do Valère (ll. 1469–72) and of course Tulle himself (ll. 1476–8). This could be one reason why, when Valère attempts to make a second intervention in the proceedings, Tulle curtly silences him (l. 1729); and although the King assures him (ll. 1731–2) that he has not forgotten his arguments he does not, in his judgement, refer to them specifically. At the same time, a clue is given to the understanding of Valère's motives when Horace, while admitting his guilt and his preparedness to die (and therefore not arguing against the consequences of Valère's case), emphasizes the nature of the personal relationships:

> Je ne reproche point à l'ardeur de Valère
> Qu'en *amant* de la *sœur* il accuse le *frère*...   (ll. 1547–8)

From neither a judicial view nor even a moral one can Valère be said to speak to a very good brief. What then of Horace when Tulle invites him to defend himself? Even before the conventional compliments of the exordium, Horace in two and a half lines signifies his inability to do so. He accepts without question the authority and justice of the King. He does not enter a plea of 'Not Guilty'. While indicating, as we have seen, that Valère's accusation springs from personal motives, he agrees with him that he ought to die, but less in order to expiate his crime than to save his honour. Within the context (ll. 1555–72) of a disparagement of the fickleness and superficiality of the judgement of the populace (a theme which his father will take up later — ll. 1711–16), Horace extols his own 'virtue' and expresses the fear that any future deeds of valour might fall short of the one he has just accomplished; and so,

> ... pour laisser une illustre mémoire,
> La mort seule aujourd'hui peut conserver ma gloire.
>   (ll. 1579–80)

Besides, his crime will detract from his glory if he must go on living with it:

. . . Pour mon honneur j'ai déjà trop vécu.
Un homme tel que moi voit sa gloire ternie,
Quand il tombe en péril de quelque ignominie . . . (ll. 1582–4)

None of this remotely represents a legal or, in any real sense, a moral argument. The fickleness of the populace is adduced only as a reason for death being preferable to a diminution of reputation. In spite of Horace's acceptance, in Act V scene i, of his father's absolute authority over him (parallel to that of the King's in his exordium) he does not repent of his crime although le Vieil Horace deeply regrets it (ll. 1416–8). Just as, before Tulle, he expresses a preference for death over dishonour, so, before his father, he has admitted that the murder of Camille sullies family honour (l. 1426) but has also claimed that it was the means of wiping out the blot already made on it by her reproaches against him and against Rome (l. 1417). In that same speech, Horace urged his father to punish him as any Roman father should, in similar circumstances, in order to save 'sa propre gloire'. As we have just seen, when he faces the King, it is his own honour which Horace seeks to ensure. In the same way as an earlier episode provides an interpretation of Valère's speech for the prosecution, Horace's interview with his father immediately before that speech reinforces our impression of his being driven solely by a sense of his 'gloire' and of that of his family.

Of course, 'la gloire' can be and has been thought of as an abstract ideal which for the hero transcends all other considerations, even those of his personal desires and aspirations and those of his family, and is completely disinterested. Yet out of three dozen mentions of the word in the play, only three, possibly four (Valère's use of it in l. 1528 being ambiguous in function), bear a general and abstract interpretation: in twenty-seven of the remaining instances (excluding the doubtful one), the concept is clearly attached to personal honour and a desire for renown, and in five it is connected with the related idea of family reputation. Horace, using the word twelve times in one of these two senses, and his father, using it nine times, show themselves to be more deeply attached to the ideal in this personal sense — the word is usually accompanied by the possessive adjective, or its equivalent — than to any notion of a moral absolute, and more deeply attached to it than are any of the other characters. (Curiace, a long way behind, is the runner-up). Similar observations can be made about allied words and expressions, such as 'honneur', 'vertu', and 'générosité'. In the light of these considerations, it is perhaps not surprising that in their speeches before the King, Horace and his father plead scarcely more than does Valère from precedent, principle, or abstract ideal.[5] (It is interesting to note that in his very last tragedy as in the first of his great series, Corneille causes Suréna to die precisely because his 'gloire' is greater than Orode's and because he boasts of it despite his awareness that it represents a danger to his life. Use of rhetoric in this way is allied to the overblown claims of petty tyrants (Ptolomée, Félix) and Machiavellian advisers.)

Before le Vieil Horace makes his plea — easily the longest and most eloquent — Sabine comes before the tribunal and makes hers. It is neither for nor against her husband: she accepts his guilt but recognizes his greatness. In asking to be allowed to die in his place her arguments are entirely emotional: her death would be a punishment for Horace; she would be released from the contrary obligations to love the man who has destroyed her family and to hate the man who has so signally served his city; her death would satisfy the gods with her blood in vengeance for Camille's. Just as in the opening scenes of the play her motives were of the most human — the desire to safeguard her happiness both in her Alban family and in her Roman marriage —, so now she seeks release from the burden of misery which is imposed upon her by its destruction. Although in the past she has endeavoured to accept Roman values and has even succeeded in putting her adopted city's glory before that of her own family (ll. 75–8), the play shows that she has reached the end of her tether in the conflict of loyalties and loves to which she is exposed (see especially the soliloquy in Act III scene i). Twice before the trial she has asked to die. Even if her plea to die in Act II scene vi is a monstrous bluff, it frightens Horace, who takes it seriously; but what is interesting about it in the present context is that she sees the whole situation in purely personal terms, on the level of personal emotion, a level beneath that of Horace and even of Curiace. The second plea comes in Act IV scene vii, after the death of Camille. She addresses Horace with deep and wounding irony, declaring that she is more guilty than his sister and deserves more to die. This plea may also be a bluff, but like the earlier one it pierces Horace's fragile moral armour: he flees from her (ll. 1391–7), in order not to lose his temper a second time and so to kill her as he had killed Camille? Perhaps, but surely his rage has passed, and Sabine, unlike Camille, is not a Roman and so is more excusable. But he loves her (l. 1355) — and she knows it (ll. 1607–10) — passionately. Her reference to his passionate love takes its place in the emotional speech she makes before the King. Nowhere does she advance a real argument, certainly not any argument based on precedent or principle.

And the plea entered by le Vieil Horace, the last of the group to speak? Its form, in the first place, does not fall within the conventions of legal rhetoric. It opens, not with the usual complimentary exordium, — but perhaps the speaker, an elderly and respected member of the aristocracy, would be excused that obligation — but with four lines addressed to the King in which the old man's lonely plight is shown to consist not only in the loss of his children but in the conspiracy of the survivors against him. Then le Vieil Horace addresses in turn Sabine, Valère, and Horace, and Tulle after each of them. The speech ends without a peroration, without even, in the last two almost laconic lines, a formal appeal to the King. The appeals to Sabine, to Valère, and to Horace could have no place in formal judicial proceedings: they are highly emotional pleadings for Sabine to do her duty by her own brothers in being loyal to her husband, for Valère to understand the horror of the glorious warrior being put to an ignominious death, and for Horace to disregard the possible attitude of the populace and to live in the confidence of his monarch's divinely-inspired judgement. While in their form these appeals do not constitute a speech for the defence, they are related to le Vieil Horace's words to the King himself, if only in the sense that they constitute a kind of living demonstration, more eloquent no doubt than a formal plea, of the reasons why Horace should be pardoned. They open (ll. 1648–54) with a statement of general principle: a spontaneous act which proceeds from a noble motive can never be criminal. This provides the foundation for the argument that Horace killed his sister with such a motive: 'Le seul amour de Rome a sa main animée' (l. 1655). And that of course is a direct echo of Camille's curses and imprecations against Rome in the murder scene itself (ll. 1301–18). In so far as the murder can be justified at all, that is the most valid argument, given the scale of values of Horace and his father. But then, in a series of rhetorical questions and exclamations, the father reminds Valère of his anger when it was reported to him that Horace had fled from the battlefield (Act IV scene ii — especially l. 1082). Cowardice and betrayal of Rome were worthy of punishment, but not the killing of an unpatriotic sister. Le Vieil Horace's bitterness in his reproaches to Valère for seeking to defend Camille when her father condemns her seems to point again to his sense of outrage against interference in a domestic matter over which he has by right complete jurisdiction and doubtless to his perception that Valère's plea is made out of self-interest. If Horace had really been guilty, his father would have punished him. All the old man's anger and his pride in his son's achievement and moral strength come to the surface here. What had begun as a discussion of principle ends in an outburst of personal emotion, expressed, however, in traditional rhetorical forms. After addressing Valère le Vieil Horace turns to the King, appealing to him again, not from principle or precedent, but from pure emotion, first asking not to be deprived of his one remaining child, then affirming that Rome needs his warrior-son for her defence.

It is evident from this analysis that the accused, and the defending and prosecuting counsels in this trial, while they may to some extent follow the forms of judicial procedures and some of their rhetorical devices, are far from presenting their arguments in a rational precedent- or principle-based manner. They speak from personal emotion and they appeal to emotion in one another, in the King, in the audience, and not to judgement. It is for that purpose that rhetoric is used.

Tulle's judgement has been prefaced in his remarks to le Vieil Horace at the beginning of Act V scene ii, before the trial proper begins. He has not prejudged the issue, but he has consoled the grieving father and honoured him and his son, thus foreshadowing his final speech and parts of that of le Vieil Horace. As we have seen, Valère's attempted second intervention in the trial is first dismissed and his plea is not subsequently referred to directly. Nor indeed does Tulle specifically reply to any of the arguments put to him. He does, however, begin (ll. 1733–9) by indirectly rejecting le Vieil Horace's plea that the spontaneity of his son's killing of Camille and the nobility of its motive remove his guilt. Legally, Horace is guilty. But Tulle also accepts that the source of his crime is at the same time the source of Rome's present greatness and of her King's. Horace's glorious service to his city places him above the law, as had that of Romulus: this is the one argument from precedent. So legal principle establishes Horace's guilt and historical precedent places him above the law (ll. 1754–8). Tulle has therefore no need to condemn or to pardon the hero, but the judgement, while seeming to weigh principle against precedent, is surely of itself of doubtful legal validity: the law can scarcely place a citizen above the law.[6] Tulle seems to be responding to emotional submissions in an essentially emotional way: he is swayed by his gratitude to and admiration for Horace and his father and has apparently taken as his cue the last two lines of the father's address to his son:

> Ne hais donc plus la vie, et du moins vis pour moi,
> Et pour revoir encor ton pays et ton Roi.   (ll. 1725–6)

As a law-giver and as being yet above the law, the King has of course every right to do this, but it does not give to the end of the trial and to the judgement any greater legality, in the strict sense, in spite of their form, than the pleas which were based on and justified by personal feeling. All are, however, entirely consistent with the centre of interest established, in Sabine's opening lines, as being in the emotions of the characters and in the actions which spring from them. No wonder that the play still appeals to theatregoers and readers, as a tragedy *entirely* concerned with patriotism, the nature of the state, the status of the ruler or the role of Providence would not, though all these things may play some part within its emotional and situational framework.[7] Junie's words (ll. 95–6) suggest the right emphasis:

>   Qu'on voit naître souvent de pareilles traverses,
>   En des esprits divers, des passions diverses!

And those words take us back, of course, to our earlier discussion of the primordial importance of situations made out of conflicting passions. D'Aubignac (*Pratique*, p. 311) puts his finger on the crux of the matter:

Mais ce que j'estime surtout nécessaire est que la délibération même soit tellement attachée au *sujet* du poème [seen as the conflict of passions within a dramatic situation] ..., que les spectateurs soient *pressés du désir de ... connaître les sentiments* [of the speakers], parce qu'alors ce n'est pas un simple conseil, mais une *action théâtrale* ... (My italics.)

Our second example of judicial procedures in Corneille's tragedies is the last act of *Cinna*. The trial here takes place entirely in private: no equivalent of Valère, as the representative of part or the whole of the Roman people, appears. Whereas in Horace, the King was not connected by family bonds nor, particularly, by friendship, with any of the other characters, Auguste has adopted Émilie after the proscription of her father and protected Cinna after the execution of his. And whereas Tulle was not personally the object of any attack, projected or realized, by those involved in the drama, Auguste of course is. A fundamental distinction can therefore be drawn between the nature of the trial in *Horace* and that of the last act of *Cinna*: in the earlier play the King is an impartial judge, in the later one the Emperor is judge, prosecutor, and plaintiff. For this reason and because the conspiracy has been discovered, no plea is addressed to the judge in any legal sense: all the conspirators admit their guilt and are willing to accept punishment. The question of establishing the truth arises no more in *Cinna* than it does in *Horace*, and that in itself distinguishes the trial scenes in these plays from those in any ordinary court of law. Auguste has no need, in the usual sense, to arrive at a verdict: what he has to decide is the nature of the sentence, if any. We should therefore expect the accused to plead mitigating circumstances in order to have the sentence lightened. Is this in fact what we find?

The trial takes place in three stages, each defendant being dealt with in turn. In Act V scene i, Auguste confronts Cinna alone, expressly forbidding him to interrupt the speech for the prosecution, and reprimanding him when he does so. All the facts are known to Auguste: his recital of them (ll. 1481–96) is brief but comprehensive. By far the greater part of the speech is taken up with other matters. First (ll. 1435–76) comes a long account of all the benefits Cinna has received from Auguste: this ends with the contrasting accusation: 'Cinna, tu t'en souviens, et veux m'assassiner.' Cinna tries to protest his innocence, but the account of the conspiracy silences him:

> Tu te tais maintenant, et gardes le silence
>
> Plus par confusion que par obéissance.   (ll. 1497–8)

Then follows a highly oratorical passage, in which the questions show the Emperor trying to understand what motives Cinna could have had for plotting against him. An ambition to rule in his place (ll. 1499–1516)? But Cinna's power and glory, such as they are, are merely borrowed ('Ma faveur fait ta gloire, et ton pouvoir en vient') and are in any case far inferior to those of many others (this is the occasion for more rhetoric, ironical in function): that Rome should accept Cinna in Auguste's place is unthinkable. What has Cinna to say to that (ll. 1517–41)?

It is clear that the accusation as such occupies little of the hundred or so lines of Auguste's speech, and Cinna does not even reply to it point by point: he is merely astonished that the conspiracy should have been betrayed (ll. 1541–4). More important, he admits his part in it, justifying it on the grounds of revenge for the death of his father and two brothers. In taking the blame entirely upon himself, and in putting forward this one motive, Cinna may, as Auguste ironically suggests (l. 1557), be playing the hero, but he is also presumably seeking to exonerate and protect Émilie: her entry immediately afterwards provokes a cry of dismay from him (l. 1564) which supports this suggestion. Auguste meanwhile ends his interview with Cinna by inviting him to pronounce his own condemnation and to choose his own punishment.

Facts and legalities enter little, then, into the 'trial' of Cinna. Auguste is much more concerned with his own problem of trying to understand how a man could be as ungrateful and disrespectful to him as Cinna evidently is. And he is of course by implication involved in an understanding and judgement of himself. The accusation takes the form not so much of a piece of juridical procedure as of a moral indictment, and the verdict and sentence are not even pronounced.

The second stage of the trial (Act V scene ii) concerns Émilie. She is led in by Livie with the words:

> Vous ne connaissez pas encor tous les complices,
> Votre Émilie en est, Seigneur, et la voici.   (ll. 1562–3)

Livie has no need to rehearse the evidence, because Émilie proudly accepts her responsibility not only for her own part in the conspiracy but also for Cinna's. She seeks not to exculpate her lover, but to be punished with him. This spontaneous plea of 'Guilty' provokes another emotional and rhetorical outburst from Auguste (ll. 1587–95) in which again he tries to understand why he should be the victim of conspiracies mounted by those whom he most trusts and on whom he has showered nothing but benefits. It is on the level of personal relationships, with his adopted daughter, that Auguste attempts to deal with the matter, and like Cinna Émilie justifies herself by referring to the revenge still to be wrought for her father's death. Judicially, she is guilty, and her 'juste courroux' (l. 1603) has nothing to do with justice in the abstract sense.[8] Émilie knows it for, after Livie has attempted to calm her and to make what is the only appeal to a political principle in the whole proceedings (ll. 1609–16), she accepts the argument and at once brushes it aside:

> Aussi, dans le discours que vous venez d'entendre,
> Je parlais pour l'[Auguste] aigrir, et non pour me défendre.
>   (ll. 1617–8)

She has no legal defence, but is animated by hatred of Auguste, just as he is angered by her ingratitude. What follows (ll. 1619 ff.) is a dialogue between Émilie and Cinna, quite inconceivable in a court of law, in which the judge listens in silence to their arguing with one another for the lion's share of the blame. It is only when (l. 1649) Émilie again addresses the Emperor, asking that she and Cinna should be united in his treatment of them, as they are in their love, their hatred of him, their 'duty', and their 'noble dessein', that he utters his resolve to unite them in their punishment. Émilie makes much of the vocabulary associated with 'la gloire' ('âmes romaines', 'devoirs', 'ce noble dessein', 'nos esprits généreux', 'l'honneur d'un beau trépas'), but it is not used to express an abstract ideal, but a common desire for revenge and, through it, for a restoration of family honour. In the same way, when the scene ends, as does the previous one, with Auguste's resolve to punish, he gives no formal verdict or condemnation, but makes an essentially emotional response to the emotional display of the preceding dialogue.

Will it go differently with Maxime (Act V scene iii)? Auguste obviously hopes so, because he welcomes him as his one remaining faithful friend. But immediately (l. 1666), Maxime launches into a self-accusation of guilt towards the horrified Auguste, but also towards the other conspirators. This time, the judge does not even threaten punishment, let alone pronounce a verdict. The revelation of universal betrayal brings about the Emperor's sudden enlightenment as to what he must do and his pardon to all three conspirators, who are as suddenly enlightened as to his moral greatness and so reconciled to him and to one another.

In what has seemed to be a trial of self-confessed criminals in which even the forms of judicial procedure and principle are not invoked, it is ultimately the judge and prosecutor who has been on trial. The conspirators' motives spring from what they see as his former injustices towards their families, injustices which have besmirched their honour and which no imperial favours could obliterate. It is only by obliterating them now in an act of pardon which is the necessary prelude to reconciliation that Auguste can arrive at the self-mastery which mastery over others cannot, in any moral sense, equal or supersede. Paradoxically, he is judged through his trial of others. And it is on the act of self-mastery, transcending his 'juste courroux' (l. 1699), that he bases his appeal to the judgement of history. The act of clemency will bring glory to Auguste, not as an example of political wisdom and expediency as Livie has suggested in Act IV scene iii, but as an example of moral strength characterized by self-mastery which will give him, as she now prophesies, real authority. The temptations of abdication (ll. 365–76), of putting happiness before duty (ll. 1227–8) and of suicide (ll. 1233–6), the temptations of evasion, have been resisted, as has that of tyrannical punishment: Auguste will continue to rule, but with authority rather than the absolute power of force. He has discovered the distinction which Guez de Balzac drew between authority and power:

Sans doute l'autorité est beaucoup plus noble que la puissance, et celle qui se forme de la révérence de la vertu beaucoup plus honnête que celle qui s'établit par la terreur des supplices...

La puissance est une chose lourde et matérielle qui traîne après soi un long équipage de moyens humains, sans lesquels elle demeurerait immobile. Elle n'agit qu'avec des armées de terre et de mer. Pour marcher il lui faut mille ressorts, mille roues, mille machines. Elle fait un effort pour faire un pas. L'autorité au contraire, qui tient de la noblesse de son origine, opère des miracles en repos, n'a besoin ni d'instruments ni de matériaux, ni de temps même pour les opérer, est toute receuillie dans la personne qui l'exerce, sans chercher d'aide ni se servir de second. Elle est forte, toute nue et toute seule: elle combat étant désarmée.[9]

No better commentary could be made on Auguste's discovery of the roots of authority within himself, or of the promptings of divine virtue, of the immediacy of the solution to his inner problem, provided that he is courageous and convinced enough to accept total exposure and total vulnerability. He has come to realize the truth of Caussin's description of ambition as a form of servitude, the ambitious man being no better than a galley-slave. Ambition is an obstacle to the spiritual freedom which, as his soliloquy shows, Auguste craves: having put himself on trial by trying others, he has learned that, as Caussin put it, ambition is 'une furieuse convoitise de tout pouvoir et de tout avoir'.[10] The trial has brought to light, not the objective facts of the case (which are either known to the judge beforehand or willingly revealed by the accused), but the truth about the nature and consequences of the judge's own actions and responsibilities. The effect is deeply ironical as is Corneille's use of what is only a quasi-judicial procedure and rhetoric.

The whole play may be said to represent, when we consider it in this light, the trial of Auguste, who has reached a point of crisis in his career which happens to coincide, although he does not at the outset realize it, with the maturing of Cinna's conspiracy against him. It is this coincidence which makes the situation so highly charged with drama and occasions the 'trépidation intérieure', as Corneille would put it, when the Emperor summons Cinna and Maxime into his presence. But he does so not in order to censure them but to seek advice in his political and moral dilemma. The scene of the great deliberation (II.i) exposes the possible solutions and, because Auguste refers to the past, invites the judgement of his counsellors on his political behaviour as on that of some of his predecessors. It is true also that their judgement is tendentious and that the audience is aware of the prejudice of each of them and so judges them while they judge Auguste. When, later, the counsellors prove to be treacherous, the dialogue-deliberation with them is impossible and becomes (IV.ii) the monologue-deliberation in which Auguste rehearses the contradictory arguments which are based on differing interpretations of the historical evidence: he has become his own judge, but only knowledge of the full extent of the defections (Act V) will allow him to arrive at a verdict. The structure of both the dialogue- and monologue-deliberations is rhetorical in the sense that the argument and counter-argument, evidence and counter-evidence are brought in in turn and weighed.[11] Moreover, the arguments from precedent, while neither the arguments nor the precedents themselves constitute legal cases, are the dramatic counterpart of juridical precedents, and from them principles for present judgement are sought and ultimately derived. But the problem which Auguste has to solve is not one of legality or of political principle: it is a personal and moral one. The issue is again, in spite of appearances to the contrary, of the kind which faces Tulle in *Horace*, and the deliberative process resembles only superficially the procedures of a court of law: its rhetorical nature is not by any means strictly legal, either.

Professor Morel has pointed out[12] that the characteristics of what he calls the 'drame-procès' of the first half of the seventeenth century and which are those of the two plays we have studied and most other tragedies by Corneille persist in many tragedies of the age of Louis XIV: he cites in particular works by Thomas Corneille, *Timocrate* (one of the greatest of all popular successes) and *Le Comte d'Essex*. But, as we have seen, the same critic suggests that, apart from what he calls the archaizing *Esther* and the comedy, *Les Plaideurs*, Racine's works do not take the form of 'drames-procès'. Our interpretation of *Cinna* as such a play, not only in so far as its last act has the appearance of a trial of the conspirators, but in that in effect in its entirety it represents the trial of Auguste (and this in particular in the great soliloquy), suggests however several instances of self-judgement, adopting a similar rhetorical procedure, in the plays of Racine. These episodes are usually treated as cases of self-doubt and of hesitation before action is taken or misgiving after it has been accomplished.

Professor Morel treats one such episode — Hermione's famous soliloquy in Act V scene i of *Andromaque* — as a scene of deliberation, and that presumably implies an attempt to decide on a further course of action. In so far as one of her questions: 'Que dois-je faire encore?' and her incomplete resolution: 'Ah devant qu'il expire...' are concerned (and these frame the whole of her speech), 'deliberation' would appear to be an adequate description of it. It contains, however, other important elements, even before the 'Que dois-je faire encore?', namely 'Qu'ai je fait?' (retrospective not prospective) and before that again, 'Où suis-je?' (in its context — 'je cours dans ce palais' — not merely connected with bewilderment as to place in any physical sense, but related to mental disorientation and uncertainty and to questioning Hermione's own moral position). The evident complexity of the passage suggests that analysis might yield interesting results:

> Où suis-je? Qu'ai-je fait? Que dois-je faire encore?
> Quel transport me saisit? Quel chagrin me dévore?
> Errante, et sans dessein, je cours dans ce Palais.   1395
> Ah! ne puis-je savoir si j'aime ou si je hais!
> Le Cruel! de quel œil il m'a congédiée?
> Sans pitié, sans douleur, au moins étudiée!
> L'ai-je vu s'attendrir, se troubler un moment?

> En ai-je pu tirer un seul gémissement?   1400
> Muet à mes soupirs, tranquille à mes alarmes,
> Semblait-il seulement qu'il eût part à mes larmes?
> Et je le plains encore? Et pour comble d'ennui
> Mon cœur, mon lâche cœur s'intéresse pour lui?
> Je tremble au seul penser du coup qui le menace?   1405
> Et prête à me venger je lui fais déjà grâce?
> Non, ne révoquons point l'arrêt de mon courroux.
> Qu'il périsse. Aussi bien il ne vit plus pour nous.
> Le Perfide triomphe, et se rit de ma rage.
> Il pense voir en pleurs dissiper cet orage.   1410
> Il croit que toujours faible, et d'un cœur incertain,
> Je parerai d'un bras les coups de l'autre main.
> Il juge encor de moi par mes bontés passées.
> Mais plutôt le Perfide a bien d'autres pensées.
> Triomphant dans le Temple, il ne s'informe pas   1415
> Si l'on souhaite ailleurs sa vie, ou son trépas.
> Il me laisse, l'Ingrat, cet embarras funeste.
> Non, non encore un coup, laissons agir Oreste.
> Qu'il meure, puisque enfin il a dû le prévoir,
> Et puisqu'il m'a forcée enfin à le vouloir.   1420
> A le vouloir? Hé quoi! C'est donc moi qui l'ordonne?
> Sa Mort sera l'effet de l'amour d'Hermione?
> Ce Prince, dont mon cœur se faisait autrefois,
> Avec tant de plaisir, redire les Exploits,
> A qui même en secret je m'étais destinée   1425
> Avant qu'on eût conclu ce fatal hyménée!
> Je n'ai donc traversé tant de mers, tant d'États,
> Que pour venir si loin préparer son trépas,
> L'assassiner, le perdre? Ah devant qu'il expire...

The rhetorical form is established from the outset, in the use of
of the questions (ll. 1393–6): they imply doubt and uncertainty,
not only, as we have seen, as to future action, and as to a deed
already accomplished (Hermione's dismissal of Pyrrhus and her
commissioning of Oreste to kill him), but also as to her present
state of mind ('Errante et sans dessein' indicating indecisiveness
and moral aimlessness). The last of the series of questions (l. 1396)
suggests an inability on Hermione's part to judge the nature of her
own passion. The self-questioning is no mere rhetorical device: she
is really seeking the answers, not simply suggesting them by implication. Because the questions are partly of a moral nature, those
answers will entail judgement as well as knowledge.

Within this context the soliloquy develops in four successive
phases, each containing juridical features, which follow a rhetorical

pattern reminiscent of formal pleas. What distinguishes it from such pleas is, however, the fact that, as the first four lines have indicated by the reiteration of the first person pronoun, Hermione seeks to establish the truth about herself, and that when she moves on from the introduction the rhetorical questions become a persuasive device (persuasion being the chief function of rhetoric) employed to persuade or reassure the speaker herself and not some interlocutor or listener.

The first phase (ll. 1397–402) comprises in effect a judgement of Pyrrhus based on the evidence of his conduct in Act IV scene v. That evidence is presented in an abridged and subjective way, the questions in themselves suggesting condemnation and a passionate response as well as an attempt on Hermione's part to persuade herself that her judgement is right.

But then (ll. 1403–8), judge and judgement turn round. In spite of the evidence just advanced, Hermione finds herself no longer prosecuting or judging Pyrrhus but pleading for him. The rhetorical form — question-marks being used in the seventeenth-century editions of the play where modern editors have usually and misleadingly preferred exclamation marks — indicates renewed self-interrogation, putting in question the judgement already arrived at and so implicitly judging and even reprimanding the judge herself. With the same swiftness (not a feature of any real deliberation)[13] as that which characterized the first judgement, Hermione now (ll. 1407–8), without reasoning or argument, having found herself guilty of momentarily pleading for Pyrrhus against the evidence, renews her resolve to sustain the death-sentence ('l'arrêt de mon courroux': the technical word revealingly linked to the emotional, the first impression of objectivity destroyed by both the possessive adjective and the passion itself).

In order to persuade herself of the validity (and objectivity) of the renewed condemnation, Hermione then (ll. 1409–20) brings in further evidence of Pyrrhus's guilt. Amongst other things, it contains a judgement of his own misjudgements ('il *croit*', l. 1411; 'il *juge*', l. 1413). Although this time the evidence does seem to be more objectively presented — one notes the total absence of the obvious rhetorical devices apparent earlier in the speech — part of that evidence is purely imaginary (ll. 1415–6), though it will be factually confirmed by Cléone in the following scene. The condemnation and sentence are repeated (ll. 1418–20), apparently with the capital evidence that Pyrrhus must have known that Hermione would seek her revenge and that his behaviour made it inevitable. The condemnation is all the more justified, therefore, by virtue not only of his having abandoned her but also of his acting in full knowledge of the consequences. All this constitutes Hermione's judgement of Pyrrhus, but it represents an implicit judgement, too, of her own earlier temptation to weaken: in that sense, of course, the judgement of Pyrrhus and the evidence against him are a necessary means of resisting the temptation, and part, therefore, of a persuasive, rhetorical process.

The actual words Hermione uses (l. 1420) in formulating the final piece of evidence bring her face to face not with Pyrrhus's responsibility but with her own (ll. 1421–2): if he dies it will be at her behest. Self-judgement is renewed and the evidence of her long-standing love for him brought in. It is not of course real evidence because, in spite of the opening ('Ce Prince', l. 1423), she does not enumerate any of his actions or even characteristics as mitigating factors, but starts from 'l'amour d'Hermione' in the present and works back (tense changing from the future to the past) to '*mon cœur*', '*je m'*étais destinée, '*j*'...ai...traversé...pour venir...'. With the re-emergence of the subjective, passionate evidence comes a return of the rhetorical language with its questions and breathless accumulation. Its persuasiveness is irresistible: Hermione judges and condemns her earlier judgement by pronouncing the reprieve (l. 1428) she had already twice rejected (ll. 1407–8, 1419–20).

One may see in all this a number of parallels with Auguste's soliloquy: the forms of pleading and counter-pleading, the production of evidence and counter-evidence (passionate and 'internal', as it were, in Hermione's case, objective and historical in Auguste's), the justification of contrary desires, the irresolution, the doubts, the judgements of others and of self, the rhetorical questions which express them. Of course Hermione's soliloquy lacks Auguste's evocation of a historical perspective and of political considerations which are used to give at least an appearance of attempting a dispassionate judgement. Yet that evocation has its counterpart in Hermione's imaginary description of Pyrrhus's behaviour at the marriage ceremony and even in her wishful recollection of her love for him. Whereas, however, Auguste directly confronts the absent Cinna in his imagination and apostrophizes him ('Quelle fureur, Cinna, m'accuse et te pardonne?' — l. 1150 — self-judgement arising out of emotion here, too, in the course of a supposed trial of another: cf. 'Mon cœur, mon lâche cœur s'intéresse pour lui'), Hermione never apostrophizes Pyrrhus in this soliloquy nor, again unlike Auguste ('Rentre en toi-même, Octave, et cesse de te plaindre', l. 1130), herself. This is very revealing, because whereas Auguste, whatever his 'fureur', tries to see himself on the same objective level as Cinna, with whom the apostrophizing represents an attempt at a direct relationship, Hermione is incapable of detaching herself from her passion but does endeavour, in order to condemn him, to detach herself from the third-person Pyrrhus. In his heart, Auguste yearns already for the reconciliation he will finally achieve: Hermione's love has, for a time at least, turned to hate. Where Hermione's initial questioning exclamation ('Ah! ne puis-je savoir si j'aime ou si je hais!', l. 1396) seems to be answered in her uncompleted final resolve, Auguste's self-examination leaves him in doubt ('Qui des deux dois-je suivre, et duquel m'éloigner?', l. 1191), and that in spite of his apparently greater ability to give expression to the nature of his difficulty (l. 1188). Within the speeches, too, the speakers stumble fortuitously on the reality, by giving it verbal expression, and are horrified by it and so reverse the resolve just determined:

> Non, non, je me trahis moi-même d'y penser,
> Qui pardonne aisément invite à l'offenser,
> Punissons l'assassin, proscrivons les complices.
> Mais quoi! toujours du sang, et toujours des supplices!
>
> (*Cinna*, ll. 1159–62)
>
> Non, non, encore un coup, laissons agir Oreste.
> Qu'il meure, puisqu'enfin il a dû le prévoir,
> Et puisqu'il m'a forcée enfin à le vouloir.
> A le vouloir? Hé quoi! c'est donc moi qui l'ordonne?
>
> (*Andromaque*, ll. 1419–22)

In the same way as Auguste's trial of Cinna and of himself has yet to be completed in a confrontation, so has Hermione's trial of Oreste and herself. At the end of his soliloquy, Auguste has still to decide on a course of action: he will be given advice and clear evidence before he does so. Hermione, on the other hand, has committed herself even before the soliloquy, to the final judgement which she now begins to doubt and regret, and it is already, whatever her misgivings, too late to revoke it. This pattern of behaviour ('les mœurs' lying at the root of the situation, as we saw earlier, and of the development of the action) repeats itself in the two scenes which follow (V. ii and iii). When Cléone returns with an incomplete account of the marriage ceremony, Hermione impulsively interrogates her, impatient to hear it, but her first question ('qu'ai-je fait?' l. 1430) is a repetition of one of those with which the soliloquy opened: it is to a sense of her own hastiness, her own guilt, to a judgement first of herself that it points. But Cléone's evidence (ll. 1431–40) represents a realization of Hermione's imaginary evidence (ll. 1414–6) and so provokes the passionately tumultuous questions through which she seeks confirmation and amplification of the truth, questions which themselves provide it.

In that sense they are rhetorical rather than interrogatory, but Hermione does require an answer. It is given, like the first evidence, simply and factually. (Throughout the dialogue, the contrast is remarkable, and of course in character, between the rhetorical, passionate language of Hermione and the straightforward, detached reporting of Cléone.) The recital of Pyrrhus's oblivious happiness provokes a new condemnation ('Le perfide! Il mourra', l. 1458), without reasoning and without awaiting any counter-evidence. It is only after the condemnation that Hermione thinks of calling for that ('Mais que t'a dit Oreste?'). Again, impatient now for the death sentence to be carried out, she leaps unreasoning to conclusions ('Oreste me trahit?' l. 1462), and despite the admitted incompleteness of Cléone's evidence ('Je ne sais') and the attempt to reassure her ('Oreste vous adore'), it is now his turn to be condemned, immediately and irrationally — and even unjustly ('Le lâche craint la mort, et c'est tout ce qu'il craint', l. 1476). The end of Hermione's speech constitutes a new resolve ('Allons. C'est à moi seule à me rendre justice', l. 1485) to take the punishment as well as the judgement into her own hands, and to join Oreste with Pyrrhus in the same condemnation; by implication, Hermione judges and condemns herself for having failed until now to inflict the punishment herself. Again, evidence — some of it drawn from the past, some of it untrue (ll. 1477–80, 1483–4) — is brought in to justify the condemnation but without reasoned argument.

Oreste, having been condemned in his absence, fares no better when he appears. He is allowed, with only one impatient interruption (l. 1525), to recount the assassination of Pyrrhus and to attempt to justify his own finally insignificant part in it, since he thinks that this will bring him adverse judgement. But the evidence in his favour, on both counts, is turned against him: he is immediately found guilty (l. 1534), this time for the opposite reason from that of the first condemnation. To the furious questions (ll. 1537–43) which constitute, apparently, by the answers implied (an obvious rhetorical device to persuade or browbeat Oreste into believing himself guilty), further evidence against him, but which are in effect an indictment of herself, he can only reply in one poor disbelieving question by way of evidence for the defence. The rhetorical questions continue, each reiterating an apparent condemnation of Oreste, but each in reality expressing Hermione's own guilt (ll. 1545–54), distorting the evidence (ll. 1555–8) and even advancing hypothesis in its place (ll. 1559–60). Hermione's rhetoric, while ostensibly directed against Oreste, is revealed as an attempt to persuade herself of her own innocence and even of that of Pyrrhus. Meanwhile, the defendant's evidence in his own favour is immediately turned against him, a not infrequent judicial procedure, perhaps, but here twisted in order to exculpate the guilty prosecutor herself.

The expression of Oreste's bewilderment and anguish takes exactly the same form, with a minor variation, as Hermione's at the beginning of her soliloquy:

> Que vois-je? Est-ce Hermione? Et que viens-je d'entendre?
> Pour qui coule le sang que je viens de répandre?
> Je suis, si je l'en crois, un Traître, un Assassin. .
> Est-ce Pyrrhus qui meurt? et suis-je Oreste enfin? (ll. 1565–8)

The sequence of lines comprising three questions, one question, one statement, two questions, closely echoes Hermione's, except that the forms of the second and fourth lines are transposed. Oreste's quatrain represents the same search for knowledge and the same foreshadowing of self-judgement, to be accomplished immediately (ll. 1570–82), as Hermione's. Like hers, it is based on a kind of evidence, but without rational or deductive argument.

From this one may conclude that, far from being deliberative, Hermione's soliloquy and succeeding scenes, while bearing many of the marks of judicial rhetoric, are a form of dramatic action in which, thanks to the imperious impulses of her passion, she leaps to conclusions on incomplete, false, or distorted evidence and without reasoned argument. Like Thésée after her, she commits herself too hastily (*hamartia*) to decision, action, judgement, condemnation. The judicial substance of Auguste's soliloquy and of his trial of Cinna, Émilie, and Maxime is, as we have seen, scarcely more real but, however suspect and prejudiced the alleged motives may be, some attempt is made to deliberate, to argue, to reason. Horace, on the other hand, acts without hesitation or deliberation in perhaps too readily satisfying his passion for personal glory by accepting without question the honour of being chosen to represent his city, and again, certainly, by killing his sister. Both he and Auguste, however, differ from Hermione in that their minds are fixed on future glory, for which they are prepared to make enormous sacrifices, in the present, while hers is fastened on the hope of immediate happiness: her judgement of Pyrrhus and Oreste condemns them because they do not procure it for her. The temptation to seek immediate happiness is what Auguste finds himself having to resist:[14]

> De tout ce qu'eut Sylla de puissance et d'honneur,
> Lassé comme il en fut, j'aspire à mon bonheur.

> (*Cinna*, ll. 1227–8)

Of the quasi-judicial episodes in Racine's tragedies, one of the most interesting is that in Act IV scene ii of *Britannicus*. After dis-

covering the young hero and Junie together (III.vii–viii) in circumstances which confirm all the suspicions of their love, Néron
concludes (III.ix) that Agrippine has been responsible for fostering
it. Act IV opens with Burrhus warning her that Néron intends to
interview her and indicating (cf. ll. 1091–2) that she is now virtually a prisoner. Burrhus makes it clear that Agrippine will be on
trial before her son and points out to her the risks she would run if
she tried to accuse him: he is the ruler (ll. 1108–9). The vocabulary suggests the judicial nature of the interview ('Oui, madame, à
loisir vous pourrez vous défendre', l. 1099) and foreshadows its
actual form ('Défendez-vous, madame, et ne l'accusez pas',
l. 1106), for Agrippine will fail to heed the admonition.

It is striking that she at once takes the initiative, assuming her
former autocratic, matriarchal manner, and making it clear to
Néron, through the imperatives, that she intends to be in control of
the situation (l. 1115). The formula, 'prenez votre place' further
suggests that, Emperor or no, he should know his place, and is
reminiscent of the defendant's being required to sit on the 'sellette'
in a seventeenth-century lawcourt. (Cf., exactly, *Cinna*, l. 1425.)
Agrippine behaves like the presiding judge, seating herself first,
and not waiting for Néron even to greet her verbally.

She immediately launches into what ought to be — and in some
ways is — a long speech for the defence (ll. 1116–96). But instead
of attempting to clear herself of blame, she actually accuses herself
of having committed a whole series of crimes and broken many
laws. The manner is not dramatically rhetorical: specific factual
evidence is presented in a straightforward narrative way. It all
speaks against Agrippine; but she has committed all those crimes
for Néron's sake, in order to seat him on the imperial throne: she
scarcely needs to say it, the nature of the recital sufficing to reveal
its drift. The account itself incriminates her, but the motive ought
to secure Néron's pardon.

Strange though this may seem as a plea for the defence, and a
perversion of legal procedures, it is stranger still that the second
part of Agrippine's speech (ll. 1196–222), much shorter than the
first, should turn from pardonable accusations against the defendant to accusation of the prosecutor — and judge (for, like Auguste, Néron is in reality both). The evidence now produced is
used to incriminate him: all the crimes committed for his sake have
been rewarded with ingratitude manifested in the natural desire of
the emperor to rule for himself. Agrippine miscalculates, as she is
soon to discover, because her evident driving passion to cling to the
reins of power by continuing to control her son incriminates her in

his eyes (ll. 1227–30). The exculpating motive pleaded by Agrippine is rejected.

Like his mother, Néron begins with a kind of self-defence. With evidence to prove his point, drawn from both what Agrippine has done and from what Rome thinks of it, he affirms that the Empire wishes him, not her, to rule. His eclipsing of her power is justified. From this base he embarks on a counter-accusation (ll. 1250–7): it is only in these few lines, in which he exaggerates (persuasively by understatement, seeming to be unrhetorical, detached, and reasonable) her political ambitions and manœuvres, that Néron really plays the prosecutor's part. By contrast with his tone, that of Agrippine's response is violently emotional: the calm recital of evidence has given way to further accusations, now couched in a series of rhetorical questions (ll. 1258–64) and chidings (ll. 1269–74). Agrippine is on the offensive again, seeking no longer to defend herself (ll. 1265–8) but to shame her son into a change of conduct and, rhetorically or in reality, to dare him (ll. 1283–6). The new accusations meet with no defence on his part (l. 1287), and the new demands (ll. 1288–92) bring forth a deceitful acquiescence (ll. 1295–304). As the dramatic sequel shows, Agrippine's emotional relief is short-lived. Her arrogance and self-assurance weave the net in which she will be trapped when, after all, Britannicus is murdered. The tactics of engaging in self-accusation in order to accuse her accuser recoil upon her. While the forms of judicial procedure (amassing of evidence, weighing up of pros and cons, explanation of motives) appear to be adopted in their first speeches by both speakers, they are used in bad faith and perverted in function as well as in form, in order to promote, not political principles, but personal ambition. As Agrippine's crimes have involved deceit, so in this first attempt to assert his power Néron·deceives her: the scales do not fall from her eyes until Britannicus is dead (see V.v). Then, like Livie, she prophesies about the Emperor's future, but unlike Livie, with foreboding and contempt. Is Racine, in this, as in the scene we have just analysed, turning Corneille's scheme of things upside down? The bitter polemics of the first Preface would lend support to such a view.

Another variant of the judicial procedure is to be found in *Iphigénie*, in the series of accusations brought against Agamemnon (IV.iv–ix). The first of the sequence of scenes is the celebrated domestic one, made up almost entirely of three speeches, one from each member of the family, Agamemnon's being significantly the shortest. Iphigénie does not actually accuse her father (ll. 1174–220), but her emotional, devoted, yet dignified plea does constitute an indictment, however gentle, respectful, and affectionate, of his behaviour and motives. In his defence (ll. 1221–48), Agamemnon produces evidence of his attempts to evade his obligations to his allies and the omnipotence of the gods, on to whom he shuffles off the responsibility for the sacrifice. Clytemnestre (ll. 1249–316) will have none of this. Unlike her daughter she vents her fury and delivers her accusation with violently emotional rhetoric, full of questions, exclamations and ironical taunts which submerge the evidence, such as it is: Agamemnon is guilty of cowardice, ambition, self-seeking, lack of paternal affection, duplicity, deceitfulness. He is given no opportunity to reply to his wife, but is left for a few moments alone (IV.v) at the end of which he again accuses the gods: in the short soliloquy he shuns self-judgement and the refusal is another symptom of evasiveness. He has no time for further reflection before an angry Achille breaks in on him with another violent accusation, though he begins merely by asking for confirmation or refutation of the rumour that Iphigénie is to be delivered to Calchas for sacrifice. Once more the king is evasive (ll. 1335–8) but to no avail. It is then that Achille bursts into angry accusations (ll. 1341 ff.) expressed in highly rhetorical questions and exclamations. Stung, Agamemnon counters equally rhetorically (ll. 1349–51), but the rhetoric is empty and expresses the petty and unworthy conceit of the autocratic *paterfamilias*. Achille returns to the attack. Characteristically, the defendant falls back on the responsibility of the gods and of his allies (ll. 1358–60). Achille being one of them, the altercation loses almost all semblance of judicial procedure, but the rhetoric continues in the young man's emotional presentation (ll. 1369–400) of the evidence of his past conduct which he adduces in order to justify his sense of injured pride. This is a kind of defence, of course, but it slips into recrimination with Achille's determination to abandon the Trojan expedition. Agamemnon's response (ll. 1401–16) takes the form of an imperious counter-accusation and an acceptance of the challenge — pure rhetoric, no doubt, for Achille's departure could bring about the collapse of the Greek alliance. His parting shot (ll. 1417–24) is again rhetorical in the sense that he defies Agamemnon and haughtily assumes the protection of Iphigénie.

That final speech represents of course a challenge to the king's authority and prestige even within his own family. As such, it may be thought of as an empty piece of rhetoric. Certainly Agamemnon sees in it a threat to his authority and, vulnerable as he is, a judgement of his intentions. Hence his angry little soliloquy (IV.vii), in which his injured pride ('Ma gloire intéressée emporte la balance', l. 1430) pushes him in the direction of accepting the challenge by ordering the sacrifice to be completed. Feeling himself already under judgement (l. 1431), he fears further judgement (l. 1432).

However, the need actually to give the fatal order, to verbalize it ('Puis-je leur prononcer cet ordre sanguinaire?' l. 1434), brings Agamemnon, like Hermione before him, face to face with the father, as well as the king, in himself. Like that of Titus, the rhetorical self-questioning ('Cruel! à quel combat faut-il te preparer?' l. 1435) is an attempt at self-persuasion, but whereas Titus succeeds, Agamemnon fails: the rhetorical accusations brought against him, not by the priest, not by the allies, but by daughter, wife, and would-be son-in-law, are still ringing in his ears. He may fear the judgement of the world at large (l. 1432), but what he most fears is that of those who, bound to him in his family, have already judged him. Once more he becomes involved in the rhetoric of self-questioning, self-judgement (ll. 1433–50), withdrawal from the brink (ll. 1451–3), recollection of the evidence (of Achille's temerity — ll. 1453–6), self-doubt (l. 1457), and finally pride, injured by Achille's judgement, which suggests the compromise of sparing Iphigénie (the father, fearful of his wife) but marrying her to someone other than Achille (the king, fearful of derisory public judgement).

Having dismissed Eurybate with this intention in mind, Agamemnon is again alone (IV.ix) and, as before, we see him returning to his doubts and charging the gods with the responsibility of confirming their demand for the life of his daughter.

This sequence of scenes represents a further adaptation, then, by Racine of the rhetoric of judicial procedures, not in a formal juridical context, but in a dramatic form of judgement and self-judgement arising out of accusation and defence with the production and interpretation of evidence.

The nearest counterpart, in Racinian tragedy, of the great self-judgement in Auguste's soliloquy, is that of Titus, in *Bérénice*. Like the consultation with Paulin in Act II scene ii, Titus's soliloquy must not be interpreted as an attempt either to arrive at a decision as to whether Bérénice must be sent away or to reverse a decision already made to do so. That Titus has resolved on the parting before the play begins is made perfectly clear:

> Pour jamais je vais m'en séparer.
> Mon cœur en ce moment ne vient pas de se rendre . . .

(ll. 446–7)

It is not the decision itself which is in doubt in the soliloquy, but Titus's capacity, less actually to carry it out than to face Bérénice with it. He has made several attempts already (ll. 471–82), but in vain. However, he says:

> Enfin, j'ai ce matin rappelé ma constance:
> Il faut la voir, Paulin, et rompre le silence.   (ll. 483–4)

Although he will charge Antiochus with escorting her back to her kingdom,

> Elle en sera bientôt instruite par ma voix;
> Et je vais lui parler pour la dernière fois.   (ll. 489–90)

It is there that the difficulty lies, and in order to try to avoid it, Titus turns to Antiochus not simply to escort the Queen after the announcement has been made to her, but actually to act as his spokesman in making it. It is harder than he thinks to carry out his resolve, though in his heart he knows that there is no alternative:

> Je connais mon devoir, c'est à moi de le suivre...   (l. 551)

Almost equally demanding is Bérénice herself, as Titus says to Antiochus:

> Elle veut qu'à ses yeux j'explique ma pensée.   (l. 740)

That constitutes the real problem, hence the recourse to Antiochus as the intermediary (ll. 741–6) which he becomes a few moments later (III.iii); but still Bérénice requires that Titus speak to her in person: it is faced with that demand and the certainty that the parting is inevitable that Titus now confronts himself (IV.iv):

> Hé bien, Titus, que viens-tu faire?
> Bérénice t'attend. Où viens-tu, téméraire?
> Tes adieux sont-ils prêts? T'es-tu bien consulté?
> Ton cœur te promet-il assez de cruauté?   990
> Car enfin au combat, qui pour toi se prépare,
> C'est peu d'être constant, il faut être barbare.
> Soutiendrai-je ces yeux dont la douce langueur
> Sait si bien découvrir les chemins de mon cœur?
> Quand je verrai ces yeux armés de tous leurs charmes,   995
> Attachés sur les miens, m'accabler de leurs larmes,
> Me souviendrai-je alors de mon triste devoir?
> Pourrai-je dire enfin, je ne veux plus vous voir?
> Je viens percer un cœur qui m'adore, qui m'aime.
> Et pourquoi le percer? Qui l'ordonne? Moi-même.   1000
> Car enfin Rome a-t-elle expliqué ses souhaits?
> L'entendons-nous crier autour de ce palais?
> Vois-je l'État penchant au bord du précipice?
> Ne le puis-je sauver que par ce sacrifice?
> Tout se tait, et moi seul, trop prompt à me troubler,   1005
> J'avance des malheurs que je puis reculer.
> Et qui sait si, sensible aux vertus de la Reine,
> Rome ne voudra point l'avouer pour Romaine?
> Rome peut par son choix justifier le mien.

Non, non, encore un coup, ne précipitons rien. 1010
Que Rome avec ses Lois mette dans la balance
Tant de pleurs, tant d'amour, tant de persévérance,
Rome sera pour nous . . . Titus, ouvre les yeux.
Quel air respires-tu? N'es-tu pas dans ces lieux
Où la haine des Rois avec le lait sucée, 1015
Par crainte, ou par amour, ne peut être effacée?
Rome jugea ta Reine en condamnant ses Rois.
N'as-tu pas en naissant entendu cette voix?
Et n'as-tu pas encore ouï la Renommée
T'annoncer ton devoir jusque dans ton armée? 1020
Et lorsque Bérénice arriva sur tes pas,
Ce que Rome en jugeait, ne l'entendis-tu pas?
Faut-il donc tant de fois te le faire redire?
Ah! lâche! fais l'amour, et renonce à l'Empire.
Au bout de l'univers va, cours te confiner, 1025
Et fais place à des cœurs plus dignes de régner.
Sont-ce là ces projets de grandeur et de gloire
Qui devaient dans les cœurs consacrer ma mémoire?
Depuis huit jours je règne. Et jusques à ce jour
Qu'ai-je fait pour l'honneur? J'ai tout fait pour l'amour. 1030
D'un temps si précieux quel compte puis-je rendre?
Où sont ces heureux jours que je faisais attendre?
Quels pleurs ai-je séchés? Dans quels yeux satisfaits
Ai-je déjà goûté le fruit de mes bienfaits?
L'univers a-t-il vu changer ses destinées? 1035
Sais-je combien le Ciel m'a compté de journées?
Et de ce peu de jours si longtemps attendus,
Ah malheureux! combien j'en ai déjà perdus!
Ne tardons plus. Faisons ce que l'honneur exige.
Rompons le seul lien . . . 1040

Self-confrontation this truly is, as is immediately evident from
the way in which, almost throughout the speech, Titus apostrophizes himself. The rhetorical questions with which he begins
(ll. 987–1000) imply an arraignment of himself before his own
moral standards. The demands of his political duty have been satisfied by his determination to send Bérénice away. Accusing himself of cruelty (l. 990) and barbarity (l. 992), he still asks whether
he has sufficient of either to carry that determination into practice.
Stoic virtue ('C'est peu d'être constant') will not be adequate without them to fulfil his 'triste devoir'.

Je viens percer un cœur que j'adore, qui m'aime. (l. 999)

Judging himself, Titus doubts whether he has it in him to cause the inevitable pain, distress, and anger. Yet, implicit in all this rhetorical probing of his will, desire, and humanity is the sense that failure to act would be weakness and cowardice.

Still the earlier wishful thinking (e.g. ll. 455–8) has not been utterly eradicated. Titus pursues his rhetorical questioning (ll. 1000–9), not now of himself, but of the feelings of the Roman people. He accuses himself of having too hastily concluded that they would be hostile to his marriage — this in spite of Paulin's confirmation (II.ii) of the hostility. But just as Bérénice has not yet heard directly from Titus, so he has not heard directly from his people that she must go. And so the entire responsibility for the decision is so far his. The wishful thinking — that Rome might yet approve of the marriage — returns in a new guise (ll. 1007–9), rekindled by self-blame (ll. 1005–6). The spurious reasoning, based not on evidence but on its alleged absence, leads to the suggestion that procrastination is the best policy, so that Rome can make up its mind ... about love (ll. 1010–13). But, in the same way as we have seen Hermione and Auguste brought face to face with reality through its verbal expression, so now we see Titus ('Rome sera pour nous ... Titus, ouvre les yeux!' l. 1013) confronted with an impossible dream and sitting in judgement on himself for believing in it, even momentarily.

The rhetorical questions continue (ll. 1013–23), but they are now used by Titus to upbraid himself for yielding to wishful thinking, self-judgement stemming from the abundant evidence of the past, in Rome's inveterate hatred of kings: the self-questioning form expresses self-contempt for his wilful pretence at ignorance. Titus has moved from judgement of his own moral weakness and cruelty to that of his succumbing so easily to idle dreams.

Just as the reality of Roman tradition cannot be shunned, neither can that of Titus's lack of achievement since his accession. He now judges himself harshly (ll. 1024–38), in the apostrophe of self-reproach and ironical rhetorical questions. Again this recalls Auguste (*Cinna*, ll. 1030–48, 1169–84) who also alternates apostrophe with judgement expressed in the first person; but an important difference may be seen between the two emperors: Auguste blames himself for crimes actually committed, as he now sees them, Titus for deeds left undone. And so (ll. 1039–40) the time has come for Titus to act, and his first action is made inevitable by the appearance, at that precise moment, of Bérénice herself.

One notes that Auguste and Titus are engaged in rhetorical self-arraignment at almost the identical point in their plays. Yet, at the end of his soliloquy, Auguste is still not resolved on a clear course of action, whereas at the end of his Titus is. Such evidence might lead to at least a partial reversal of the commonly-held view that Corneille's characters are resolute and Racine's vacillating, but I do not propose to investigate that possibility here. What is important for our discussion is that we should see the distinction in relation to what we have already discovered about suspense in the works of the two playwrights, and about the place and nature of the dénouement. If, in *Cinna*, Auguste's soliloquy does not, even in Act IV, result in a clear decision, it is evident that Corneille is keeping the audience in a state of curiosity as to what it will ultimately be and, therefore, as to the fate of the conspirators; and he will maintain the curiosity to the very end. In *Bérénice*, on the other hand, it is in the heroine's fate that we are primarily interested: it is finally sealed at the end of Titus's soliloquy. Is the play then over? No, because we still need to be satisfied as to how the characters will respond to the emperor's decision and, above all, we have to see Bérénice discovering, not that Titus must send her away, but that he does so despite his continuing love.

The connexion between judgement and discovery is nowhere more apparent than in *Phèdre*. The heroine's discovery, in Act IV scene iv — a quite fortuitous discovery that Hippolyte is in love with Aricie, occasioned by Thésée's brushing his son's revelation aside as merely 'un frivole artifice' to protect himself against accusations of illicit passion — leads to judgement of Œnone, of Aricie and of Phèdre herself. It is in the midst of her disbelieving, jealous rage that Phèdre begins her accusation of Œnone (ll. 1233–6), who is, at that point, simply blamed for being a bad informant; but the anger is apparent in the rhetoric of the questions. Deprived of the knowledge of the young lovers' meetings, Phèdre forgets Œnone for the time being and fills the void with imagined scenes of innocent bliss which she contrasts with her own guilty misery. Apart from one brief plea for pity (l. 1258), Phèdre becomes completely absorbed in her own thoughts until Œnone interrupts (l. 1295) to express horror at her mistress's terrifying vision of her guilt and to try to comfort her by belittling the moral importance of her 'feux illégitimes' (ll. 1295–6, 1304–6), the adjectives clearly indicating that nothing could (in spite of the earlier temptation (ll. 349–50): 'Vivez: vous n'avez plus de reproche à vous faire...') make her passion innocent. This time the temptation to minimize her moral guilt is angrily repulsed: Œnone is arraigned, held responsible for Phèdre's falling into sin, dismissed with a call for divine retribution.

In her jealousy, Phèdre judges and condemns Aricie, too: 'Le crime de la sœur passe celui des frères' (l. 1262). In order to satisfy her passion for revenge, Phèdre resolves momentarily to incrimin-

ate Aricie before Thésée by turning her real 'crime' (that of being
loved by Hippolyte) into a political one.

The impulse is, however, short-lived, because the actual expression of it brings Phèdre face to face (like Hermione and Titus) with
the horrifying reality of her thoughts and intentions (ll. 1263–4).
The rhetoric is turned upon herself:

> Que fais-je? Où ma raison se va-t-elle égarer?
> Moi jalouse! Et Thésée est celui que j'implore!
> Mon époux est vivant, et moi je brûle encore!   (ll. 1264–6)

The 'crime' of Aricie becomes 'mes crimes':

> Je respire à la fois l'*inceste* et l'*imposture*.
> Mes *homicides* mains promptes à me venger
> Dans le sang *innocent* brûlent de se plonger.   (ll. 1270–2)

Immediate self-condemnation — again without rational argument
but with the production of emotional evidence — brings desire for
escape in the all-concealing darkness of death; but that leads only
to the evocation of judgement before Minos, her own father, in the
halls of Hell and to a renewed vision of the horror of her crimes
(ll. 1284, 1291), of punishment (l. 1287–8), of pardon (l. 1289), of
judicial torture and eternal suffering ('les tourments', l. 1294). The
vocabulary itself discloses the panoply of judgement, condemnation, and punishment. And yet the crime has never been committed save in intention and desire (ll. 1291–2).

Having begun with a real judgement of herself, Phèdre has
moved, through certainty of her guilt, to imagined judgement by
her father. The factual discovery of the love of Hippolyte and Aricie led her, by way of her jealous desire to punish the girl for having
stolen the young man's heart, to judgement of her own outrageous
intent. In its turn, that judgement brings discovery of the depths of
her moral depravity, of the true nature of her passion and its
effects, and it is that that in her imagination she sees Minos judging and punishing. Once discovered, that depravity appears unpar-
donable: it is because she advances mitigating circumstances that
Œnone in her turn brings Phèdre's judgement on her head. But it
is on her own head that Phèdre must bring it, in act, not in word,
judgement being implicit in the self-inflicted punishment of suicide.
The reasons she finally gives, in her last, dying appearance
(ll. 1617–19, 1622–44), represent an exculpation of Hippolyte, a
condemnation of Œnone, but a judgement of herself in her expression of remorse (l. 1635). In this completion of the dramatic process, Thésée is at last enlightened as to the facts of Phèdre's guilt
and Hippolyte's innocence and as to the real nature (l. 1647) of
what he himself has done. Thésée finally judges himself, as Auguste does, and — although this is seldom noticed — resembles

him too in the magnanimous gesture (made too late this time) of reconciliation with Aricie, the sole survivor of his supposed enemies. Of Hippolyte, he says, in the last lines of the play,

Rendons-lui les honneurs qu'il a trop mérités.
Et, pour mieux apaiser ses mânes irrités,
Que, malgré les complots d'une injuste famille
Son amante aujourd'hui me tienne lieu de fille.

In the light of the foregoing discussion, it is evident that Professor Morel's distinction between Corneille and Racine with which this chapter opened calls for considerable qualification. In the first place, while it may be true that in Racine's tragedies no formal judicial framework is set up (although he approaches it in *Britannicus*), judicial procedures and rhetoric are to be found in them. On the other hand, although such a framework may be set up in Corneille's plays, the arguments, the nature and use of evidence, and the way in which verdicts are arrived at are by no means of a legal character. Professor Morel himself has pointed out [15] that the arguments ostensibly advanced may not be genuine, but that they are often an attempt — vain, as far as the audience is concerned, and deliberately so — to conceal the real motives or feelings which they are intended to promote or fulfil. This kind of ambivalence is of course related to what we have seen of the audience's ironical view of the dramatic action. It is also to be found in episodes which, while not constituting judgement or self-judgement as in the examples we have studied, are parallel to them, deliberations and interrogations.[16] The armoury of rhetoric is used on two levels, one apparently political and rational, the other personal and passionate. In Auguste's consultation with Cinna and Maxime (II.i), in Ptolomée's with Photin, Achillas and Septime (I.i), in Pompée's discussion with Sertorius (III.i), in the dialogues of the first act of *Sophonisbe*, in Oreste's embassy to Pyrrhus (I.ii), in Hippolyte's conversation with Phèdre (II.v), the advice or information given is plausible and acceptable enough in itself, but its significance lies in the unspoken feelings behind it.[17] Those feelings may or may not be known to or suspected by the interlocutor and sometimes the development of the plot may largely take the form of confirmation of suspicions or pursuit of the truth. As in the parallel interrogatory episodes (e.g. *Mithridate*, III.i–vi; *Suréna*, IV.iii–iv; *Phèdre*, III.v; IV.ii; V.iii), subterfuge may be attempted on either side, and this involves not necessarily deliberate lying, but the use of arguments and evidence which are real enough in their own terms but calculated to deflect the interrogator from discovering the real intentions and motives of the speaker.

In the second place, it will be evident that to think of Corneille's

drama as the place where punishment is inflicted and Racine's as
that where the crime is committed is a considerable over-
simplification. In a judicial sense, Rodrigue, Horace, Cinna and
his fellow-conspirators, and Nicomède are not punished, although with the exception of the last-named they are convicted of
crimes. On the other hand, Ptolomée (in *Pompée*), Cléopâtre (in
Rodogune), Sertorius, Sophonisbe, Attila, and Suréna are indeed
punished, but not necessarily from a judicial point of view. But of
course we have seen that the proceedings of the 'trials' are far from
being directed in any recognizable juridical manner towards the
establishment of guilt or innocence. In fact, in cases like that of
Suréna, only the interrogatory episodes bear any resemblance to
legal procedures and they take place without even the semblance of
a court of law. He and Eurydice bring their punishment on themselves by their obstinate silence, as do Ptolomée by his rash and
foolish revolt against Roman power, and Cléopâtre (in *Rodogune*)
by her overweening and unscrupulous personal ambition. In
Racine's plays, punishment — self-inflicted though it may be, as
with Cléopâtre — abounds as does judgement, though some
(Agamemnon, Thésée) escape the ultimate penalty. In this matter,
therefore, real resemblances exist between the works of the two
dramatists, but, as we have seen, so do distinctions, which are revealed as much at least by the place of the 'trials' in the development of the plays as by their nature and expression. Indeed their
place virtually indicates their nature and their nature determines
their expression.

Set in their context, the judicial elements in the plays, with all
their rhetorical devices, are seen to have little to do with legal processes or even legal offences. Charged with emotion, they concern
deceit, self-seeking, ambition, passionate love or hate or jealousy,
pride (whatever it may be called), glory: what is really on trial,
and what the audience perceives even when the characters themselves do not, is the overwhelming destructive power of the passions. In one form or another, trials, deliberations, interrogations,
and judgements are seen to be part of a whole pattern which includes situation, plot, and dénouement, and the rhetoric which
accompanies them performs its function neither in a juridical way
nor as pure ornament but as a revelation of what really drives the
characters to their tragic ends.

CHAPTER VIII
Tragic Quality

Tous les hommes recherchent d'être heureux. Cela est sans exception, quelques différents moyens qu'ils y emploient. Ils tendent tous à ce but. Ce qui fait que les uns vont à la guerre et que les autres n'y vont pas est ce même désir qui est dans tous les deux accompagné de différentes vues. La volonté [ne] fait jamais la moindre démarche que vers cet objet. C'est le motif de toutes les actions de tous les hommes, jusqu'à ceux qui vont se pendre.

Pascal, *Pensées* (Lafuma, 148)

The end of tragedies or serious plays, says Aristotle, is to beget admiration, compassion, or concernment.

Dryden, *Essay on Dramatic Poesy* (1668)

In the course of this enquiry I have concentrated on dramatic technique in the tragedies of Corneille and Racine in an endeavour to suggest comparisons and distinctions between the two dramatists in matters which, in the many parallels drawn hitherto, have received little attention. Although the problem is fraught with many difficulties, it is perhaps important to consider briefly the tragic quality of some, at least, of the plays. Some critics have come to the conclusion that Racine is the only French dramatist to have achieved tragedy in any real sense; others that Corneille's art is the more tragic. It may of course all be a matter of terminology, of the meaning one gives to the word 'tragic', and the concept has certainly been diversely understood and interpreted and found therefore in many different kinds of play. It may well be that no satisfactory single definition can be devised in intellectual or rational terms and that all we can say is that we may be able to recognize intuitively tragic quality in a dramatic work and perhaps identify some but not all of its conditions and ingredients. It is in that tentative spirit that I propose to approach the question, and as far as possible without a preconceived and hermetic definition in mind. I shall also attempt to arrive at conclusions on the basis of the kind of analysis made in the preceding chapters, and to avoid large generalizations, deeming it more prudent to discuss the question in terms understood by Corneille and Racine than to impose on their works concepts which belong to more recent times, and within the framework of their art as we have analysed it so far.

Even a cursory investigation reveals that the word 'tragic' itself did not, for the writers of the sixteenth and seventeenth centuries, convey a qualitative concept, the idea of an essence, but the characteristics of a genre. It was applied equally to the now recognized masterpieces of that genre and to the scores of inferior plays to which we might now hesitate to attach it. The dictionaries of the late seventeenth century give the word 'tragique' as adjective and as noun. Furetière states that the noun can be applied only to a poet who writes tragedies or to the genre of tragedy. Richelet, relying on its derivation from the Latin 'tragicum' (itself, like its French and English equivalents, basically an adjective), says that it denotes 'le genre tragique' and quotes the example also given in the Academy Dictionary in support: 'Ce poète s'applique au tragique, et ne réussit pas dans le comique.'

If definitions of the noun are on these generic lines only, those of the adjective are scarcely more revealing of a quality or an essence. In two of his prefaces, Racine uses the adjective to denote, not plays of a particular genre, nor their authors, but the characters who appear in them. The instance in the second preface to *Bajazet* (1676) seems straightforward: 'Les personnages tragiques doivent être regardés d'un autre œil que nous regardons d'ordinaire les personnages que nous avons vus de si près.' This sentence is embedded in Racine's discussion of the respect due to characters in tragedy and of the possibility of equating geographical and cultural distance with temporal and historical remoteness: 'maior e longinquo reverentia'. Yet the association of the genre with that respect does seem to imply something qualitative about the nature of the characters of tragedy: they must be worthy of the respect of the audience. That idea is obviously connected with the exemplary nature of those characters which, as we have seen, it is part of the function of their historical dimension to achieve. The other instance of the term being used in this way by Racine is to be found in the first preface to *Andromaque*, written several years earlier (1668): 'Aristote, bien éloigné de nous demander des héros parfaits, veut au contraire que les personnages tragiques, c'est-à-dire ceux dont le malheur fait la catastrophe de la tragédie, ne soient ni tout à fait bons, ni tout à fait méchants.' Here, it appears that Racine is not applying the epithet to all characters in tragedy indiscriminately but only to those whose misfortune constitutes the catastrophe. The distinction may at first sight seem to be based on the quality, status, or dignity of such characters, but in fact it is more closely connected, in this context, with the action in which they are involved: 'dont le malheur fait la catastrophe'. Taken together, however, these two sentences from the prefaces suggest that Racine uses the word 'tragique' with the idea of characters appropriate to tragedy, both from the point of view of their exemplary nature and from that of their particular kind of role. By implication, this suggests that tragedy itself is concerned with an action of some elevated importance and, in a catastrophe, bringing misfortune upon some, at least, of the characters.

Both Richelet and Furetière, in seeing the word 'tragique' as being applied to the style and diction (Richelet: 'élevé, sublime, touchant et qui sent la tragédie') appropriate to tragedy, imply the same quality of elevation and dignity in the genre. The idea of catastrophe is evident, however, in their further definition of the term as a synonym for 'funeste' or 'fâcheux': in this Corneille obviously joins them, and his definitions of tragedy, with its emphasis on the kind of action appropriate to it, characterize that action as both elevated and catastrophic. Without Corneille's somewhat polemical angle, the Academy Dictionary adapts, as he does, Aristotle's definition of tragedy: 'Pièce de théâtre qui représente une action grande et sérieuse entre des personnes illustres, et propre à exciter les grandes passions, comme la terreur et la compassion, et qui finit d'ordinaire par quelque événement funeste.'

Such definitions simply sum up seventeenth-century usage. In his *Discours de la tragédie* (1639), Sarasin writes that plot and dénouement are the two parts of a play which 'contiennent toute l'action tragique': in the context, 'l'action tragique' clearly means 'the action of the tragedy' and is not indicative of a quality or essence.[1] Ogier, in his preface to Schelandre's *Tyr et Sidon* (1628), had written almost tautologously of 'les affaires sérieuses, importantes et tragiques' appropriate to tragedy,[2] but the later, Cornelian emphasis on elevation of subject-matter is already perceptible. D'Aubignac was to make the emphasis more explicit when he sought to correct those who 'se sont imaginés que le mot de *tragique* ne signifiait jamais qu'une aventure funeste et sanglante', and explained that its use should be confined to tragedy, described as 'une chose magnifique, sérieuse, grave et convenable aux agitations et aux grands revers des prines'.[3] But it need not have an unhappy ending, Corneille's 'péril de vie', etc., suffices.

The implications of Racine's use of the word 'tragic' are, as we have seen, rather more subtle. Two more of his prefaces contain the term. In one, the preface to *Bérénice*, a further dimension is introduced: '... le dernier adieu qu'elle [Bérénice] dit à Titus, et l'effort qu'elle se fait pour s'en séparer, n'est pas le moins tragique de la pièce: et j'ose dire qu'il renouvelle assez bien dans le cœur du spectateur l'émotion que le reste y avait pu exciter'. Here, in addition to ideas of dignity and catastrophe, is that of the arousal of particular emotions characteristic of tragedy: the remainder of the preface allows us to see what they are; but at this point of course Racine is associating them specifically with the catastrophe of his play. The preface to *Iphigénie* takes up Aristotle's remark in the thirteenth chapter of the *Poetics*, to the effect that Euripides was 'the most tragic certainly of the dramatists'. Racine's paraphrase, 'entre les poètes Euripide était extrêmement tragique', exchanges the superlative for an adverb of intensity, and when he then adds by way of explanation or justification, 'c'est-à-dire qu'il savait merveilleusement exciter la compassion et la terreur, qui sont les véritables effets de la tragédie', he is developing the idea of the emotions again and specifying those which characterize tragedy. In the corresponding passage of the *Poetics*, however, the emotions are discussed in terms of the kind of action appropriate to tragedy (a change from happiness to misery), of the source of the catastrophe (not depravity, but some great error), and of the moral nature of the character involved (the intermediate kind of personage, not pre-eminently virtuous and just). It is these characteristics that

Aristotle finds in Euripides whose plays, 'on the stage, and in the public performances...', are seen to be the most truly tragic'. Aristotle's preference — or that of the Athenian audiences, to which Racine also refers — is based in part on Euripides's presentation of characters as the agents or victims of 'some deed of horror' and of plays which usually have unhappy endings. These things feature in Racine's allusions only by implication, if at all: in this context, the word 'tragic' is associated with the emotions aroused by tragedy, with those emotions in themselves and not, as in the *Poetics*, with their sources. Racine's view is consonant with that of D'Aubignac and, like his, it does not insist on the unhappy ending: as the preface to *Bérénice* puts it, 'ce n'est point une nécessité qu'il y ait du sang et des morts dans une tragédie'. On the contrary, in fact, the essential ingredients of tragedy are the elevation of the action, the status of the characters and the arousal of the right emotions.

It is quite clear that, seen in this light, Racine's *concept* of tragedy depended, as did all others in his time, on these three ingredients. Does this mean that he and Corneille inherited a tradition in which some kind of tragic essence, the tragic, was already taken for granted and therefore simply not discussed? Such a tradition would have originated in the Renaissance, and theorists of tragedy were certainly not lacking, not only in France but in Italy and the Low Countries as well. These theorists all repeat the Aristotelian formula about arousing pity and fear, but without exception they find it impossible to arrive at a proper understanding of the catharsis of these emotions. They can scarcely be blamed, since they almost all — before Corneille, Castelvetro was the only notable exception[4] — regarded tragedy as having a moralizing function and since that function was interpreted in a Christian sense. The use of the word 'purgatio' in the many Latin translations for Aristotle's 'catharsis' complicated matters further. One effect of Christian morality was to emphasize pity at the expense of fear and to interpret it as compassion, an effect which continued to be felt throughout the seventeenth and eighteenth centuries. In discussing the emotions, both Italian and French theorists tended to neglect Aristotle's insistence on the distinctive nature of the action which tragedy was supposed to represent. Instead, they stressed the rank and status of the characters: in tragedy they are noble and high-born, in comedy, of the common people. The arousal of compassion for such characters together with their nobility makes for the third condition of tragedy, an appropriate style. Since, in the sixteenth century, tragedy was thought of as an essentially rhetorical and declamatory art, it is perhaps not surprising that Du Bellay, for example, discussed style to the exclusion of all else.[5] In his dedication of a

translation of Euripides's *Hecuba* (1554), Guillaume Bouchetel puts
a characteristic emphasis not only on 'haulteur de style, grandeur
d'argumens, et gravité de sentences', but also on the moral purpose
to be achieved: 'pour remonstrer aux roys et grans seigneurs l'in-
certitude et lubrique instabilité des choses humaines, afin qu'ils
n'ayent confiance qu'en la seule vertu'.[6]

Even in works with such promising titles as *Diffinition de la tragédie* (1537), by Lazare de Baïf, or *L'Art de la tragédie* (1572) by Jean
de la Taille, this kind of statement is echoed. The subjects and
moral function of tragedy may be discussed, but not the nature of
the action. So Baïf writes that 'tragédie est une moralité composée
des grandes calamitez, meurtres et adversitez survenues aux nobles
et excellentz personnages...',[7] while La Taille states that 'son vray
subject ne traicte que de piteuses ruines de grands seigneurs, que
des inconstances de Fortune, que bannissements, guerres, pestes,
famines, captivitez, execrables cruautez des tyrans, et bref que
larmes et miseres extremes...'.[8] Laudun d'Aigaliers, in 1579, repeats the list: 'les commandements des rois, les batailles, meurtres,
violemens de filles et de femmes, trahisons, exils, plaintes, pleurs,
cris, faussetés et autres matières semblables...'.[9] The sequence is
incongruous, including as it does not only the subjects (in the
proper sense) fit for tragedy (battle, war, murder, etc.) but their
effects ('plaintes, pleurs, cris', 'larmes et miseres extremes'). It is
in these that the rhetoric and declamation come into their own: the
works of Jodelle, Garnier and others reveal their virtuosity in composing speeches calculated to arouse compassion but a complete
lack of interest or skill in creating an interesting and moving action. The endless lamentations will need to be supplanted by such
an action before French tragedy can become a reality, and indeed
tragedy will have to be born again, of tragi-comedy and the *pastorale*, before it can become a living reality. Nowhere in the definitions
I have cited can one find the idea of the tragic essence, because
nowhere do their authors consider the tragic action as a change
from happiness to misery as a result of some great error. When we
come to Heinsius's treatise, *De tragoediae constitutione* (1611), we find
him writing of plot as the essence of tragedy, and of the emotions of
'commiseratio' and 'horror' being aroused by the action 'non nar-
ratione simplici aut verbis, sed inprimis ipsa actionis dispositione'
('not by a simple narrative or by words, but chiefly by the very
disposition of the action').[10] Although this represents a distinct and
explicit change of emphasis, away from rhetoric and poetry and
towards the ordering of plot and action as the essence of tragedy
and as the source of the tragic emotions, it is not until tragedy has
been converted into an appropriately dramatic genre that the

theory becomes a theatrical reality. That transformation is effected, of course, by the growth of tragi-comedy which ousted tragedy from popular pre-eminence and eventually, about 1640, becomes the basis for a new kind of tragic drama.

Meanwhile, neither in the sixteenth-century tradition of tragedy nor, naturally, in early seventeenth-century tragi-comedy, do we find any development of the essence as described by Heinsius. Might a key to its nature be found in the notion of the catharsis of the emotions? As early as 1561, Scaliger had discussed the question of the dénouement and its effect: 'Catastrophe', he wrote, 'conversio negotii exagitati in tranquillitatem non expectatam' ('The catastrophe effects a change from agitated activity to unexpected tranquillity').[11] The unexpected tranquillity which is achieved in the catastrophe after the troubling of the emotions may be described as a cathartic effect. Heinsius likewise thought that the tragic action reduces or calms the emotions of pity and fear in the human soul. Sarasin believed that Aristotle's view of the aim of tragedy was that it should calm these emotions and restore tranquillity to the soul, reducing pity and fear to 'une médiocrité raisonnable'.[12] It has been shown that Racine probably owed something directly to both Heinsius and Sarasin[13] and, perhaps through them, to Scaliger. In his annotations on the *Poetics* (*Principes*, pp. 11–12), he translated part of the sixth chapter ('a tragedy...is in a dramatic, not a narrative form with incidents arousing pity and fear, wherewith to accomplish its catharsis of such emotions') by saying of tragedy: 'Elle ne se fait point par un récit, mais par une représentation vive qui, excitant la pitié et la terreur, purge *et tempère* ces sortes de passions' (editor's italics). Evidently dissatisfied with the idea of the purgation of emotions, fear and particularly pity, which from the point of view of Christian morality were far from unhealthy, Racine added not only the italicized words in an endeavour to translate more adequately what he sensed to be Aristotle's meaning, but a whole new sentence b[y] way of explanation: 'C'est-à-dire qu'en émouvant ces passions, el[le] leur ôte ce qu'elles ont d'excessif et de vicieux, et les ramène à [un] état modéré et conforme à la raison.' The link with Sarasi[n's] 'médiocrité raisonnable' is obvious as is the connexion with Sca[li]ger's and Heinsius's tranquillity.

Now if Racine's understanding of Aristotelian catharsis is [cor]rect, as most modern scholars would agree, and if it provides [us] with a key to the tragic essence, which experience of his p[lays] might lead us to believe, that does not of course mean that ca[thar]sis and the tragic can be simply equated. Are the tragic emo[tions] an end in themselves? The example of the fall of the statue of [P] in the ninth chapter of the *Poetics* seems to suggest otherwise:

Tragedy . . . is an imitation not only of a complete action, but also of incidents arousing pity and fear. Such incidents have the very greatest effect on the mind when they occur unexpectedly and at the same time in consequence of one another; there is more of the marvellous in them than if they happened of themselves or by mere chance. Even matters of chance seem most marvellous if there is an appearance of design as it were in them; as for instance the statue of Mitys at Argos killed the author of Mitys's death by falling down on him when a looker-on at a public spectacle; for incidents like that we think to be not without a meaning. A plot, therefore, of this sort is necessarily finer than others.

This passage calls for comment in several directions. In the first place, one notes that chance is not excluded from Aristotle's concept of tragedy, based as it was on his own experience of plays performed in the theatre. What is crucial is that the events of tragedy should *appear* to happen by design and in consequence of one another. The inner coherence of the plot, at which *vraisemblance* is chiefly directed, far from excluding the marvellous, can in fact be a source of the marvellous if the unexpected is seen to be the logical outcome of its antecedents, and if the combination of apparently chance events appears to serve some higher purpose. The fall of the statue was a chance event. The fact that it fell on the murderer of the man whom it commemorated may also be thought of as a chance event, but, properly presented, it can be seen as a representation of divine justice and retribution. Part of that proper presentation concerns the arousal of pity and fear, and part of it the suggestion of a vision which makes sense of the 'chances and changes of this fleeting world' by giving to the audience, in retrospect at least, the feeling that, for all the surprises and unexpectedness of the sequence of episodes in a play, the outcome was inevitable.[14] That is of course another aspect of *vraisemblance*. Vauquelin de la Fresnaye touched upon this in 1605 when he said that the poets of antiquity could 'monstrer par leurs vers combien la vie des hommes estoit fresle, debile, et infortunée, au respec de la bienheureuse felicité de Dieu'.[15] The Christian dimension is of course only Vauquelin's way of expressing the sense of some transcendent power behind human events.[16] Aristotle seems to suggest that the best kind of tragic plot conveys a sense of fitness and inevitability behind the sequence of even chance incidents. The awareness one feels of that ultimate design is not unconnected with the tranquillity of mind we have seen to be associated with catharsis.

Corneille's extended discussion of the problem, which opens his second *Discours* (*Writings*, pp. 28–37, and see my notes), reveals that he does not understand catharsis and does not believe that it is ever achieved. The whole passage revolves around the difficulty of choosing one of a dozen or more possible interpretations, particularly in view of the supposedly simultaneous arousal and purgation of pity and fear, and of the inappropriateness, as Corneille sees it, of the examples provided by Aristotle. Behind this lies no doubt his own concept of tragedy as being concerned less with characters of middling virtue than with those possessed of 'la grandeur d'âme' whether in terms of moral goodness or of moral evil. Polyeucte and Cléopâtre (in *Rodogune*) are examples of such characters. A lack of theoretical understanding, however, does not mean that the kind of tragic awareness I have mentioned is absent from one's experience of his plays, but rather that it is perhaps achieved by means different from Racine's.

Our analysis of plot development suggests a similar concern for *vraisemblance* as inner coherence in both dramatists, and a similar concern· for creating surprise and arousing curiosity. Modern theories of tragedy frequently exclude these last as detracting from the sense of tragic inevitability and so of transcendence. Classical scholars do not believe that the emotions of surprise and curiosity were excluded from Aristotle's concept of tragedy. One of them, basing his conclusions in a recent article[17] on an analysis of the *Poetics* and other evidence, has said that its author's remarks (chapter IX) about the playwright's not being bound to a 'rigid adherence to the traditional stories' and about 'even the known stories [being] known only to a few' are to be taken literally: the familiar stories were by no means universally familiar to the audiences of fourth-century Athens. So 'Aristotle could have conceived of dramatic tension of the most basic kind'. His 'general emphasis upon suspense, excitement, surprise does show an acute awareness of one level at which even the greatest of tragedies work upon the emotions of the audience, a level increasingly recognized by modern scholarship'. Not only does the possible ignorance of the audience allow the dramatist to exploit the creation and arrangement of episodes in such a way as to arouse curiosity and suspense: it encourages him also to alter the familiar stories, in their details at least, and even to invent them. The representation of characters of middling virtue can be seen to be calculated to permit the audience to identify itself with them in their own suspense or surprise as well as in their terror or false hopes. Hope springs eternal: we hope against hope, even when we know we delude ourselves — like and with the stage characters. With them, too, we feel apprehension as to the future. Scudéry wrote that 'la terreur et l'espérance, qui sont les deux ressorts qui font mouvoir tous les esprits, sont aussi les deux pivots sur quoi tourne toujours la scène'.[18] Hope, apprehension: the inseparable contraries on which Descartes's analysis of the passions is based are in themselves a principle of tension, a kind of tension dramatically exploited by Corneille in Camille, for example — the hope, kindled early in the play, of marriage, after long separation, on the morrow to Curiace, and the apprehension occasioned by the uncertainties of the battle —, or by Racine in Bérénice — the hope of union with Titus and the apprehension aroused by his evasions. The creation or adoption of dramatic situations, put to new purposes, the creation or alteration of plot developments as sequences of episodes: these, as we have seen in the works of both dramatists, generate suspense, tension, curiosity, surprise, emotions which can enhance the tragic effect. It would be mistaken to think that either of them scorned these emotions or that, if he did not, he was unworthy of the great tragic tradition. It is unlikely, from this point of view, that either playwright paid much heed to whether the audience was or was not familiar with the story of his play. (Corneille does not seem to have had a very high opinion of his spectators' historical knowledge — see the *Avertissement* to *Rodogune: Writings*, p. 187 — and Racine pours scorn on the competence even of his critics.) In terms of its dramatic qualities, the story was indeed less important than the largely or entirely invented situation to which it was adapted. All the arguments about fidelity to 'sources' deliberately avoid the real issue: they were intended for rivals, critics, and theorists, not for the spectators to whom both Corneille and Racine appealed as the ultimate arbiters.

At the same time, of course, the contrary emotions and the tensions to which the *péripéties* subject the characters bring them to breaking-point: then, and in the succession of the episodes, they attract our pity and fear. The tension created for Camille between her human love for the dead Curiace and the inhuman demands her brother makes on her provoke her ultimate outburst against him, the catastrophe. In the same way, Hermione is eventually provoked by contrary emotions into angrily dismissing Pyrrhus and sending him to his death. So it goes with Roxane. With less spectacular but no less moving effect, the same is true of Bérénice. None of this is achieved without the ordering of a plot in which the episodes, causing and resulting from the passions with *vraisemblance* and yet with surprise, produce dramatic alternations of fear and hope which eventually drive the characters to a fatal miscalculation, an error of judgement. In all these factors, which coexist, it is possible to see that the audience experiences complex emotions, the so-called dramatic and tragic, simultaneously and indeed inseparably.

The dramatic qualities of the plays bring us face to face with what has been considered to be one of the paradoxes of seventeenth-century French tragedy, a paradox allegedly resolved sometimes if not always by Racine (but by no one else) by emphasizing 'poetic' qualities to convey 'une vision sans cesse approfondie de la misère de l'homme'[19] and eliminating all hope of a better destiny: the paradox of combining a sense of the tragic with the dramatic form and function of tragi-comedy. It is of course true that, with the exception of a few glimmers here and there, the plays called tragedies written by the lesser contemporaries of Corneille and Racine fail to convey that sense: they offer to the public a form of theatrical entertainment in which suspense, curiosity, and surprise were paramount, but that does not imply that they were lacking in psychological depth or penetration — the tragedies of Rotrou, Quinault and others testify to the contrary. Nor does it imply that at least occasionally the secondary authors, as we now see them, failed to write moving and evocative verse. But these writers, it has been said,[20] used tragedy 'as a mere Form [i.e. a subdivision of Drama], with an unsuitable filling'. It has been found possible, following Lanson's definition — *Historique–Royale–Sanglante–Élevée de style*[21] — to deduce from works by Corneille and Racine a 'standard specification' of tragedy as it was understood by them and, presumably, their public. That specification, 'a minimal definition', provides a framework based not on character or emotion but on content and style. If one substitutes for Lanson's *Sanglante* Corneille's 'péril de vie, de pertes d'État, ou de bannissement', it is possible to see that, as they both claim, Corneille and Racine write plays in which *grandeur* is inherent in the subject, endowing it with seriousness, as the *Poetics* has it (chapter VI) and in that way drawing it 'towards the tragic pole'.[22]

All the tragedies written in seventeenth-century France were founded on 'history'; all their authors took liberties with it. When it suited their dramatic purposes Corneille and Racine, like their lesser rivals, had scant respect for many of the facts of history — chronology, relationships, motives could all be changed. Yet one of the hallmarks of the works of the two great authors was that, in spite of the liberties taken, something authentically historical informed their plays. Tragedy is not history: it is concerned not with accurate particularities but with an imaginative grasp of some of the forces behind them. That imaginative grasp eluded the minor playwrights or, if it did not, they had not the poetic gifts to convey it to us. Their cavalier treatment of history was not compensated for by an understanding of its inner and universal significance. Corneille, on the other hand, was capable, in *Pompée* for example, of seizing upon the remarks of Dio Cassius about the fall of a great man and his ignominious end, 'whereby was proved once more the weakness and the strange fortune of the human race', and of seeing in them a tragic vision of history.[23] Similarly, in *Iphigénie*, Racine drew into his very original treatment of the legend elements from the Homeric account of the rivalry between Agamemnon and Achilles and of the hostility between Agamemnon and his wife which was to culminate in their destruction (in the Electra plays),[24] and so infused his play not only with an authentic sense of 'history' but with a sense of tragedy as well.

So tragedy concerns historical figures, but often with an invented psychology, and involved in historical events, but with new connexions between them. The original aspects of the plays formed an essential characteristic of the situation with which the dramatist began and which he developed through a plot of his own creation in order to arrive at a historical catastrophe. The historical figures were necessary, however, not only for the sake of a plot which was in part contrived in such a way as to suggest the tragic vision, but also because, of their very nature as historical, they are the personages whose memory has survived in the annals of our race. They are the exceptional men and women who have come to be regarded as exemplary of human nature and behaviour and caught up in great events: public figures having to shoulder grave responsibilities, but exemplars also of humanity and propelled by passions common to us all. The minor dramatists filled the form of tragedy not with issues patently of serious import: they trivialized such issues by submerging them in romanesque and sentimental intrigues in which curiosity, suspense, and surprise were often generated with skill, but which allowed of no identification of spectator with stage character and of no sense of grandeur in either events or personages. Certainly these characters are placed in harrowing dilemmas, having to choose between what are to them equally demanding alternatives: Quinault and Thomas Corneille were outstandingly successful in devising such situations, and their plays were immensely popular — popular, too, for the frequently convincing portrayal of passion. The dilemma was far from being absent from the tragedies of the two masters: from Chimène and Rodrigue to Eurydice and Suréna, from Andromaque to Hippolyte, their characters are faced with impossible choices. Hence the 'combats intérieurs', the 'généreuses irrésolutions' which lie at the heart of all their situations and plots: in devising them, they were willing to learn, as we have seen, from their lesser rivals, and Racine was willing to learn from Corneille.

A crucial passage from the latter's first *Discours* (*Writings*, p. 8) deserves to be remembered in connexion with his concept of tragedy:

Lorsqu'on met sur la scène un simple intrique d'amour entre des rois, et qu'ils ne courent aucun péril, ni de leur vie, ni de leur État, je ne crois pas que, bien que les personnes soient illustres, l'action le soit assez pour s'élever jusqu'à la tragédie. Sa dignité demande quelque grand intérêt d'État, ou quelque passion plus noble et plus mâle que l'amour, telles que sont l'ambition ou la vengeance, et veut donner à craindre des malheurs plus grands que la perte d'une maîtresse. Il est à propos d'y mêler l'amour, parce qu'il a toujours beaucoup d'agrément, et peut servir de fondement à ces intérêts, et à ces autres passions dont je parle; mais il faut qu'il se contente du second rang dans un poème, et leur laisse le premier.

It is clear that for Corneille the distinguishing mark of tragedy is an 'elevated' action, something more 'serious', as Aristotle would put it, than a love plot between even 'illustrious' characters.[25] One may of course see here a definition of Corneille's own type of tragedy, with its emphasis on great public events or issues and the passions that go with them (ambition, 'gloire', revenge, etc.), but love does play its part, usually an essential part, and often as the spring of the great interests at stake. Cinna conspires against Auguste, reluctantly, but because if he refuses he will lose Émilie: the great public issue is made dependent on his love. César's conquests are inspired, so he says, by his passion for Cléopâtre (the Egyptian). Rodogune and Cléopâtre (the Syrian) manipulate and test the love of Antiochus and Séleucus for their own political ends. Nicomède and Laodice draw inspiration from their love to resist the manœuvrings of Prusias and Arsinoé, as will Suréna and Eurydice much later and in a different key. Corneille's plays of the 1660s are characterized by the sacrifice of love and the exploitation of marriage as means to achieving alliances and other political ends. If, therefore, love provides the basis of the great issues it does tend to be the secondary — if sometimes more appealing — dramatic interest. Corneille's emphasis is on the 'action illustre, extraordinaire, sérieuse' (ibid. p. 9), but the love plot is never frivolous or trivial for it always springs from a powerful passion which has its own dignity and seriousness, because it raises moral problems for the characters graver than 'la perte d'une maîtresse'.

Corneille's definition of the tragic action in these terms has often been used as a basis for contrasting his plays with those of Racine, in which love has been assumed to preponderate to the extent of eliminating more 'serious' questions. Certainly the emphasis is different, at least in all the secular plays; but *Athalie*, in particular, entirely devoid of love interest, has an action which is perhaps — at least from the point of view of Racine and his contemporaries —

of greater moment and of more universal import than any of Corneille's tragedies. As far as the present discussion is concerned, the
tragic quality of that play is hardly in doubt. But the contrast between the two dramatists is essentially one of emphasis, not of
black-and-white distinctions. In Racine's secular plays, the action
is never one which involves only 'un simple intrique d'amour entre
des rois'. The royal personages are indeed royal in the sense that
they have not merely their personal dignity or that of their office,
but also their responsibilities, often of a dynastic nature. The risks
they run in pursuing their amorous passions are those of loss of life
or kingdom, or of banishment. Pyrrhus threatens the Greek
alliance and the safety of his own state by insisting on marriage to
Andromaque and saving the life of her Trojan son, the heir of Hector. Néron and Britannicus may indeed be rivals for the hand of
Junie, but their rivalry is essentially a factor both in the legitimacy
of the emperor's rule and in Néron's desire to throw off the tutelage
of his mother. The 'elegy' of Bérénice and Titus is far from being a
simple love-story: the emperor's rule is at stake and its importance
prevails. In every one of Racine's tragedies, problems of authority,
royal succession, usurpation of kingdoms or political rebellion are
inextricably bound up with what is doubtless the central action
constituted by the conflict of loves, hates, and jealousies to which
the characters are subject. In effect, it is only the purport of the
last sentence of the passage quoted from Corneille's first *Discours*
which clearly distinguishes his tragedies from those of his rival
who, after all, could write in defence of his 'elegy' that in tragedy 'il
suffit que l'action en soit grande, que les acteurs en soient héroïques,
que les passions y soient excitées...'. One notes the emphasis
on action and its grandeur, on the characters as agents, and their
noble status and stature, and on a multiplicity of passions. Besides,
the passions which drive Racine's characters are in themselves serious and not frivolous, all-consuming forces and not ephemeral
whims, as well as being serious by being bound up with public
issues and events.

The characters of the tragedies of both dramatists are conscious
of the import, the seriousness of the action in which they are engaged, and of their own status and dignity, which none of them
ever loses. Both the seriousness of the action and their own dignity
find expression in noble discourse and diction. Their language itself (like the alexandrine verse) enhances the dignity of their status
and is the sign of their being raised above the level of common
humanity. They are raised above it also in the sense of their being,
as Aristotle puts it (*Poetics*, chapter XIII), 'in the enjoyment of
great reputation and prosperity'. If it is to impress as being tragic,

as being poetic, 'more philosophic and of graver import than history' (*Poetics*, chapter IX), their fall must be from a great height both of status and authority and of happiness or prosperity. The connexion between the depth of the fall and the arousal of tragic emotions is clear in La Mesnardière's formulation: '... Les infortunes qui arrivent à ces puissances qui semblaient être au-dessus de toutes ces calamités frappent notre imagination avec plus de véhémence, d'étonnement et de terreur que si elles s'attachaient à des personnes vulgaires.'[26] Horace builds up his heroic stature by the exercise of his own will to the point where his glory is for ever assured; and then, to save it from dishonour, as he sees it, he falls into the 'great error' of killing his sister. 'How are the mighty fallen!' If he had not achieved high glory and prosperity, his fall would not be felt to be tragic. And 'he that is down need fear no fall'. By often ignoble means, Auguste has reached the pinnacle of political power; his power being threatened, and his self-esteem wounded by the conspiracy of his trusted friends and advisers, he becomes aware of his imminent fall, less from power itself than from future glory, in time to avoid it by taking the risk of refusing to exercise that power itself; but the menace of the fall is felt as acutely as though it had actually taken place. César and Rodogune are similarly sayed from falling — they look over the precipice and we with them — but Ptolomée and Cléopâtre are not. Pyrrhus as one of the vanquishers of Troy and, as he imagines, the possessor of Andromaque, falls from those heights to his doom; Oreste, deluding himself that the assassination will bring him the happiness of marriage to Hermione, is precipitated into the depths by her passionate refusal. Bérénice, prospering with her ever-growing kingdom and happy in the thought that Titus is now free to marry her, is thrown into despair when she learns that he is not free at all. Without their 'heroic' stature, the fall of such characters would be reduced to the level that has often enough been attributed to Racine's, if not to Corneille's, dramatic action, that of disappointment, disillusionment, 'crime passionnel'. While of course these things are present they are raised, as is not usually the case in the plays of the minor dramatists, to the tragic pitch.

In considering the causes of the tragic fall, we need to look at the ways in which, in the plot of the plays, it is brought about and represented. The tragedies of both dramatists may be seen to be constructed on a series of tests to which their characters are subjected. Distinctions emerge, however, when we consider the nature and outcome of the tests. Horace, Polyeucte, César, Cléopâtre (in *Rodogune*), and Nicomède, for example, are characters capable of making up their minds immediately to pursue what they see as the path to glory, and of acting without hesitation. But the incidents of the plot test the strength of their will to carry out their decisions and are often occasioned by external factors or events over which they have no control: Horace faced with Curiace as friend and brother-in-law, taunted by Sabine, threatened by the death of his brothers, taunted again by Camille, 'tried' by his king; Polyeucte tested by Néarque and going early to baptism, his will endangered by the presence of Pauline and the arrival of Sévère... It is unnecessary to enlarge upon this question. Some characters survive the tests with their will unscathed and their glory undiminished (César, Nicomède), though not necessarily without suffering or a threat of doom to come.

Some pass one set of tests only to fail in another: Horace's self-control falters fatally when he is faced with his lost humanity in Camille's passionately angry outburst. Some must pay the price of death for what they believe to be their glory, Polyeucte in goodness and virtue, Cléopâtre in evil and vice. Other characters hesitate between alternative courses of action because a move in either direction has equal attractions or dangers. The tests then search out their wills and eventually lead to decision and action. Cinna hesitates to carry out the conspiracy because he is torn between love for Émilie and gratitude and respect for Auguste; once he has made up his mind and is discovered, his resolve is tested by the interrogation to which he is subjected; Auguste himself hesitates early in the play between abdication and pursuit of his reign by conventional political (and often cruel and unjust) means, and later, once the conspiracy is revealed, between punishment and pardon; both these characters pass through irresoluteness to decision, Cinna to commit an error and a crime, Auguste to see the path to true glory and a kind of personal salvation. But others (Sertorius, Eurydice) hesitate too long and, although they do eventually make decisions, they make them too late. None of these characters has any direct control over the critical incidents which mark the stages in the development of the plot (often as *péripéties*) and which demand decision and will-power. The possibility of error or failure is always present and sometimes realized.

The linear development of the plot, one test succeeding another, raising tension and arousing curiosity, gives to Corneille's plays an appearance of complexity and a sense of precipitation, particularly as the tests frequently arise from external causes. They give to the characters an urgent sense of purpose and an impression that survival of the tests and solutions to the problems are really possible. Sometimes the will is ultimately frustrated; sometimes its exercise proves fatal. Even when it does not, however, the play is not necessarily deprived of its tragic quality. Although in most of Corneille's plays mistaken identity plays no part and so does not lead to recognition either before or after the fatal deed has been committed, Auguste for example does seem to be the psychological and moral equivalent of the heroes of Greek plays cited by Aristotle (*Poetics*, chapter XIV) as featuring in the 'best of all' tragedies, in which the deed is not actually done because its victim is recognized in time. In *Cinna*, the emperor knows the identity of the conspirators: what saves them, and him, is his recognition of himself and his true ideals. Corneille would presumably deny this, since he wishes to substitute for Aristotle's favourite type of tragedy — and indeed for the others he describes — 'une espèce nouvelle'. Interestingly enough, *Cinna* (along with *Le Cid*, *Rodogune*, *Héraclius*, and *Nicomède*) is used as an example. This new kind of tragedy is based on the one which Aristotle found least satisfactory: 'the worst situation is when the personage is with full knowledge on the point of doing the deed, and leaves it undone. It is odious and also (through the absence of suffering) untragic.' In his analysis of *Cinna* (second *Discours: Writings*, p. 40) — and it is accompanied by remarks on the other plays just listed —, Corneille thinks of the uncommitted deed as the assassination of Auguste, not as the punishment of the conspirators. It is curious that he should confine himself to this angle on the play to which he gave the sub-title of *La Clémence d'Auguste*; but of course from the point of view of plot he is right, as his own brief analyses in the first *Discours* (*Writings*, pp. 10, 13) show. What is important to him is that the withdrawal from precipitating catastrophe should not be the result of 'un simple changement de volonté'. But, he says, when the characters

...font de leur côté tout ce qu'ils peuvent, et qu'ils sont empêchés d'en venir à l'effet par quelque puissance supérieure, par quelque changement de fortune qui les fait périr eux-mêmes, ou les réduit sous le pouvoir de ceux qu'ils voulaient perdre, il est hors de doute que cela fait une tragédie d'un genre peut-être plus sublime que les trois qu'Aristote avoue... (Ibid., p. 40.)

The application of this rule to *Cinna* is obvious:

Cinna et son Émilie ne pèchent point contre la règle en ne perdant point Auguste, puisque la conspiration découverte les en met dans l'impuissance, et qu'il faudrait qu'ils n'eussent aucune teinture d'humanité, si une clémence si peu attendue ne dissipait toute leur haine.

These comments are interesting, not only because they reveal something of their author's conception of consistency of character and behaviour (which is linked to *vraisemblance*), particularly when the characteristic involved is strength of will ('vertu'), but also because the legitimate ways of frustrating the will indicate the presence in his plays of powers outside themselves over which the personages have no control. We see here, then, confirmation of our own analysis of the type of incident on which Corneille's plots are largely based. (The 'simple changement de volonté' which he condemns is, however, a frequent feature of the plays of the lesser dramatists.)

His discussion of the problem moves, as does Aristotle's, to the question of which kind of plot is the most capable of arousing the tragic emotion of pity. He thinks that the third of Aristotle's categories, which comprises the kind of situation in which agent is unaware of victim's identity but discovers it in time to draw back, does not arouse pity. Assuming that the audience does know the identity of the victim, such a situation can give rise only to 'un certain mouvement de trépidation intérieure': the audience fears that the deed may be done, and hopes that it will not be done, before that identity is discovered. Corneille pursues the matter further in writing that when the discovery is not made until after the death of the unknown victim, 'la compassion qu'excitent les déplaisirs de celui qui le fait périr ne peut avoir grande étendue, puisqu'elle est reculée et renfermée dans la catastrophe' (ibid., p. 41). The importance of this argument is twofold. First, Corneille himself conceives of tragedy as arousing at least one of the two authentic tragic emotions, though one does note how easily he slips from 'pitié' to 'compassion' and, in the following sentence, to 'commisération', so that one may wonder whether, without the Christian overtones of these last two words, pity might not be felt for the agent in the kind of situation described. Be that as it may, Corneille certainly believed that he was discussing authentic tragic pity. Second, because he believes that pity must be aroused, it must be aroused within the working-out of the plot for, as we have seen, the dénouement is in his plays delayed as long as possible in order that suspense and curiosity be maintained throughout the play. So if the victim's identity is discovered after his death, the discovery will necessarily come too late in the play for pity for the agent to be developed to any great effect. But to say this is not at all to say that Corneille ignores or dismisses the need to arouse pity, though it is a commonplace of criticism to affirm that he substitutes 'admiration' for it and that, unlike Racine's, his characters are too 'heroic' in their actions to excite the traditional tragic emotions.

It is perfectly clear, on the contrary, that Corneille is intent on arousing pity. As he continues his argument in favour of the kind of situation in which agent knows victim but is prevented by some extraneous power from killing him, he insists on the need to arouse pity in the course of the development of the plot. The implications of one sentence are, for all its celebrity, rarely considered either in terms of the context we have been examining or in relation to Corneille's own dramatic practice:

> Mais lorsqu'on agit à visage découvert, et qu'on sait à qui on en veut, les combats des passions contre la nature, ou du devoir contre l'amour, occupent la meilleure partie du poème, et de là naissent les grandes et fortes émotions qui renouvellent à tous moments et redoublent la commisération.
>
> (Ibid., p. 41.)

The plot then is so devised as not only to produce what we have called tests of the will, but also to allow those same incidents to give rise to the 'combats intérieurs' experienced by the characters. The source of pity lies in their being torn by contrary desires and imperatives, and as crisis follows crisis that pity is renewed and intensified. The development of the plot does more, therefore, than increase the dramatic excitement; it intensifies the tragic emotions at the same time, And if the excitement is to be maintained to the very end, the tragic emotions are sustained along with it. Implicit in Corneille's thinking about his art is the belief that the dramatic and tragic emotions are complementary, not incompatible.

Without being entirely subjective it is of course difficult to discuss emotional arousal and response on the part of the audience. Is our response to any play of Corneille or to any incident in it one of pity? All one can say with certainty — because pity, if experienced, is not necessarily authentic — is that down the last three centuries audiences appear to have confirmed the experience to which Corneille refers in the passage we have been studying. While he admits that some of his characters, Œdipe for example, arouse less pity than others, he seems to imply that the arousal of pity is not occasioned simply by a single character but by 'le poème entier': Aristotle writes, not of characters exciting pity, but of the situations in which they are placed. It is the situations that give rise to the 'combats intérieurs', 'les grandes et fortes émotions' in the characters, and it is the spectacle of those emotions that renews and intensifies our pity with each successive crisis, *péripétie*, or 'coup de théâtre'. Here again it is evident that the dramatic features of the plot need not preclude our experience of the tragic emotions — quite the contrary, in fact. It is also evident that the minor dramatists contrived what D'Aubignac (*Pratique*, pp. 70–1) called 'sujets d'incidents, intrigues ou événements': concentration on the 'accidents imprévus' brought about by conflicts of interests produces only 'une attente agréable et un divertissement continuel'. At the opposite pole are the 'sujets de passions', 'quand d'un petit fonds le

poète tire ingénieusement de quoi soutenir le théâtre par de grands
sentiments': the episodes, which appear as integral to those subjects, are the occasion for changes of feeling in the characters and
maintenance of interest in the audience.

Now this kind of distinction has often been applied to Corneille
and Racine. It may be true that in some of his plays written in the
1660s the elder dramatist did, as D'Aubignac seeks to prove in his
Dissertations, concentrate unduly on incident and that Racine took
his cue from the abbé's criticisms and produced what are usually
called simple plots. In fact, however, both Corneille and Racine,
with varying emphases, wrote plays with that D'Aubignac calls
'sujets mixtes', 'lorsque par des événements inopinés, mais illustres, les acteurs éclatent en des passions différentes, ce qui contente
infiniment les spectateurs, quand ils voient tout ensemble des accidents qui les surprennent et des mouvements d'esprit qui les ravis-
sent'. These 'mouvements d'esprit' correspond of course to Corneille's 'commisération', and the 'passions' to his 'grandes et fortes
émotions'. Without the dilemmas arising out of the situation, these
passions and emotions would not be produced; without the successive tests and crises, they would not be intensified.

Corneille's plots may be thought of in terms of a linear development in which the hero progresses from height to height, but not
without suffering and temptation, through incidents which challenge him along the way, to turn aside — hence the pity —, nor
without the danger of falling — hence the fear. He is sustained in
his march by his conviction that in itself it will bring him honour
and that his goal is really attainable, fulfilment of duty (however
that is understood) or realization of glory. He works towards the
image of himself that he wishes to see projected on the future. The
image may ultimately be tarnished (Horace, Cinna) and even
where it is not (César, Nicomède) the hero's hold on it may be
precarious. Sometimes the image is false and is shattered
(Cléopâtre).

If one may use another geometric metaphor, Racine's plots are
circular. Far from being open on to the future they enclose the characters within the present. We have seen how these characters aspire,
not to future glory, but to present happiness. The plots are not
lacking in incident, but the incidents themselves spring from within
the initial situation which renews itself through them and revolves
upon itself. It is because Andromaque cannot make a choice that
Pyrrhus, Hermione and, ultimately, Oreste hesitate between equally imperative and equally repelling alternatives. They are faced
with contrary passions because they are placed in a certain
situation thanks to the arrival of Oreste at the court of

Pyrrhus. His embassy on behalf of the Greek allies precipitates the need for Pyrrhus to arrive at a decision over the fate of Astyanax. But Pyrrhus is emotionally dependent, in trying to reach it, on Andromaque, as is Hermione on him, and Oreste on her. And since Oreste's mission is motivated, not by political ends, but by his passion for Hermione, the circle is complete, and it turns on its own axis, each hesitation or change of mind pushing it round towards the next. The plot is Racine's 'machine infernale'. With variations its characteristics appear in all his plays. While Corneille condemns the type of escape-dénouement which results from 'un simple changement de volonté', Racine so creates characters and places them in such situations that changes of passion — from love to jealousy or hate, from one form and object of love to another — are a perpetual feature of the whole plot, not simply of its end. The characters' being caught in a situation in which, as Marmontel was to put it, 'l'alternative n' [a] point de milieu',[27] and their vain endeavours to surmount obstacles which lie within themselves, attract our pity, particularly when, as we have seen, we are aware, as they are not, of the futility of their efforts, and when they resort to lying and subterfuge, deceiving not only others but themselves. Of course many of Corneille's characters have delusions of grandeur — Horace perhaps, Cinna certainly (Auguste tells him so), Ptolomée, Cléopâtre (the Syrian queen), Félix, Prusias, Sophonisbe, Sertorius, Suréna — but they are not the same thing (although in their pursuit they may lie) as the deliberate self-deception of Oreste, Agrippine, Bérénice, Roxane, Mithridate, Agamemnon, or Phèdre.

None of this implies, however, that Racine's characters are reduced to mere passivity. Even when, like Andromaque, they are in effect powerless, they still actively search for an escape from their dilemma which may not be an actual solution to it. Bérénice, Roxane, Mithridate, Achille, Thésée, and Athalie all display great vigour. Oreste, tired of the burden of his 'innocence', is determined to carry off Hermione:

> Il faut que je l'enlève, ou bien que je périsse.
> Le dessein en est pris; je le veux achever.
> Oui, je le veux.   (ll. 714–6)

Right to the end, up to his last encounter with Hermione after the assassination, Oreste is still endeavouring to achieve that one purpose in spite of frustrations, hesitations, and the knowledge which he will not admit to himself, except momentarily (ll. 537 ff.), that she can never return his love. It is precisely the absence of that underlying self-contradiction which distinguishes the Cornelian hero: once he has come to his resolve, often after doubts and hesitations, he puts them behind him even to the extent, as with Horace, of trying to kill them as they still lurk within.

If, then, Racine's plots appear to be simpler than Corneille's, it is not because they lack incident or *péripéties* but, as we have seen, because the incidents are not externally occasioned. If, in Corneille's tragedies, the incidents test the will of the hero, in Racine's they are created by the characters themselves: in seeking to inflict suffering on others they entail it for themselves. Even when, as with Titus, the suffering is reluctantly inflicted the agent suffers in doing it. Another way of expressing the distinction would be to say that in Corneille's plays it is the externally-produced changes in the situation which give rise to the 'combats intérieurs' which then become the cause of pity, while in Racine's it is the unresolvable 'combats intérieurs' which occasion the incidents, the inner conflicts still being in themselves a source of pity, the more so as the incidents intensify the suffering. If we are to make use of D'Aubignac's distinction between 'sujets d'incidents' and 'sujets de passions' in applying it to the tragedies of Corneille and Racine or, more appropriately, if we are to see in both 'sujets mixtes' with different emphases, the distinction needs to be qualified to take account both of the nature and origins of the incidents and of their relationship to the inner conflicts.

Another aspect of our earlier discussion of simplicity emerges in the context of the arousal of pity. We have seen Corneille suggesting that pity should be excited, renewed, and intensified as the plot develops and as one test or crisis succeeds another. There is no doubt that in Racine's tragedies, too, pity is aroused by the emotions occasioned in the characters by the successive incidents or giving rise to them. But where Corneille (third *Discours: Writings*, pp. 63, 66, 67, 74), anxious to maintain to the end the interest of the audience not only in the feelings of his characters but in the dramatic developments in themselves, does not see how either intellectual curiosity or tragic emotion can continue beyond the end of those developments,[28] Racine writes plays in which the plot is virtually completed in a dénouement at the end of the fourth act. The pity is indeed at its most intense when the 'deed of horror' has been done and the characters are brought face to face with the reality of what they have done and of what they are. Racine's tragedies are constructed as they are not simply because he needs to accommodate the discovery itself, but because the discovery is, as Corneille saw that it might be, a powerful source of tragic emotion.

Analysis of plot brought us to some important distinctions between the plays of the two dramatists. They are of course not simply a matter of technique, but have a deeper significance in that they reveal different ways of expressing a tragic vision. The absence of discovery on the part of the characters who have committed the fatal deed does not necessarily mean that the play is not tragic, and it is only with rare exceptions (*Héraclius, Iphigénie, Athalie*) that discovery of personal identity is involved.

In our discussions of the trial-and-judgement episodes in some of the works of Corneille and Racine, we established some parallels. It is possible also to see divergences. With the dénouement in Corneille's plays being delayed for as long as possible and with the absence of the kind of discovery analysed by Aristotle, it is clear that any deliberations must occur as the plot develops and progresses towards the dénouement. Horace and Curiace deliberate on their responses to being nominated each to be his city's champion (II.i, iii): they both fear the outcome, but neither hesitates to accept the honour and the challenge; each judges himself and his adversary; each sees his own weakness but, overcoming it, is ready for action. Although tempted to yield to it when faced by the women, neither turns back. When at the end, although subjected to a kind of trial, as we have seen, Horace has the opportunity to do so he makes no discovery about himself or about the nature of what he has done. He remains to the end true to the ideal expressed to Curiace when they both know their destiny:

> La solide vertu dont je fais vanité
> N'admet point de faiblesse avec ma fermeté,
> Et c'est mal de l'honneur entrer dans la carrière
> Que dès le premier pas regarder en arrière.
> Notre malheur est grand, il est au plus haut point;
> Je l'envisage entier, mais je n'en frémis point.
> Contre qui que ce soit que mon pays m'emploie,
> J'accepte aveuglément cette gloire avec joie...   (ll. 483–91)

There is to be no looking back: the inner conflict is over — almost as soon as it has begun — and the decision, once taken, will resist all attempts to undermine it. But Horace already admits that he takes it blindly. He is never enlightened — or at any rate never admits to being enlightened — as to its real nature and consequences.

Cinna and Émilie deliberate on the action to be taken and, in spite of doubts and hesitations, decide on the assassination of Auguste. At the same time, he deliberates with Maxime and Cinna — trusted counsellors as well as treacherous conspirators — as to whether he should abdicate. Having decided to continue his imperial rule, he discovers the conspiracy. Deprived of his advisers, he deliberates with himself and again appears, after another deliberation with Livie (IV.iii), to decide on punishment. That impression is confirmed in the 'trial' of the conspirators which occupies the final act. Auguste has overcome what he calls the 'rigoureux combat d'un cœur irrésolu' (l. 1188), and it is only in the last scene that the true light causes him to relent; but by then he has in fact carried out his decision to try the criminals and has already condemned them.

Any of Corneille's tragedies will reveal a similar structure, the deliberations occurring as the plot develops and resulting in further developments: Ptolomée deliberates with Photin, Achillas, and Septime right at the beginning of the play, with César in the third act, with Photin and Achillas and with Cléopâtre in the fourth. Each of these deliberations results in a decision which carries the plot forward; the last one proves fatal to him; nowhere is he given the opportunity either to turn back or to judge himself. Sertorius deliberates with Viriate, Perpenna, and Pompée, as does Suréna with Eurydice, Orode, Palmis, and Pacorus. The deliberations bring decisions but no subsequent judgement.

The pattern is not the same in Racine. In his tragedies, certain kinds of deliberation do exist. In *La Thébaïde* and *Alexandre* they are not markedly different, in form or in function, from Corneille's; but thereafter a change takes place. Whereas Auguste, Ptolomée, Sertorius and others genuinely seek advice in a predicament, Pyrrhus and Titus, for example, do not. When Pyrrhus 'consults' Phœnix (II.v) he merely invites advice that he wants to hear, and when it turns out to be different he rejects it while pretending (to himself as well as to his confidant?) to accept it. When Titus 'consults' Paulin (II.i) he is seeking simply to fortify his own resolve, to confirm what he already knows about Roman tradition and prejudices.

The shape of Racine's plots is such, however, as to allow the characters time to be involved in catastrophe as well as in the dénouement, to face the consequences of the final decision. In place of deliberation before the decision is reached they often go through self-judgement after it and, however fleetingly, come to enlightenment. Our analysis of *Britannicus* and *Mithridate* led to this conclusion. The self-judgement is much more extended in certain other cases: Hermione and Oreste, Bérénice, Phèdre, Athalie. While close parallels exist, as we have seen, between the self-judgement of Auguste and that of Hermione, those parallels chiefly concern the rhetorical structure of the scenes in question and the inward searching in which the characters are engaged. But, placed in their context within the development of the plot, they are seen to be quite different in function. Auguste is seeking a way forward and fails at that moment to find it; Hermione has already acted

and is reflecting on what she has done. Racine's version of that part
of Aristotle's *Poetics* (chapter XIV) which we have examined in
connexion with Corneille's comments on it is very revealing, partly
because of the liberty he takes with the text, inserting the italicized
phrase, and partly because this passage is followed by a misunderstanding of the text, since he confuses two of the types of situation
and assimilates them. 'Mais le·meilleur *de bien loin*, c'est lorsqu'un
homme commet quelque action horrible sans savoir ce qu'il fait, et
qu'après l'action il vient à reconnaître ce qu'il a fait.' (*Principes*,
p. 25.) The phrase 'de bien loin' exaggerates the preference, presumably because (unconsciously?) Racine had in mind his own
type of tragedy. The same impression is created by his substitution
(suggested perhaps by Vettori's '*imprudentem* quidem egisse...
agnovisse') of the nature of the deed for identity of the victim.
Racine's characters, no less than Corneille's, act 'à visage
découvert' except in so far as their own selves are masked from
them. The doing of the deed faces them with their own reality.

The nearest Corneille comes to this is in his discussion, in the
same context (*Writings*, p. 40), of *Nicomède*: Prusias, he writes, 'est
forcé de reconnaître son injustice après que le soulèvement de son
peuple, et la générosité de ce fils qu'il voulait agrandir au depens
de son aîné, ne lui permettent plus de la faire réussir'. Essentially,
however, Prusias recognizes less the nature of his own error than
the presence of *force majeure*. In fact, although Corneille acknowledges that Aristotelian discovery may be 'un grand ornement'
(p. 42), he maintains that it causes the loss of 'beaucoup d'occasions de sentiments pathétiques qui auraient des beautés plus considérables', and those beauties are to be found in the pathos of 'les
oppositions des sentiments de la nature aux emportements de la
passion ou à la sévérité du devoir' (p. 38). Yet, no less than
Racine, Corneille seems to accept D'Aubignac's advice that the
dramatist should choose to begin his play as close to the catastrophe (or dénouement) as possible. But whereas Racine presumably does this in order to accommodate the moment of self-
discovery and indeed to extend it considerably in some cases, Corneille does it partly in order to avoid difficulties over observing the
unity of time, partly to precipitate the action and maintain the excitement, and partly to incorporate the maximum of those beauties
which arise from the pathetic portrayal of characters repeatedly
caught in crisis or dilemma. Did Racine perhaps find his way to his
particular form of tragedy by following negative clues in the writings, dramatic and critical, of his elder?

But the problem raises another question: if the characters themselves (or some of them) do not make the discovery, is the play

necessarily untragic? Eugène Vinaver seems to come to the conclusion that it is; but he quotes with approval a passage from Professor T. B. L. Webster's comments on Greek tragedy which suggest
something rather different:

The process [of tragedy] has three stages: first *hybris*, the initial act of
pride, violence or folly; secondly, *ate*, infatuation, sent by the gods to lead
the sinner to his ruin; and thirdly, enlightenment whether of the sinner
himself *or of the world through his example.*[29] (My italics.)

We may take this formulation as a framework against which to set
some of Corneille's plays to which many have denied tragic quality, though as far as *hybris* is concerned some qualifications will be
entered later. Of the three stages set out by Professor Webster it is
the nature of the last, of course, which most concerns us here.

Let us begin with *Horace*, so often regarded as the touchstone of
Corneille's art as a writer of tragedies. The initial act of pride or
folly — *hybris* — on the part of the hero is no doubt his unquestion-
ing acceptance of the honour of being chosen as the champion of
his city in its conflict with Alba Longa. His pride is such that,
without hesitation, he suppresses the claims of his more human
and instinctive feelings for family and friends. Eventually he is
victorious in the combat. If the play were to end at that point,
which is what Corneille's critics would have preferred, both on the
grounds of morality (*bienséance*, etc.) and on those of technique
(unity of action), we should no doubt agree that it would scarcely
be tragic: it would be a piece of heroic drama, exciting and morally
uplifting, perhaps, but no more. Corneille, however, never yielded
to criticism of the morality and construction of his play. He persisted in having Horace overtaken by *ate* when he meets Camille
and is taunted by her for his inhumanity. Infatuated with his
'gloire' and vulnerable still in his 'barbarity', he turns on her and
kills her. This and all that follows may be a technical blemish, but
the end of Act IV and the whole of Act V are necessary, from the
tragic standpoint, to show Horace isolated and standing a kind of
trial, more perhaps before the audience than before his King, and
in that sense moral rather than legal. The fact that he is not actually punished by Tulle is not a sign of pardon, and the fact that he
now has to go on living with his crime is for him a worse punishment than the death he asks for. Horace does not, as we have seen,
admit to being other than legally guilty: in wishing to die he seeks
to preserve the 'gloire' which is put in jeopardy by his survival. So
he is not enlightened as to the true nature of his error, but his very
silence on that score, coming at the end of the whole dramatic
process, enlightens the audience. That enlightenment about the

error which leads the hero to think of himself literally as a super-
man, superior to his own humanity as well as to that of others, is
what fulfils the tragic purpose of the play. The technical defect,
which Corneille was always unwilling to correct, draws attention to
his belief in his perception of what constitutes tragedy.

Two of his own favourite plays, while from this point of view not
technically defective, confirm these findings. In *Rodogune*,
Cléopâtre, queen of Syria, is depicted, as Corneille himself says, as
entirely evil. Bent on the destruction of Rodogune, the Parthian
princess who is virtually her captive, and of her own sons who are
both the latter's suitors, she engages in Machiavellian manœuvres
in order to trap them into self-incriminating attitudes. The trick-
ery, however, recoils upon her: instead of separating her adversaries she succeeds only in uniting them and isolating herself. Reduced to despair, she drinks the poison intended for them. Like
Horace, but on a very different moral level, Cléopâtre single-
mindedly pursues the goal of her own glory which takes the form of
political power. Like Horace, too, yielding to *hybris* and *ate*, she
knows the solitude of apparently reaching the summit of her ambition. Brought to ruin, she fails, like him again, to recognize the
vanity of power pursued for its own sake and at the expense of all
maternal feeling, although in the last scene the opportunity could
be made. The audience is, however, enlightened and brought to
the realization that a would-be superhuman will, especially when
directed to morally worthless ends, can cause men and women to
transgress beyond the limits of their place in the moral universe.
The terrible nature of the inevitable retribution shows where the
limits lie, not by argument but by working on the imagination
through the emotions, particularly, in this instance, through fear.

Cinna is in many ways the converse working-out of the same
problem as in *Rodogune*. Where Cléopâtre brings upon herself solitude ·and destruction and dies without the tragic privilege of enlightenment, Auguste eventually binds others to himself. By passing through isolation he comes to the enlightenment which saves
him from moral destruction. As in his solitude he recalls the past in
which he succumbed to *hybris* in capturing and maintaining his own
imperial power, no matter how inhuman the means, he resists the
temptation of *ate* which would by the same methods perpetuate his
despotic and unscrupulous hold on the empire. But the pardon he ac-
cords to the conspirators is refused until he discovers the true motive
for granting it: it must not be for political ends, in order to maintain
his power over others, but to achieve, at the expense of that power,
mastery over himself as the source of true authority. The development of the play as a whole ensures that that enlightenment,

briefly expressed as it is (ll. 1693–1700), and the instantaneous recognition and reconciliation it brings from the conspirators, is ours as well as his and shows where human dignity really lies.

These few examples show that if Professor Webster's formula is applied to them some at least of Corneille's plays, even when discovery is not present or is not developed for the tragic victim, are nevertheless to be considered as true tragedies. The pattern is different from Racine's but the ultimate effect not necessarily dissimilar. Although enlightenment is universally — whether for the tragic character or for the audience — discovery of *hybris*, this word is often incorrectly or inadequately understood. In an examination of the Athenian law relating to *hybris*, Professor W. M. MacDowell has drawn upon a wide variety of documents, including tragedies, to show that it usually has less to do with an actual assault on a victim than with crowing over him. It is associated with the high spirits of youth, with disobedience to the commands of kings or gods, with bold-speaking at the expense of others. But the word is also to be found in Euripides and used to describe the adultery of Aegisthus and, by Theseus, to describe Hippolytus's alleged attempt to seduce Phaedra. In Aeschylus, its commonest meaning is 'lust'. Its general meaning in this context is connected with a desire for sexual possession, by force if need be. '*Hybris* is therefore having energy or power and misusing it self-indulgently.' It is always bad, voluntary (i.e. without the self-restraint of its antonym, *sophrosyne*), and usually involves a victim.[30]

It will be obvious that Horace, for example, is not guilty of *hybris* only in his rash pursuit of his 'gloire' whatever the cost but in his boasting to Camille at the beginning of the murder scene (IV.v). Already in Act II scene iv, he had given his orders in a sequence of imperatives (ll. 517 ff.) which is followed now by others; but the substitution of the second person singular for the plural marks the self-indulgent pride of the victor and a kind of boastful verbal assault on his sister before she has even spoken to him. The appropriateness of the *nemesis* which now overtakes him is redoubled when one knows that the boasts and orders are addressed to the fiancée of one of those he has just killed. In the same way, Cléopâtre continues to boast and threaten even when she has drunk the poison. This is precisely what Auguste, in the end, refrains from doing. What finally brings retribution on Suréna is his refusal to refrain: imprudently, to say the least, he constantly reminds Orode of his superior 'gloire' and military prowess and, at the same time, disobeys his King. Nicomède, on the other hand, for all his greatness, strives only to sustain the royal authority vested in Prusias: he is not self-indulgent nor merely boastful.

In Racine, *hybris* can be thought of as associated not only with political or military power but with an unrestrained passion for sexual possession. Pyrrhus sees himself — and boasts of it — as capable of defeating the combined forces of the Greek allies: it is their soldiers, under Oreste, who kill him. His passion to possess Andromaque is such that he endeavours to force her into marriage. In so doing he boasts of his power to destroy her and Astyanax: this is undisguised threat and blackmail. When he thinks that he has succeeded, he boasts of his success to Hermione and throws taunts at her. She, in her injured pride and frustrated passion to possess him, responds in kind. Although thereafter we see no more of Pyrrhus, Hermione at least comes to a realization that her possessive passion, while still overwhelming, has, once disappointed, been so unrestrained as to cause her to destroy its very object. And of course we do hear, as she does, of how *nemesis* has overtaken Pyrrhus as it overtakes her and then Oreste who has also allowed his passion to possess her to blind him to the reality of her feelings towards him. The web of unrestrained and unreciprocated passion for possession is complete and inescapable.

Agrippine boasts of what she mistakenly believes to be her political triumph over Néron, and not for the first time: her self-indulgent, autocratic treatment of her son is an act of *hybris* which brings about the unintended result, the death of Britannicus, and through it her own downfall which she ultimately recognizes. But if Néron at the end of the play appears to have succeeded in his political struggle with his mother, he too has committed *hybris* both in that sphere and in his cruel attempt to take Junie into his own possession and to wrest her from Britannicus. This attempt recoils upon him, and he fails, in spite of the assassination: the overconfident and unscrupulous certainty of success finally eludes him.

It is Bérénice's overwhelming desire to possess Titus that blinds her to the reality of the situation in which his imperial duty is all-commanding. Like Néron, she is over-confident and imprudent, as her first dialogue with Phénice shows, and she is self-indulgent in her conversations with Antiochus. She boasts of her love for Titus only to discover that he cannot be hers, and for reasons which, in spite of his passion for her, ought to be known to her in advance and which are not those which she deludes herself into believing or imagining.

The possessive passion of both Roxane and Atalide for Bajazet is *hybris*, too, and so is that of Mithridate and Pharnace for Monime. Roxane and Mithridate attempt an impossible compulsion and each loses the object of passion. In order to possess Achille, Ériphile resorts to the cruellest means to eliminate her successful

rival, only to find that her confidence in Calchas's interpretation of
the oracle in her favour is misplaced. It is Phèdre's possessive passion for Hippolyte that drives her to her jealous refusal to exculpate him in the eyes of Thésée and then, in remorse and guilt,
brings her to destroy herself. Thésée, too, commits *hybris* in his
initial certainty that his son is guilty and in his over-hasty call for
Neptune's punishment: like his wife, he discovers too late the
enormity of the error into which it has led him.

In the politico-religious sphere Athalie and Mathan fall into the
same error. Their boasting expresses their unwarranted confidence
in their ability to overcome right with might — although for Athalie, as for Thésée, disquieting doubts present themselves to her
mind — and it takes the spectacular form of their entry into the
sacred and forbidden precincts, only to be faced with the terrifying,
unsuspected truth.

It is not surprising that Racine should comment on the *Ajax* of
Sophocles (ll. 760–5) in these terms (Picard, ii, p. 860): 'Raison de
la colère des Dieux contre Ajax. — Son orgeuil, sa confiance sur lui
seul, et le mépris de leur secours.' 'Rome est tout où je suis', boasts
Sertorius, but the boast will not save him. What the discovery of
hybris teaches is the folly of man's assumption that he can outreach
or suppress his own humanity or that of others, that he can rely
upon himself alone, that he is certain of being right.[31]

Any discussion of enlightenment or discovery presupposes prior
blindness. The tragic paradox of Oedipus is that of a man seeking
the truth and not seeing it where it is, and of his moral sight being
clear only when his physical sight is lost and he is isolated and able
to look only into himself, and when his certainties have been destroyed. He has leapt to conclusions and acted on them in the
haste and pride of those certainties. Although the distinctions are
obvious enough, we find in both Corneille and Racine adaptations
of the Oedipus paradox; we have, however, seen that enlightenment is not usually granted to the elder playwright's characters
themselves. In general, indeed, we find them resisting the truth to
the very end even if it is made plain to us. It has often been said
that Corneille's heroes act in full knowledge of what they are doing
and see their way forward in the clear light of day. This clarity of
vision is supposed to be the mark of the 'généreux'. But the consequence of their action usually turns out to be the reverse of what
they intended or expected. Yet they profess to believe in certain
ethical values ('gloire', 'honneur', 'vertu', 'devoir', etc.) which
guide their choices. If, however, we look closely at what they say
we can see that those values and their consequences are blindly
and unquestioningly accepted. Once again, Horace exemplifies

these beliefs and this behaviour. When (II.i) he has been nominated as his city's champion, he is not certain of victory, but

La gloire de ce choix m'enfle d'un juste orgueil:
Mon esprit en conçoit une mâle assurance . . .　　(ll. 378–9)

It is his 'gloire', not that of his city, which gives him pride; he is already sure that he is right; that certainty gives him confidence. In his dialogue with Curiace after the announcement of the names of the Alban champions (II.iii), the certainty itself takes the form of blindness:

J'accepte aveuglément cette gloire avec joie,　　(l. 492)

and that 'gloire' supersedes all other feelings and values:

Celle de recevoir de tels commandements
Doit étouffer en nous tous autres sentiments.　　(ll. 493–4)

Horace in his certainty blinds himself to all except that 'gloire'. When Camille challenges him after his victory (IV.v) he persists in his blindness and expects her to follow him in suppressing her instinctive human feelings for the sake of his own 'gloire':

Tes flammes désormais doivent être étouffées,
Bannis-les de ton âme, et songe à mes trophées:
Qu'ils soient dorénavant ton unique entretien.　　(ll. 1275–7)

Camille treats ironically 'cette gloire si chère à ta [Horace's] brutalité', and her irony is for him 'ce mortel déshonneur'. In killing his sister Horace tries not simply to suppress those other feelings which he possessed and which Camille, to his shame, still possesses and proclaims, but to kill the humanity they represent. His glory is all that occupies his mind, the glory blindly accepted, the glory he will seek to preserve by asking to die before it is besmirched, the glory that blinds him to all else and does not allow him in his pride and certainty to be enlightened as to its folly.

The conversions and reconciliation of the end of *Polyeucte* may be indicative of a worthier concept of glory, but even there the hero pursues it blindly in the sense that his vision of it is so bright that he sees nothing else. Cléopâtre's 'gloire' is doubtless hardly estimable, but she too pursues it relentlessly, suppressing all other feelings, blinding herself to every other consideration and to the consequences of her actions. Auguste on the other hand realizes in time to save himself that he has likewise been blind to the nature of his political activities and to the moral destruction they have wrought in him. But he is an exception in Corneille's tragic drama. It is this kind of blindness which occasions *hamartia*, literally, missing the target aimed at:[32] how can the target be struck if the eye is blinded? And the eye is blinded because of *hybris*. Dramatically, however, *hamartia* is not a flaw in the character but the error which precipitates the catastrophe.

Now Racine's characters also speak the ethical language of 'gloire', 'devoir', etc., but their use of it is different.[33] Sabine's dilemma, expressed emotionally as it is (I.i), does involve duty in a real sense: where in fact does her duty lie? in Rome, or in Alba? to her husband, or to her family? But Hermione argues from duty, not in order to find a way forward, but in order to shame Pyrrhus (shame also, ironically enough, the antonym of glory) for what he has done to her in deciding to marry Andromaque. In her last interview with him (IV.v), she attacks him with his failure to do his duty:

> Quoi! sans que ni serment ni devoir vous retienne,
> Rechercher une Grecque, amant d'une Troyenne?
> Me quitter, me reprendre, et retourner encor
> De la fille d'Hélène, à la veuve d'Hector? (ll. 1317–20)

In her anger and sorrow over the lost illusion of love and happiness, and still feeling that she loves him, she returns to the same language:

> J'ai cru que tôt ou tard à ton devoir rendu,
> Tu me rapporterais un cœur qui m'était dû. (ll. 1363–4)

But it is all in vain: Pyrrhus heeds not the call of duty but the call of love, just as he had done in his first dialogue with Andromaque (I.iii), when he had brushed aside her attempt to recall him to his obligations to his allies — but here too, of course, the idea of duty was only used as a means of deflecting him from his amorous purposes and not with any moral compulsion in mind:

> Seigneur, que faites-vous, et que dira la Grèce?
> Faut-il qu'un si grand cœur montre tant de faiblesse?
> Voulez-vous qu'un dessein si beau, si généreux,
> Passe pour le transport d'un esprit amoureux? (ll. 297–300)

The idea of 'générosité' is used only in ironical vein, and that of 'devoir' and 'gloire', associated with it, only for a negative purpose.

Nowhere is this negative function clearer than in the case of Titus, who uses this vocabulary more frequently than any other of Racine's characters and even than many of Corneille's heroes. But where Horace actively pursues his 'gloire', where he strives to realize a vision of the glory to come — that of Rome as well as his own —, or where Auguste apostrophizes the future, projecting into it the true glory of his new-found authority and self-mastery, Titus sees glory and duty as a burden (l. 462), not as an ideal to be striven after but as a fate-given obstacle to happiness (ll. 719–22, 736). He speaks of his 'triste devoir' (l. 997). Where the Cornelian hero pursues his own vision of glory — however mistakenly, we may think —, for Titus it is the reverse: 'Ma gloire inexorable à toute heure me suit' (l. 1394). At best, acceptance of it for the fu-

ture is urged only as a poor consolation for the loss of present joy
(ll. 1051–3, 1058, 1095–8, 1172–4, 1365, 1375–6). This negative,
sorrowful form of glory, which is equated for Titus with fate
(l. 715), is expressed in the form and action of the play. The resignation he has already arrived at must be understood and
accepted by Bérénice: she must pass from illusory hope to discovery of the force of the duty and glory which are the Emperor's
destiny. Where, for Corneille's Bérénice, love and personal happiness are obstacles to her fulfilment as a queen, for Racine's, it is
her queenship and the imperial glory of Titus that stand in the way
of her fulfilment as a woman. But in Corneille's play, the sacrifice
of immediate personal happiness ensures the compensating
achievement of glory, while in Racine's the imposition of glory
means the destruction of that happiness. The end of the play:

> Adieu. Servons tous trois d'exemple à l'univers
> De l'amour la plus tendre et la plus malheureuse
> Dont il puisse garder la mémoire douloureuse.
> Tout est prêt; on m'attend. Ne suivez point mes pas.
> Pour la dernière fois, adieu, Seigneur . . .   (ll. 1502–6)

with its pain and separation is very different from that of *Cinna*,
where Livie's prophecy, following the reconciliation, looks forward
eagerly to future glory.

Now if Bérénice is ultimately enlightened, and if the whole play
is needed to bring her to discovery of the truth, it is clear that the
painful length of the process is due in part to her resistance to
knowledge. It has been observed[34] that in Sophoclean tragedy in
particular — and it is obviously true of Oedipus — two contrary
desires exist simultaneously: to know the truth and to blind oneself
to it. Bérénice is unable to account for Titus's coldness and evasiveness: she needs to know the reason. But before the play opens,
she has already deluded herself (*hamartia*) into believing that he is
all-powerful, like the Oriental despots with whom she is familiar,
and that as emperor he can impose his will on his people. If he
no longer expresses his love for her as he has done in the past, then
either he is jealous of Antiochus who also loves her or he has
ceased to love her. She draws the wrong conclusions because she
resists the truth which is staring her in the face. Yet she relentlessly
pursues it, too, but only in directions compatible with her own
experience, whether of politics or of men. The play represents to us
the gradual overcoming of the resistance, the discovery being
made, 'de parole en parole', of the bitterest truth of all. To Titus
she says:

> J'aimais, Seigneur, j'aimais, je voulais être aimée.
> Ce jour, je l'avouerai, je me suis alarmée:

J'ai cru que votre amour allait finir son cours.
Je connais mon erreur, et vous m'aimez toujours . . .

(ll. 1479–82)

To Antiochus,

Je l'aime, je le fuis; Titus m'aime, il me quitte.   (l. 1500)

The *hybris*, the pride of passion with which she began, the *ate*, the
infatuation with which she persisted, have yielded at last to enlightenment, not only for us, but for herself. The progressive search
for the truth and continuing resistance to it characterize the plot of
Athalie, too, and the behaviour of the heroine of that play, from her
first questioning of the meaning of her terrible dream (II.v),
through her determination to find certainty ('Mais je veux de mon
doute être débarrassée' — l. 611) and her interrogation of Josabeth
and Joas, to the discovery in which the resistance is finally overcome: 'Oui, c'est Joas; je cherche en vain à me tromper' (l. 1769).

Corneille's heroes, once they have arrived at a decision, do not
look back:

    . . . C'est mal de l'honneur suivre la carrière
    Que dès le premier pas regarder en arrière,

says Horace (ll. 487–8); but Racine's characters, in their pursuit of
the truth and even when they have decided or have put the decision into effect, are always looking back. We have seen how Hermione, having faced the truth of Pyrrhus's abandonment and sent
him to his death, can still be the victim of her own illusion and is
still on the point of relenting, of once again resisting the truth.
Oreste, too, nurtures the illusion that she can love him, even to the
point of committing murder and regicide for her, while knowing
that it is impossible for her to love him. Thésée returns home in the
hope of a happy reunion with wife and son: they shun him. Like
Bérénice, he thinks he has found the truth, even confirmed it, and
so he brings down his curses and the divine wrath on Hippolyte.
The *hybris* of hot-tempered judgement blinds him to the real truth,
but immediately misgivings arise in his mind (ll. 1161–2). After
Hippolyte has reappeared with Aricie, and she has tried to deflect
Thésée's wrath, they are reinforced (V.iv), and still more so when,
immediately afterwards, he hears of the suicide of Œnone. The
light, imperfectly perceived, begins to dawn ('Qu'on rappelle mon
fils . . .', l. 1481), but too late. Yet at each step, Thésée has tried to
resist the truth, to justify his own hubristic action. With Théramène's account of Hippolyte's fatal chariot-drive and Phèdre's
ultimate confession, he is finally enlightened. Like Bérénice before
him, he recognizes his *hamartia*:

    Allons de mon erreur, hélas, trop éclaircis
    Mêler nos pleurs au sang de mon malheureux fils. (ll. 1647–8)

The much-discussed and frequently-criticized *récit de Théramène* is a vital part of this process. Thésée listens, without interrupting, throughout its length: it works on his imagination, as on ours, painting a picture of horror which terrifies, arousing in Thésée anger and remorse, sorrow and pity and in us, as we witness his enlightenment, the true tragic emotions.

Thésée has to face the truth alone as Théramène speaks. He has felt himself isolated and deserted on his home-coming and baffled in his soliloquies. It has been remarked[35] that such solitude is a mark of the tragic character: the moment of truth is often the moment of solitude, hence the importance of the soliloquy. It is then that the appearances and illusions fall away. And even when the tragic character fails to be enlightened, as in the last act of *Horace* or of *Rodogune*, his moral isolation, represented before us, symbolizes for us the significance of what has taken place.

It is not difficult to see how in *Horace* the hero is gradually isolated from all the other characters, and feels himself eventually cut off even from his father. In one sense, of course, this can be seen as his progressive ascent as a hero, as he lifts himself above the moral level of ordinary men. But in his fall he drops below it through trying to outreach the limits of humanity. Among the members of his family, Nicomède is isolated, too; but he constantly strives less for his own glory than for that of royal authority as vested in his father. The play ends in reconciliation as does *Cinna* when Auguste, having suffered isolation in the form of desertion and treachery, and incomprehension on the part of Livie, sees in his solitude where true glory lies: when he embraces it he draws all others to him. Before that, however, he has become aware of his solitude:

> Ciel, à qui voulez-vous désormais que je fie
> Les secrets de mon âme et le soin de ma vie? (ll. 1121–2)

The plot in *Andromaque* is devised in such a way as eventually to bring Hermione to isolation when, the dénouement being reached (IV.v), all the other characters, even Cléone, are at the temple, for the marriage or for murder. That is the moment of truth for her, as his isolation (V.iv) will shortly be for Oreste. Titus's solitude (IV.iv) marks the dénouement also, for at that point not only is the truth finally discovered but action has become inevitable. Similarly, it is when they are alone that Roxane (IV.iv), Mithridate (IV.v) and Phèdre (IV.v) — the parallel pattern is striking — have to face the truth and that Thésée (IV.iii; V.iv and even V.vi and, after Phèdre has expired, V.vii) comes 'de parole en parole', and more and more alone, to realize the truth which he resists:

> Mes entrailles pour toi se troublent par avance. (l. 1162)

Quelle plaintive voix crie au fond de mon cœur?   (l. 1456)

O mon fils! cher espoir que je me suis ravi!   (l. 1571)

... De mon erreur, hélas! trop éclaircis ...   (l. 1647)

The tragic vision, essentially imaginative although conveyed through the emotions, as the emotions are aroused by the *mimesis*, the imaginative enactment, comes from a sense, in retrospect at least, that what has taken place could not have been otherwise. The initial situation, the relationships between the characters and their nature, the circumstances in which they are caught up are all products of the dramatist's imagination. Given these, he so works out his plot, obeying the laws of *vraisemblance* and *bienséance*, logic and fittingness and consistency, as to produce at the end a feeling of inevitability. Without that feeling, and it is more than mere intellectual understanding (the plays are neither arguments nor syllogisms), we should experience the emotions, dramatic (and tragic?), only as ephemeral, which is perhaps why even the second-rate plays of the seventeenth century attracted the same spectators time after time. But Corneille and Racine, thanks to this feeling of inevitability, arouse in us what have been called aesthetic emotions:[36] these arise in part out of our fellow-feeling with the characters being balanced by our being spectators watching them detachedly. The sense of fittingness and of inevitability give rise to what Corneille suggested as a new tragic emotion, admiration: admiration on the aesthetic level for the coherence of the play as a work of art, but admiration, too, as Descartes understood it, awe and wonder both at the spectacle of heroic and terrible deeds and at the revelation of an abiding moral order. It is here that Professor Kitto sees the source of what he calls 'Awe and Understanding':

... when we have seen terrible things happening in the play, we understand, as we cannot always do in life, *why* they have happened; or if not so much as that, at least we see that they have not happened by chance, without any significance ...[37]

No better explanation could be given of the importance attached to the need for *vraisemblance*, the probable and necessary sequence of episodes within the plot which does not, as we have seen, diminish effects of surprise and unexpectedness. The marvellous, as it is called in the *Poetics*, and as it was taken up by the contemporaries of Corneille and Racine as the Latin 'admiratio',[38] arises in part out of these aspects of their tragedies, out of the experience of the unexpected being shown to be inevitable. The gods of old and all their powers may have disappeared, but it is the power of the passions within the characters (and it is in their terms that the dramatic situation is created and developed) which now cause what must be to be. All things fall into place.

What is experienced in this way (it has been said) must be particularised in individual situations ... They are situations which we can grasp and understand because their structures are laid bare before us, in art works which mimetically present whatever they have to say in a whole connected 'according to probability or necessity'.[39]

One last word. While it is true that tragedy brings us to an awareness that one false step (*hamartia*), taken with the best of intentions but in a moment of passion (*hybris*), can precipitate us into an abyss of destruction or despair (*nemesis*), and perhaps that, in spite of apparent light and certainty, it is bound to be taken in the dark, it does also, in its highest forms, bring us not only to peripety but to discovery. The currently fashionable idea of Providentialist endings in Corneille's plays — conversions, reconciliations, interventions by judge or ruler, even when they seem to be contrived — has perhaps a value in terms not of Corneille's belief in a particular theological or political system but of their being simply guides to our interpretation of the plays as tragedies. Tragedy disturbs the order of the moral universe, but it also restores it. If Horace is not condemned to die, he is being integrated anew into the world of Rome and of men. That kind of re-acceptance, of the restoration of order, is to be found in many of Corneille's plays, and often on a higher plane than before: in *Horace*, Rome has absorbed Alba and is set on her course to greatness; in *Polyeucte*, martyrdom has changed the lives of the surviving characters; in *Cinna*, the discovery and pardon of the conspirators lifts Auguste — and indeed the conspirators themselves — to a higher moral level; in *Pompée*, Rome's hold over a subject-nation is assured; in *Nicomède*, imperilled royal authority is safeguarded; in *Sertorius*, the unity of the empire is restored. Even in *Rodogune*, the world is cleansed by the punishment of an evil queen. When we turn to Racine, we find that in *Andromaque* the heroine and her infant son survive and are released. In *Bérénice*, the threat to Roman law is removed; in *Bajazet*, the life of the seraglio returns to its 'ordre accoutumé'; in *Mithridate*, the dissension within the royal house is closed; in *Iphigénie*, the Greek expedition to Troy can at last go forward.

All these things are wrought through suffering, another source of wonder and awe and, in the end, of joy and relief — perhaps the true catharsis — as we experience them when in her final despair Athalie cries: 'Dieu des Juifs, tu l'emportes!' or Phèdre breathes her last words:

Et la mort, à mes yeux dérobant la clarté,
Rend au jour qu'ils souillaient toute sa pureté.

Those lines are a strange echo of a couplet (ll. 1425–6) spoken by Horace to his father in their interview after the death of Camille and before the entry of the King:

Reprenez tout ce sang de qui l'a lâcheté
A si brutalement souillé la pureté.

Where Phèdre's imminent death brings relief from suffering and restores the world to its former purity, Horace is granted no release and he must go on living with his foul deed which remains for ever a stain on his family's honour. While it is true that those plays of Corneille which end in reconciliation, however dearly bought — *Cinna, Polyeucte, Nicomède, Œdipe, Sertorius —*, give a final impression of optimism, others, including *Horace*, produce the opposite effect. For the characters, certainly, for the audience, too, the endings of *Bérénice*, of *Bajazet*, and above all of *Britannicus*, are dark and hopeless, but unlike Racine's farewell to both secular and sacred tragedy, the end of Corneille's dramatic career, in *Suréna*, is marked by a sense of waste, desolation, and disintegration. Who shall say which is the more tragic?

Notes

CHAPTER 1
The Dramatic Subject: its nature and its disposition

1. In *Writings on the Theatre* (hereinafter *Writings*), pp. 12–13.
2. Cf. D'Aubignac (*La Pratique de théâtre*, pp. 83 f.) on the unity of action, which involves the dramatist in the choice of a 'point d'histoire' for his subject.
3. See Racine, *Andromaque*, ed. R. C. Knight and H. T. Barnwell.
4. See, for example, his letter to the Abbé de Pure, 3 Nov. 1661 (*Iutégrale*, p. 861).
5. In the introduction to his edition of the play, Dr P. H. Nurse seems to exaggerate the significance of the changes in the last act (pp. 52 ff.).
6. Chapelain, *Sentiments de l'Académie sur le 'Cid'*, in *Opuscules critiques*, pp. 165, 160.
7. *Pratique*, p. 68.
8. In the first and third *Discours: Writings*, pp. 10, 62.
9. *Pratique*, pp. 66–7, 133–4.
10. See the *Examen* (*Writings*, p. 128).
11. See D'Aubignac's second *Dissertation* and Corneille's *Examen* (*Writings*, pp. 163–8).
12. This is the preference consistently shown by Chapelain and D'Aubignac, among others.
13. Second *Discours: Writings*, pp. 46–7.
14. Second preface to *Andromaque*, ed. cit., pp. 205–6.
15. *Principes de la tragédie*, p. 13.
16. e.g. second *Discours: Writings*, p. 46.
17. For a detailed examination of this question, see *Andromaque*, ed. cit, It is interesting to note that Voltaire uses the word 'ordonnance' in the sense of 'disposition'.
18. See R. C. Knight, '*Cosroès* and *Nicomède*'; and the same scholar's edition of *Nicomède*.
19. See Gustave Rudler, 'Une source d'*Andromaque? Hercule mourant* de Rotrou', and the discussion in the introduction to *Andromaque*, ed. cit.
20. Cf. Rudler's edition of the play for a detailed analysis.
21. Le Moyne, *La Gallerie des femmes fortes*, pp. 128–42.
22. See R. C. Knight, *Racine et la Grèce*, pp. 336 ff., and 'Hippolyte and Hippolytos'; J. Pommier, *Aspects de Racine*, pp. 311 ff.
23. Stegmann, *L'Héroïsme cornélien*, vol. ii, pp. 501, 612–4. Cf. P. J. Yarrow, *Corneille*, pp. 234–5.
24. The suggestion for the friendship of Titus and Antiochus could well have come from that (also invented) between Vespasian and Vologèse in the obscure *Aricidie* of Le Vert (1646).
25. The etymology of the word suggests not simply 'creating' but 'finding', and its use in the seventeenth century was certainly ambiguous.
26. Second *Discours, Examen* of *Héraclius*, and preface to *Sophonisbe: Writings*, pp. 46–7, 138–9, 164. Cf. much earlier (1647) the original prefaces to *Rodogune* and *Héraclius*, ibid., pp. 186–91. It has been suggested that Corneille in effect produced his own version of the Electra story in *Rodogune*, in which Antiochus-Orestes does not actually kill Cléopâtre-Clytemnestra, though she does die at the end. See M. Fumaroli, 'Tragique païen et tragique chrétien dans *Rodogune*', pp. 599–600.

27. See Corneille's insistence on it in many of his critical writings, e.g. the *Épîtres* accompanying *L'Illusion comique, Le Menteur*, and *Don Sanche*, the *Examens* of *Mélite, Clitandre, Rodogune*, and *Héraclius*, the preface to *Clitandre*, and the *Au Lecteur* of *Nicomède* and *Agésilas*.
28. La Mesnardière, *La Poétique*, p. 28.
29. Peter France, *Racine's Rhetoric*, p. 31.
30. Ibid., p. 13.
31. *Pratique*, p. 228; Jean de la Taille, *De l'art de la tragédie*, pp. 26–7; Saint-Évremond, *Œuvres*, vol. iv, p. 429; Poussin, letter to Chambray, 1 Mar. 1665, *Lettres*, p. 310. It is not without interest that, in his preface to *Samson Agonistes*, Milton wrote of the 'economy and disposition of the fable'.
32. Poussin, *Lettres et propos sur l'art*, pp. 172–3; Pascal, *Pensées*, Lafuma 22; Poussin, *Lettres*, pp. 138, 155, 240, 241.
33. See Louis Racine's *Mémoires*, Picard, vol. i, p. 42; the prose draft, ibid., pp. 948–51, and Picard's comment, p. 947 (he does however admit that 'Racine avait déjà modelé la légende à son gré' and in the same ways as in *Iphigénie* and *Phèdre*); Picard, vol. ii, p. 397, and cf. Racine, 28 Mar. 1662, *Lettres d'Uzès*, p. 61; Poussin, *Lettres*, p. 298. Cf. Aristotle's comment (*Poetics*, chap. IX) on the poet being the maker of his plot before being the maker of his verses.
34. Félibien, *Entretiens*, vol. iv, pp. 155–6. (All references are to vol. iv.)
35. By Lawrence Gowing: see A. Blunt, *Nicolas Poussin* (text vol.), p. 244 and n. 72. Cf. the earliest account of Poussin's methods of work, given by Sandrart, cit. in *Lettres et propos d'art*, pp. 176–7.
36. Cf. R. Picard, 'Le Brun–Corneille et Mignard–Racine', pp. 178–81; W. McC. Stewart, 'Charles Le Brun et Jean Racine', p. 218.
37. Cf. Félibien, pp. 123, 158.
38. Racine, *Lettres d'Uzès*, p. 104; Poussin, *Lettres*, pp. 244, 282, 284, 292, 295; *Lettres et propos sur l'art*, pp. 188, 179; Félibien, pp. 99–100.
39. Second *Discours: Writings*, p. 57.
40. Félibien, pp. 14, 21, 37; Corneille, first *Discours: Writings*, pp. 1–2; Racine, second preface to *Bajazet*.
41. Bénichou, *L'Écrivain et ses travaux*, p. 228, n. 52; cf. Knight, *Racine et la Grèce*, p. 270 (this idea having been developed more recently — see *Andromaque*, ed. cit.).
42. See Professor A. H. T. Levi's survey in *French Moralists...*
43. Cf. M. Fumaroli, 'Rhétorique et dramaturgie: le statut du personnage dans la tragédie classique', pp. 247–9.
44. Descartes, *Les Passions de l'âme*, Articles 53 ff.
45. Poussin, *Lettres et propos sur l'art*, pp. 160, 161; Félibien, p. 81; Corneille, first *Discours (Writings*, pp. 8–9); Racine, preface to *Bérénice*.
46. Félibien, pp. 10, 91, 26.
47. Notes on Sophocles (*Electra*, ll. 1326 and 77; *Ajax*, l. 387; *Trachiniae*, ll. 488–9, 497), and Euripides (*Medea*, ll. 1–45, 263–6): Picard, vol. ii, pp. 852–3, 846, 858, 869, 871, 873.
48. The words italicized represent a misinterpretation of Aristotle's text influenced by seventeenth-century French dramatic practice (see Vinaver's note, pp. 62–3). The nature of the misinterpretation need not detain us here: as it stands, the passage reveals Racine's own thought as much as that of Aristotle.
49. Descartes, *Les Passions de l'âme*, Article 52. Cf. *Traité de l'homme*, pp. 166–7.
50. This kind of statement seems to be ignored by those critics who make direct comparisons between Racine and, say, Euripides, without reference to any intermediaries or to the conventions of seventeenth-century tragedy. See, for instance, Evelyne Méron, 'De l'*Hippolyte* d'Euripide à la Phèdre de Racine'.

CHAPTER II

Sources: tragedy and history

1. Third *Discours: Writings*, p. 62. The unity of action is in tragedy constituted by the unity of peril.
2. See the *Examen* and the passage it refers to in the first *Discours*, where Corneille explains the mechanical structure of the play (*Writings*, pp. 123, 10).
3. Lucan, *Pharsalia*, trans. R. Graves, Penguin Classics, 1956. (See p. 14.)
4. *Œuvres*, ed. Marty-Laveaux, vol. v, pp. 107–9.
5. Cf. Corneille's comments, on Aristotle's statement, in the first *Discours: Writings*, p. 18.
6. See Caesar, *Civil Wars*, III. 108.
7. Pompée was assassinated on 28 Sept. 48 BC, and Ptolomée killed on 25 Mar. 47 BC.
8. Dio Cassius, *Roman History*, XLII. 41.
9. The war was chiefly caused by César's levies on the population of Alexandria in respect of unpaid debts, and by the Alexandrians' hatred of Cléopâtre — hence the historians' criticism of César's enslavement to her. See Dio Cassius, *Roman History*, XLII. 34; Plutarch, *Life of Caesar*, XLVII; Appian, *Roman History*, II. xii. 90.
10. Dio Cassius, *Roman History*, XLII. 3; Caesar, *Civil Wars*, III. 103.
11. Plutarch, *Life of Pompey*, LXXX; Appian, *Roman History*, II. xii. 85.
12. Dio Cassius, *Roman History*, XLII. 7.
13. This is a departure from at least one of the historical narratives, that of Plutarch (*Life of Caesar*, XLIX) who recounts the romantic story of love at first sight.
14. For the historical account, see Hirtius, *The Alexandrian War*, XXXI.
15. Plutarch, *Life of Pompey*, LXXVII.
16. Corneille perhaps omits Theodotus because, according to Plutarch (*Life of Pompey*, LXXX), he escaped Caesar's vengeance. Racine may have followed Corneille's example in omitting Seneca from the number of Néron's advisers in *Britannicus*.
17. Dio Cassius, *Roman History,*, XLII. 3, 4; Appian, *Roman History*, II. xii. 84.
18. Plutarch, *Life of Pompey*, LXXVIII — LXXIX; Caesar, *Civil War*, III. 104–6.
19. L. M. Riddle, *The Genesis and Sources of Corneille's Tragedies* ... This critic thinks that Antoine, in Corneille's play, is an 'incongruous' confidant.
20. Ibid., pp. 89–93.
21. Cf. L. Herland, 'Les Éléments précornéliens...'. See also H. L. Cook's introduction to Scudéry's play, pp. 10 ff.
22. *Examen* of *Rodogune* (*Writings*, p. 131).
23. According to Professor Stegmann, *L'Héroïsme cornélien*, ii, p. 131. The play is Pazzi's version of the Dido story. Of Boisrobert's *Didon*, Professor Stegmann writes: '... Son Jarbas [Énée's rival] prend figure de héros courtois; au lieu de rival forcené [as he was in the tradition], le voici devenu amant distingué, qui prône la vertu de Didon, malgré l'amour qui la consume pour l'insensible Énée.' This representation of Jarbas is clearly a predecessor of Racine's Antiochus.
24. See, for example, C. L. Walton's introduction to the play, pp. 24–5.
25. Louis Racine, *Remarques sur Bérénice*, in *Œuvres*, vol. i, p. 520.
26. Josephus, *Jewish Antiquities*, XX, 7, 1.
27. On the peculiarity of his role, see R. C. Knight, 'The Rejected Source in Racine', pp. 162–4.
28. First *Discours:Writings*, p. 18.
29. *Principes*, p. 27.

30. Plutarch, *Life of Caesar*, XLIX; Suetonius, *Life of Caesar*, LII.
31. See Montaigne, *Essais*, II, xxxiii, and cf. Dio Cassius, XLII. 34; Suetonius, *Life of Caesar*, L–LII; Plutarch, *Life of Caesar*, XLVIII.
32. The key passage is in Plutarch, *Life of Caesar*, LVIII (Amyot, LXXV). Cf. ibid., III, IV, and Dio Cassius, XL. 54, 56.
33. Cf. A. S. Gerard, 'Baroque and the Order of Love', pp. 118–31, 210–11; and R. C. Knight, '*Andromaque* et l'ironie de Corneille', p. 24.
34. Cf. Plutarch, *Life of Pompey*, LXXX; Dio Cassius, XLI. 62 and XLII. 6; Plutarch, *Life of Caesar*, XLVI; and even the hostile Suetonius, *Life of Caesar*, LXXV. (Both Dio Cassius — XLII. 8 — and Lucan — *Pharsalia*, IX — make out that Caesar's grief was a sham.)
35. First *Discours: Writings*, p. 8.
36. For details, see my edition of the play, notes to ll. 473 ff., 1707, 1743–4.
37. Cf. P. Bénichou, *L'Écrivain et ses travaux*, p. 226.
38. Tacitus, *Histories*, II, 81.
39. Josephus, *Jewish Antiquities*, XX. 7, 3.
40. Dio Cassius, LXXV. 4–5; Herodian, IV. 1–2.
41. Suetonius, *Life of Titus*, V.
42. Dio Cassius, LXVI. 15, 3–4 (cf. 18, 1–2); Aurelius Victor, *Caesares*, X. 4. Cf. Suetonius, *Life of Titus*, VI: 'Egitque aliquanto incivilius et violentius [...] ut non temere quis tam adverso rumore magisque invitis omnibus transierit ad principatum' ('And he acted somewhat too unjustly and violently, so that no one came to power with so much hostility and public disapproval').
43. *Bérénice*, ll. 301 ff., 1484 ff.; Suetonius, *Life of Titus*, I.
44. Suetonius, ibid., IV, VII.
45. Ibid., VII.
46. Though he did not mention all the possibilities, e.g. 'In puero statim corporis animique dotes explenderunt...forma egregia...' ('In his youth his physical and mental endowments were outstanding...He was extremely handsome...'). Cf. *Bérénice*, l. 311.
47. Suetonius, ibid., III, V.
48. Herman Bell, in his introduction to Magnon's *Tite*.
49. Dupleix, *Histoire Romaine*, XXXIII, ix (vol. iii, pp. 305–7). Dupleix's account is an adaptation of course of those of the historians, notably Dio Cassius, Suetonius, and Tacitus.
50. This seems to give the lie to Professor Stegmann's contention that in Corneille the eponymous characters are a 'couple tout entier racinien' (*L'Héroïsme cornélien*, ii, p. 548).
51. Suetonius, ibid., VIII, IX (cf. *Bérénice*, IV. iv and *passim*).
52. Ibid., IX.
53. Tacitus, *Histories*, II. 2.
54. Mesnard, ed. cit., p. 427, n. 2. See also Mesnard's notes on individual lines in this scene.
55. Coëffeteau's account (*Histoire romaine*, pp. 508–9) may have been important in this respect: Titus 'donna aussi congé à la Reine Bérénice, quoiqu'avec un regret égal de part et d'autre, mais l'amour de la gloire fut plus puissant que les attraits de la volupté'. Incidentally, the same historian suggests something of Titus's resignation to his fate.
56. Guez de Balzac, letter to Corneille (17 Jan. 1643, cited in *Cinna*, ed. D. A. Walts pp. 55–7) who seems to have remembered it when writing the first paragraph of his *Examen* of *Pompée*.
57. Le Moyne, *La Gallerie des femmes fortes*, pp. 130–1.
58. Cf. R. Zuber, '*Les 'Belles Infidèles'*...'
59. Villiers, in Granet, *Recueil...*', vol. i, pp. 40–1.

60. Gouhier, 'Remarques sur le "théâtre historique"', pp. 22–3. Cf. W. D. Howarth, 'History in the Theatre...', p. 159.
61. e.g. Dio Cassius on the death of Pompey (XLII. 5) or Dionysius of Halicarnassus on the combat of the Horatii and Curiatii (III. xviii. 1 and xix. 3).

CHAPTER III
'La Vraisemblance': its dramatic function and significance

1. Chapelain, *Préface de l'Adonis*, in *Opuscules critiques*, pp. 86–8, 93–4.
2. In Gasté, *La Querelle du 'Cid'*, pp. 365–6.
3. Barnwell, in 'Some Reflections on Corneille's Theory of *Vraisemblance* as Formulated in the *Discours*'.
4. By Sarcey, in *Quarante ans de théâtre*.
5. Professor Knight ('Hippolyte and Hippolytus') has stressed the importance of the dynastic theme in *Phèdre*. Yet little attention has been paid to it in that or any other of Racine's plays.
6. Schaper, *Prelude to Aesthetics*, p. 96.
7. Cf. M.–O. Sweetser, *La Dramaturgie de Corneille*, p. 86.
8. Vossius, *Poeticarum Institutionum Libri Tres*, Bk. II, p. 64; Mambrun, *De epico carmine*, p. 17. Cf. E. Kern, *The Influence of Heinsius and Vossius...*, pp. 123–4; R. Bray, *Doctrine classique*, p. 319; and Aristotle's reference to the marvellous in *Poetics*, chapter IX.
9. Racine's annotations on the *Poetics* probably date from 1662 or later, i.e. two years after the publication of Corneille's *Discours*. See Professor Knight's conclusions (*Racine et la Grèce*, pp. 152–3).

CHAPTER IV
Plot: drama in tragedy

1. See Eric Bentley's remarks, in the opening pages of *The Life of the Drama*, on those who are little concerned with 'the living experience of the play' and with the 'elementary' problems which are perhaps the hardest to solve.
2. Longepierre, Preface to *Médée* (1694), ed. T. Tobari, p. 28.
3. See R. C. Knight's argument that Corneille's *Horace* is a different (and new) kind of play as compared with those of Mairet and Rotrou: '*Horace*, première tragédie classique'.
4. Boileau, *Art poétique* III, ll. 31, 55–6.
5. Sainte-Beuve, *Portraits littéraires*, i, p. 79 (1843 edn.).
6. See, in particular, E. Vinaver, *Racine et la poésie tragique*, and G. May, *Tragédie cornélienne, tragédie racinienne*, which is followed up approvingly by Professor Odette de Mourgues (*Racine or, the Triumph of Relevance*, pp. 36–8). Cf. Dr P. H. Nurse's criticism of Professor May in 'Quelques réflexions sur la notion du tragique dans l'œuvre de Pierre Corneille', p. 171.
7. Lanson, *Esquisse d'une histoire de la tragédie française*, p. 103.
8. In particular, Mornet and Lancaster, who provide detailed catalogues of the devices used and re-used. See also Lapp, *Aspects of Racinian Tragedy*, Chapter I, on the question of themes.

9. Similar instances are to be found in *Sertorius* (ll. 1857–64, 1909–12), *Othon* (ll. 1708–9), *Agésilas* (ll. 1994–5, 2071). Less obvious ones are in *Rodogune* (ll. 383–5), *Théodore* (ll. 946–7), *Attila* (ll. 71–2).
10. For this kind of technique more generally, see Scherer, *La Dramaturgie classique*, pp. 206–8, and May, op. cit., pp. 78–87.
11. *Pratique*, pp. 129–30.
12. Rapin, *Les Réflexions sur la poétique*, p. 114.
13. Cf. R. C. Knight, *Racine et la Grèce*, p. 278.
14. Chapelain, cit. Ch. Arnaud, *Étude sur . . . C'abbé d'Aubignac* pp. 348, 350.
15. La Mesnardière, *La Poétique*, pp. 55; Sarasin, *Œuvres*, vol. ii, p. 20; La Bruyère, *Caractères*, 'Des ouvrages de l'esprit', §§ 51, 54.
16. Donatus, in H. W. Lawton, *Handbook . . .*, pp. 14–16.
17. P. Bénichou, *L'Écrivain et ses travaux*, pp. 230 ff.
18. Picard, in his edition of Racine, vol. i, pp. 663 ff.
19. May, op. cit., pp. 149–51.
20. For a detailed comparison, see Knight, *Racine et la Grèce*, pp. 298 ff.
21. Girdlestone, in his edition of the play, p. 81.
22. Scherer, op. cit., p. 129.
23. See also Rudler's detailed comparison in his edition of the tragedy, pp. xxiii ff.
24. May, op. cit., p. 152.
25. Corneille, letter to M. de Zuylichem, 28 May 1650. In *Intégrale* edition, p. 854.
26. Marmontel, *Éléments de littérature* in *Œuvres complètes*, vol. xii, pp. 108–9.
27. Some of the similarities between the playwrights in this connexion have been examined by Professor Knight in 'Andromaque et l'ironie de Corneille'.
28. Cf. Knight, *Racine et la Grèce*, p. 214; M. P. Haley, *Racine and the 'Art poétique' of Boileau*, p. 115; and, much less cogently, J. A. Stone, *Sophocles and Racine*. Aspects of the question other than those I study here are discussed by Professor W. McC. Stewart (see Bibliography).
29. All the references are to line numbers of the Greek work and page numbers in Picard, vol. ii.
30. Cf. Diderot's remark in *De la poésie dramatique* (*Writings on the Theatre*, p. 154): 'Si j'ignore une seule raison pourquoi Néron écoute l'entretien de Britannicus et de Junie, je n'éprouve plus la terreur.'
31. Scherer, op. cit., p. 207.
32. Vinaver, in his edition of Racine, *Principes*, p. 43.
33. Ibid., p. 223.
34. G. Germain, *Sophocle*, p. 152; T. B. L. Webster, *An Introduction to Sophocles*, pp. 118 ff.
35. Bentley, op. cit., pp. 18 ff.
36. e.g. J. Jones, *On Aristotle and Greek Tragedy*, ch. I; H. D. F. Kitto, *Form and Meaning in Drama*, p. 233.
37. Racine's debt to Corneille is of course particularly evident in his first play, *La Thébaïde*: see my article, 'Intrigue et pathétique dans le théâtre de Racine'.
38. Cf. Knight, 'Sophocle et Euripide ont-ils "formé" Racine?'
39. Granet, *Recueil . . .*, vol. ii, pp. 37–8.

CHAPTER V

Simplicity: situation and dénouement

1. See *Writings*, pp. 26–7, 79, and my notes 156, 157 and 158 on pp. 221–2, and note 92 on p. 241.

2. The quotations from the *Dissertations* are taken from Granet's *Recueil* and the page references are to that collection.
3. *La Dramaturgie classique*..., pp. 92–4.
4. By R. C. Knight, in 'The evolution of Racine's *Poétique*'.
5. See Rudler, 'Une source d'*Andromaque? Hercule mourant* de Rotrou'. For the restrictions to be brought to Rudler's thesis, see the introduction to *Andromaque* (ed. Knight and Barnwell), pp. 11–12.
6. Cf. R. C. Knight, 'Le sens de *Pertharite*'.
7. See D. A. Watts's introduction to his edition of the play.
8. Cf. Lancaster, *History*, part III, vol. ii, pp. 440–1.
9. Saint-Évremond, *Dissertation sur le Grand Alexandre*, in *Œuvres en prose*, ed. R. Ternois, vol. ii, p. 84.
10. Cf. Scherer, op. cit., pp. 63 ff.
11. Donneau de Visé, *Défense du 'Sertorius' de M. Corneille*, in Granet, vol. ii, pp. 314, 317.
12. Quoted by Scherer (op. cit., p. 96) from B.N. MS 559.
13. Saint-Évremond, loc. cit., pp. 81–2, 93.
14. See Scherer, op. cit., pp. 125 ff.
15. Third *Discours*, Examen of *Nicomède* (*Writings*, pp. 73–4, 153).
16. *Andromaque*, ed. Knight and Barnwell, ll. 1494 *d, k, l*.
17. Scherer, op. cit., p. 96.

CHAPTER VI
Simplicity: peripety and discovery

1. Scherer, *La Dramaturgie classique*, pp. 83–5.
2. Heinsius, *De tragoediae constitutione liber*, chap. VI. See the 1643 edition, p. 48: 'Quemadmodum in neutro, subito et praeter expectationem ulla in contrarium mutatio apparet, quam peripetiam maximus virorum Aristoteles vocavit.' Cf. p. 66 of the same edition.
3. La Mesnardière, *La Poétique*, p. 55: 'Par le mot de péripétie, le philosophe a voulu dire un événement imprévu qui dément les apparences, et, par une révolution qui n'était point attendue, vient changer la face des choses...Ce soudain renversement est la plus grande beauté du sujet de la tragédie...'.
4. Sarasin, *Discours de la tragédie*, published as a preface to Scudéry's *L'Amour tyrannique* (1639) and reprinted in the various editions of Sarasin's *Œuvres*. In the 1663 (Paris) edition the relevant passage is on p. 323. There Sarasin gives this definition of peripety ('la péripétie'): '...Un changement inopiné de l'action, et un événement tout contraire à celui que l'on attendait et que l'on s'était proposé ...'.
5. Dr P. R. Sellin has rightly drawn attention ('Le Pathétique retrouvé: Racine's Catharsis Reconsidered') to the common misinterpretation of Heinsius's use of the word 'perturbationes' as the equivalent of 'péripéties'. The matter is discussed at length in chapters VIII and IX of Heinsius's treatise, where 'perturbationes' is used quite clearly in the sense of 'emotions disturbed by the peripety' See also Sarasin (loc. cit.) who equates 'perturbatio' with the French 'trouble' and the Greek 'pathos'. Vossius, on the other hand, does seem to use the word with the meaning of 'disturbances in the course of events' ('rerum perturbationes': *Poeticarum Institutiones*, pp. 42, 43).
6. Vinaver, *Racine et la poésie tragique*, and his edition of Racine's *Principes de la tragédie*.

7. House, *Aristotle's 'Poetics'*, p. 96. Cf. Walter Lock, 'The use of *peripeteia* in Aristotle's *Poetics*', p. 251: '...any event in which any agent's intention is overruled to produce an effect which is the direct opposite of [his] intention'; and Mary Philip Haley, 'Peripeteia and Recognition in Racine', p. 426: '...the character bends his efforts to accomplish a certain purpose only to find in the end that his own action has turned back upon him·in such a way as to bring about the very opposite of the effect intended.' Cf. D'Aubignac, *Pratique*, p. 95.

8. Cf. D'Aubignac, *Pratique*, p. 340: '...il ne se faut pas contenter d'émouvoir une passion par un incident notable, et la commencer par quelques beaux vers; mais il la faut conduire jusqu'au point de sa plénitude.' See, also, pp. 89 and 284.

9. Vinaver, *Racine et la poésie tragique*, p. 156; cf. *Principes*, pp. 50–1.

10. *Principes*, pp. 50–1.

11. See R. C. Knight, *Racine et la Grèce*, pp. 150–3.

12. First *Discours*: (*Writings*, p. 10). Cf. D'Aubignac, op. cit., p. 139: 'Il faut... prendre garde que la catastrophe achève pleinement le poème dramatique, c'est-à-dire qu'il ne reste rien après, ou de ce que les spectateurs doivent savoir, ou qu'ils veuillent entendre...'.

13. First *Discours*: (*Writings*, p. 12). Cf. D'Aubignac, op. cit., p. 140: '...Il ne faut pas ajouter à la catastrophe des discours inutiles et des actions superflues qui ne servent de rien au dénouement...' (this remark is followed by a criticism of the last part of Corneille's *Horace*).

14. See *Poetics*, chapter VI: 'A tragedy...is the imitation of an action that is... complete in itself...'.

15. Caussin, *La Cour sainte*, vol. ii, chap. III, pp. 865 ff.

16. Scherer, op. cit., p. 88.

17. See the account of Oedipus's past life in ll. 771–809 of Sophocles's play.

18. Cf. R. C. Knight, op. cit., pp. 388 ff.

19. Professor Weinberg has compared Mithridate's quest with that of Oedipus (*The Art of Jean Racine*, p. 197).

20. See Scherer, op. cit., pp. 145–6; R. C. Knight, 'Corneille's *Suréna*: A Palinode?', p. 179.

21. For a detailed discussion see my article, 'Some Reflections on Corneille's Theory of *Vraisemblance*...'.

22. According to Brossette (*Correspondance*, p. 566), '...la *péripétie* et l'*agnition* se doivent rencontrer ensemble dans la tragédie: et c'est ce qui arrive dans la *Phèdre* de M. Racine...'.

CHAPTER VII
Rhetoric: trial and judgement

1. J. Morel, 'Rhétorique et tragédie au XVIIᵉ siècle', p. 95.

2. Ibid.

3. Clemency is obviously a mark of magnanimity, an aristocratic virtue completely devoid of the kind of tyranny Auguste has practised in the past and is now tempted to practise again. Magnanimity is the root of all other virtues. See Du Bosc, *La Femme héroïque*, pp. 12–15, 21–2.

4. Much recent research has been devoted to the importance of rhetoric in seventeenth-century literature (see for example the Bibliography under the following names: H. M. Davidson, P. France, M. Fumaroli, A. Kibédi Varga, P. Kuentz, I. D. McFarlane, J. Morel), but little attempt has so far been made to relate it

to plays (or other works) in their totality. While I am indebted to work of this kind, my purpose here is not to identify rhetorical devices but to examine some aspects of their organic function within the framework of the whole play. P. Kuentz's article draws attention to the very large number of works on rhetoric (in French alone, 114, and excluding school text-books) published between 1610 and 1715.

5. Professor W. H. Barber ('Patriotism and "Gloire" in Corneille's *Horace*') argues the same point, showing that 'la gloire' is essentially connected, not with patriotism, but with self-fulfilment, and that this is consonant with the ethic of 'la gloire' in the first half of the seventeenth century. See also H. C. Ault ('The Tragic Genius of Corneille', pp. 164–5) and P. H. Nurse ('A propos de l'héroïsme cornélien'). In Racine, of course, 'la gloire' is always used in a highly personal way, often negatively, and not as an ideal to be aspired to. (See my articles on 'La Gloire dans le théâtre de Racine'.). In his *Rhétorique* (p. 247), however, Lamy suggests that the persuasive speaker should establish a general principle as the basis for discussing a particular case.

6. The problem of morality and legality in *Horace* is interestingly examined by C. J. Gossip ('Tragedy and Moral Order in Corneille's *Horace*').

7. In his chapter on deliberations, D'Aubignac (*Pratique*, pp. 304 ff., and cf. pp. 341–2) points out that abstract discussions of this nature are essentially untheatrical and can only be made interesting to the audience thanks to 'des événements qui de moment en moment se contredisent et s'embarrassent', or by 'des passions violentes qui de tous côtés naissent du choc et du milieu des incidents...' This principle underlies part of his adverse criticism of Corneille's *Sophonisbe*. See his *Dissertation* in Granet, vol. i, p. 143. Cf. his criticism of the famous deliberation scene in *Sertorius* (*Dissertation*, ibid., pp. 265–7). La Mesnardière, (*La Poétique*, p. 9) and Longepierre (*Parallèle*) discuss the same point.

8. Professor Peter France (*Racine's Rhetoric*, p. 230 and n. 2) has pointed out the frequency of the specious use of the words 'juste' and 'justice' in Racine's plays, where we might more readily expect to find it in Corneille's.

9. Guez de Balzac, *Le Romain*, in *Œuvres* (1658 edition), pp. 15–17.

10. Caussin, *La Cour sainte*, vol. i, chap. I, p. 348.

11. See Dr D. A. Watts's analysis in his edition of the play (p. 165).

12. Morel, loc. cit., p. 96.

13. D'Aubignac (*Pratique*, p. 305) discusses the danger, from the dramatic point of view, of allowing deliberations to 'languish' for want of passion.

14. Cf. A. Stegmann, *L'Héroïsme cornélien*, vol. ii, p. 500: 'Le tragique réside...dans le renoncement au bonheur immédiat.'

15. Morel, loc. cit., pp. 100 ff. Cf. R. C. Knight, '*Andromaque* et l'ironie de Corneille'.

16. See M. Fumaroli, 'Rhétorique et dramaturgie: le statut du personnage dans la tragédie classique', p. 240.

17. Cf. P. France, op. cit., pp. 230, 237–8.

CHAPTER VIII
Tragic Quality

1. Sarasin, *Discours de la tragédie*, in *Œuvres* (1926), vol. ii, pp. 19–23.
2. Ogier, in Mantero, *Corneille critique*, p. 57.
3. *Pratique*, p. 143.

4. Castelvetro, *La Poetica d'Aristotele vulgarizzata e sposta*, 1576.
5. Du Bellay, *Deffense et illustration de la langue françoise*, bk. II, chap. IV.
6. Bouchetel, in Weinberg, *Critical Prefaces*, p. 108.
7. Baïf, ibid., p. 73.
8. La Taille, *De l'art de la tragédie*, p. 24.
9. Laudun d'Aigaliers, *Art poétique*, V, iv, in Lawton, *Handbook...*, p. 94.
10. Heinsius, *De tragoediae constitutione*, pp. 46–7. Cf. p. 19: 'iisque actionibus horrorem pariter et miserationem movet.'
11. Scaliger, *Poetices libri Septem*, I, lx, p. 33.
12. Sarasin, op. cit., pp. 4–5.
13. Cf. E. Vinaver in Racine, *Principes*, p. 49; P. R. Sellin, 'Le Pathétique retrouvé ...'
14. Cf. E. Krantz, cit. G. May, *Tragedie cornélienne...*, pp. 213–14: it is at the end of the play (not before) that 'la marche de l'action et l'enchaînement des scènes prennent un aspect déductif qu'ils ne pouvait avoir tant que la conclusion... demeurait cachée...'.
15. Vauquelin de la Fresnaye, Preface to *Satyres françoises*: Weinberg, op. cit., p. 272.
16. Cf. H. Gouhier, *Le Théâtre et l'existence*, pp. 33 ff., and 'The Tragic: Transcendence, Freedom and Poetry'.
17. J. Moles, 'Notes on Aristotle, *Poetics* 13 and 14'.
18. Scudéry, *Apologie du théâtre* (1639), in Mantero, op. cit., p. 84.
19. Vinaver, *Racine et la poésie tragique*, pp. 45–6.
20. R. C. Knight, 'A Minimal Definition...', pp. 298, 305.
21. Lanson, *Esquisse d'une histoire de la tragédie française*, p. 5.
22. This is one of the points developed by Marmontel in his *Discours de la tragédie* (*Poétique française*, vol. ii, pp. 95–228).
23. Dio Cassius, *Roman History*, XLII. 5. See my discussion of this aspect of *Pompée* in my edition of the play, pp. 200–5. See also Dupleix, XXIX, p. 36.
24. Cf. R. Pfohl's detailed study in *Racine's 'Iphigénie'*.
25. An illustrious day for the enactment of the tragic event was also advocated by Corneille (*Examen de Rodogune: Writings*, p. 132) and by D'Aubignac (*Pratique*, pp. 232–3). Cf. Scaliger, III, xcv.
26. La Mesnardière, *La Poétique*, p. 17.
27. Marmontel, *Discours de la tragédie* in *Poétique française*, vol. ii, p. 199.
28. D'Aubignac advises beginning the play as near to the catastrophe as possible, so as to allow the maximum of time for the development of the passions, but he also thinks that once curiosity is satisfied the audience's interest is lost (*Pratique*, pp. 125–6, 138).
29. Webster, *Greek Art and Literature*, p. 53.
30. MacDowell, '*Hybris* in Athens', passim.
31. Cf. J. Jones, *On Aristotle and Greek Tragedy*, pp. 166 ff.
32. Cf. G. Germain, *Sophocle*, pp. 85–90.
33. See my articles on 'La Gloire dans le théâtre de Racine'.
34. By Karl Reinhardt, in his book on Sophocles. (See, for example, pp. 114 ff.)
35. O. Nadal, *Le Sentiment de l'amour dans l'œuvre de Pierre Corneille*, pp. 195–6.
36. E. Schaper, *Prelude to Aesthetics*, pp. 116–17.
37. Kitto, *Form and Meaning in Drama*, p. 235.
38. e.g. Vossius, *Poeticarum Institutiones*, II. 19 (p. 64); Mambrun, *De epico carmine*, pp. 17–19; cf. Dryden and Goulston (see Mary Gallagher, 'Goulston's *Poetics* and tragic "admiratio"').
39. E. Schaper, loc. cit.

Bibliography

The bibliography contains only those works which are actually referred to in the text or which have some direct bearing on it. The following abbreviations are used:

FMLS : *Forum for Modern Language Studies*
FS : *French Studies*
MLR : *Modern Language Review*
RSH : *Revue des Sciences Humaines*
LCL : Loeb Classical Library

Place of publication is Paris unless otherwise stated.

A. The Works of Corneille and Racine

Pierre Corneille

Œuvres, ed. Ch. Marty-Laveaux. 12 vols. 1862–8.
Œuvres complètes, ed. A. Stegmann. 1963.
Le Cid, ed. M. Cauchie. 1946.
Le Cid (1637), ed. P. H. Nurse. London, 1978.
Cinna, ed. D. A. Watts. London, 1964.
Horace, ed. P. H. Nurse. London, 1963.
Nicomède, ed. R. C. Knight. London, 1960.
Polyeucte, ed. R. A. Sayce. Oxford, 1962.
Pompée, ed. H. T. Barnwell. Oxford, 1971.
Rodogune, ed. J. Scherer. 1945.
Suréna, ed. R. Sanchez. Bordeaux, 1970.
Writings on the Theatre, ed. H. T. Barnwell. Oxford, 1965.

Jean Racine

Œuvres, ed. P. Mesnard. 8 vols. and 2 albums, 1865–73.
Œuvres complètes, ed. R. Picard. 2 vols. 1966.
Andromaque. Texte de 1668, ed. R. C. Knight and H. T. Barnwell. Geneva, 1977.
Athalie, ed. P. France. Oxford, 1966.
Bajazet, ed. C. Girdlestone. Oxford, 1964.
Bajazet, ed. M. M. McGowan. London, 1968.
Bérénice, ed. C. L. Walton. Oxford, 1965.
Britannicus, ed. W. H. Barber. London, 1967.
Britannicus, ed. P. F. Butler. Cambridge, 1967.
Lettres d'Uzès, ed. J. Dubu. Uzès, 1963.
Mithridate, ed. G. Rudler. Oxford, 1960.
Phèdre, ed. R. C. Knight. Manchester, 1955.
Principes de la tragédie, en marge de la 'Poétique' d'Aristote, ed. E. Vinaver. Manchester, 1944.

B. *General Bibliography*

Amyot, J., see Plutarch.

Appian, *Roman History*, ed. H. White. 4 vols. London, 1912–13. (LCL)

Aristotle, *Poetics*, ed. I. Bywater. Oxford, 1909.

Aristotle, *Rhetoric*, ed. J. H. Freece. London, 1926. (LCL)

Arnaud, Ch., *Étude sur la vie et les œuvres de l'abbé d'Aubignac*. 1887.

Aubignac, F. Hédelin, abbé d', *Dissertations* on Corneille's *Sophonisbe, Sertorius* (vol. i) and *Œdipe* (vol. ii). See Granet, *Recueil*.

Aubignac, F. Hédelin, abbé d', *La Pratique du théâtre*, ed. P. Martino. Algiers and Paris, 1927.

Ault, H. C., 'The Tragic Genius of Corneille'. *MLR*, XLV (1950), pp. 164–76.

Aurelius Victor, *Caesares*, ed. P. Dufraigne. 1975.

Baïf, L. de, *Diffinition de la tragédie*. See Weinberg, *Critical Prefaces*, pp. 73–5.

Baillet, A., *Jugemens des sçavants sur les principaux ouvrages des auteurs*. 13 vols. 1685–90.

Balzac, J.-L. Guez de, *Lettres diverses*. 2 vols. 1659.

Balzac, J.-L. Guez de, *Œuvres diverses*. 1658.

Barber, W. H., 'Patriotism and "Gloire" in Corneille's *Horace*'. *MLR*, XLVI (1951), pp. 368–78.

Barcillon, J., '*Le Cid*: une interpretation psychanalytique'. *Studi Francesi*, 57 (1975), pp. 475–80.

Barnwell, H. T., 'From *La Thébaïde* to *Andromaque*: a View of Racine's Early Dramatic Technique'. *FS*, V (1951), pp. 30–5.

Barnwell, H. T., 'La Gloire dans le théâtre de Racine'. *Jeunesse de Racine*, 1961 (pp. 21–31), Jan.–Mar. 1962 (pp. 5–27), Apr.–June 1963 (pp. 21–40), Oct.–Dec. 1964 (pp. 61–9), 1966 (pp. 51–61), 1970–1 (pp. 65–84).

Barnwell, H. T., 'Histoire romaine et tragédie classique: transformations et survivances poétiques'. *Mélanges à la memoire de Franco Simone*, ed. L. Sozzi and others. Geneva, 1981. Vol. ii, pp. 309–22

Barnwell, H. T., 'Intrigue et pathétique dans le théâtre de Racine'. *Actes du Ier congrès international racinien*. Uzès, 1962. Pp. 67–75.

Barnwell, H. T., 'Peripety and Discovery: A Key to Racinian tragedy'. *Studi Francesi*, 26 (1965), pp. 222–34.

Barnwell, H. T., 'Racine and Sophoclean Tragedy'. *FMLS*, II (1966), pp. 48–52.

Barnwell, H. T., 'Racine as Plot-maker: Seventeenth-Century and Sophoclean Techniques'. *The Classical Tradition in French Literature. Essays presented to R. C. Knight*, ed. H. T. Barnwell, A. H. Diverres and others. London, 1977. Pp. 103–14.

Barnwell, H. T., 'Seventeenth-Century Tragedy: A Question of Disposition'. *Studies in French Literature presented to H. W. Lawton*, ed. J. C. Ireson and others. Manchester and New York, 1968. Pp. 13–28.

Barnwell, H. T., 'The Simplicity of Racine — yet again'. *FS*, XXXI (1977), pp. 394–406.

Barnwell, H. T., 'Some Notes on Poussin, Corneille and Racine'. *Australian Journal of French Studies*, IV (1967), pp. 149–61.

Barnwell, H. T., 'Some Reflections on Corneille's Theory of *Vraisemblance* as Formulated in the *Discours*'. *FMLS*, I (1965), pp. 295–310.

Barnwell, H. T., *The Tragic in French Tragedy*. Belfast, 1966.

Bénichou, P., *L'Écrivain et ses travaux*. 1967.

Bénichou, P., *Morales du Grand Siècle*. 1948.

Bensserade, I. de, *Cléopâtre*. 1636.

Bentley, E., *The Life of the Drama*. London, 1965.

Bidar, M., *Hippolyte*. Lille, 1675.

Blunt, A., *Nicolas Poussin*. Text and Plates. 2 vols. New York, 1967.

Boileau-Despréaux, N., *Epîtres. Art poétique. Lutrin*, ed. Ch.-H. Boudhors. 1952.

Boileau-Despréaux, N., *Satires*, ed. Ch.-H. Boudhors. 1952.

Boisrobert, F. le Métel de, *La Vraie Didon, ou la Didon chaste*. 1643.

Bonnard, A., *La Tragédie et l'homme*. Neuchâtel, 1950.

Boorsch, J., 'Remarques sur la technique dramatique de Corneille'. *Yale Romanic Studies*, 1941, pp. 101–62.

Bouchetel, G., *Dedication to Euripides's 'Hecuba'*. See Weinberg, *Critical Prefaces*, pp. 103–9.

Boyer, Cl., *Chresphonte, ou le retour des Héraclides dans le Péloponèse*. 1659.

Boyer, Cl., *Clotilde*. 1659.

Boyer, Cl., *Oropaste, ou le faux Tonaxare*. 1663.

Boyer, Cl., *Porus*. 1648.

Bray, R., *La Formation de la doctrine classique en France*. 1927.

Bray, R., *La Tragédie cornélienne devant la critique classique*. 1927.

Brossette, Cl., *Correspondance entre Boileau-Despréaux et Brossette, publiée sur les manuscrits originaux*, ed. A. Laverdet. 1858.

Butler, P. F., *Classicisme et baroque dans l'œuvre de Racine*. 1959.

Caesar, *Civil Wars*, ed. A. G. Peskett. London, 1914. (LCL)

Castelvetro, L., *La Poetica d'Aristotele vulgarizzata e sposta*. Basle, 1576.

Caussin, N., *La Cour sainte*. 6 vols. 1645.

Champigny, G., *Le Genre dramatique*. Monte Carlo, 1965.

Chapelain, J., *Opuscules critiques*, ed. A. C. Hunter, 1936.

Chaulmer, Ch., *La Mort de Pompée*. 1638.

Coëffeteau, N., *Histoire romaine, contenant tout ce qui s'est passé de plus mémorable depuis le commencement de l'empire d'Auguste jusqu'à celui de Constantin le Grand*. 1647.

Corneille, Th., *Poèmes dramatiques*. 5 vols. 1722.

Corneille, Th., *Camma, reine de Galatie*, ed. D. A. Watts. Exeter, 1977.

Couton, G., *Corneille*. 1958.

Couton, G., *La Vieillesse de Corneille*. 1948.

Davidson, H. M., 'Pratique et rhétorique du théâtre, étude sur le vocabulaire et la méthode de d'Aubignac'. *Critique et création littéraires en France au XVIIᵉ siècle*, ed. M. Fumaroli. 1977. pp. 169–75.

Dawson, S. W., *Drama and the Dramatic*. London, 1970.

Descartes, R., *Les Passions de l'âme*, ed. G. Rodis-Lewis. 1966.

Descartes, R., *Traité de l'homme*, in *Œuvres*, vol. xi, ed. Adam and Tannery. 1909.

Diderot, D., *Writings on the Theatre*, ed. F. C. Green. Cambridge, 1936.

Dio Cassius, *Roman History*, ed. E. Cary. 9 vols. London, 1914–27. (LCL)

Dionysius of Halicarnassus, *Roman Antiquities*, ed. E. Cary. 7 vols. London, 1939. (LCL)

Donatus, *De tragoedia et comoedia*. See Lawton, *Handbook*, pp. 2–21.

Donneau de Visé, J., *Défense de la 'Sophonisbe' de M. Corneille*. See Granet, vol. i.

Doubrovsky, S., *Corneille et la dialectique du héros*. 1963.

Du Bellay, J., *La Deffence et illustration de la langue françoyse*, ed: H. Chamard. 1948.

Du Bosc, J., *La Femme héroïque*. 1645.

Du Bosc, J., *L'Honneste Femme*. 1632.

Dupleix, Scipion, *Histoire romaine depuis la fondation de Rome*. 3 vols. 1638–43.

Dupont, J.-B., *L'Enfer d'amour*. Lyons, 1603.

Edwards, M., *La Tragédie racinienne*. 1972.

Euripides, *Works*, ed. A. S. Way. 4 vols. London, 1912 (LCL)

Félibien, A., *Entretiens sur les vies et les ouvrages des plus excellents peintres*. 5 vols. Trevoux, 1725.

Florus, *Epitome of Roman History*, ed. E. S. Forster. London, 1929. (LCL)

France, P., *Racine's Rhetoric*. Oxford, 1965.

Fumaroli, M. (ed.), *Aspects de l'humanisme jésuite au début du XVIIᵉ siècle. RSH*, 1975, pp. 245–93.

Fumaroli, M., 'Le *Crispus* et la *Flavia* du P. Bernardino Stefonio S.J.,' *Les Fêtes de la Renaissance*, III (ed. Jacquot and Kerigan). 1975, pp. 505–24.

Fumaroli, M., 'Rhétorique et dramaturgie: le statut du personnage de la tragédie classique'. *Revue de l'Histoire du théâtre*, XXIV (1972), pp. 223–50.

Fumaroli, M., 'Tragique païen et tragique chrétien dans *Rodogune*'. *RSH*, 1973, pp. 599–631.

Gallagher, M., 'Goulston's *Poetics* and tragic "admiratio".' *Revue de littérature comparée*, XXXIX (1965), pp. 614–19.

Garnier, R., *Cornélie*, ed. R. Lebègue. 1973.

Gasté, A., *La Querelle du 'Cid'. Pièces et pamphlets*. 1898.

Gerard, A. S., 'Baroque and the Order of Love'. *Neophilologus*, XLIX (1965), pp. 118 ff.

Germain, G., *Sophocle*. 1969.

Gilbert, G., *Arie et Petus, ou les amours de Néron*. 1660.

Gilbert, G., *Hipolyte, ou Le Garçon insensible*. 1647.

Goldmann, L., *Le Dieu caché. Étude sur la vision tragique dans les 'Pensées' de Pascal et dans le théâtre de Racine*. 1955.

Gossip, C. J., 'Tragedy and Moral Order in Corneille's *Horace*'. *FMLS*, XI (1975), pp. 15–28.

Gouhier, H., 'Remarques sur le "théâtre historique"'. *Revue d'esthétique*, 1960, pp. 16–24.

Gouhier, H., *Le Théâtre et l'existence*. 1952.

Gouhier, H., 'Tragédie et transcendance. Introduction à un débat général'. *Le Théâtre tragique*, ed. J. Jacquot. 1962, pp. 479–83.

Gouhier, H., 'The Tragic: Transcendence, Freedom and Poetry'. *Cross Currents*, Winter 1960, pp. 17–28.

Granet, F., *Recueil de dissertations sur plusieurs tragédies de Corneille et de Racine*. 2 vols. 1740.

Grenaille, F., *L'Innocent malheureux, ou la mort de Crispe*. 1639.

Guérin de Bouscal, *Cléomène*. 1640.

Haley, Sister M. P., 'Peripeteia and Recognition in Racine'. *Publications of the Modern Languages Association*, LV (1940), pp. 426–39.

Haley, Sister M. P., *Racine and the 'Art poétique' of Boileau*. Baltimore, Oxford and Paris, 1939.

Hall, H. G., 'Pastoral, Epic and Dynastic Dénouement in Racine's *Andromaque*'. *MLR*, LXIX (1974), pp. 64 ff.

Heinsius, D., *De tragoediae constitutione liber*. Leiden, 1643.

Herland, L., 'Les Éléments précornéliens dans *La Mort de Pompée* de Corneille'. *Revue d'histoire littéraire de la France*, L (1950), pp. 1–15.

Herland, L., 'La Notion de tragique chez Corneille'. *Mélanges de la Société toulousaine d'études classiques*, I (1948), pp. 265–84.

Herodian, *Works*, ed. C. R. Whittaker. 2 vols. London, 1969–70. (LCL)

[Hirtius], in Caesar. *Libri tres de bello civile: cum libris incertorum auctorum de bello Alexandrino*, etc., ed. R. du Pontet. Oxford [1900].

House, H., *Aristotle's 'Poetics'*. London, 1956.

Howarth, W. D., 'History in the Theatre: the French and English Traditions'. *Trivium*, I (1966), pp. 151–68.

Jacquot, J. (ed.), *Le Théâtre tragique*. 1965.

Jasinski, R., *Vers le vrai Racine*. 2 vols. 1958.

Jones, J., *On Aristotle and Greek Tragedy*. London, 1968.

Josephus, *Jewish Antiquities*, ed. L. H. Feldman. 9 vols. London, 1965. (LCL)

Justin, *Epitoma Historiarum Philippicarum Pompeii Trogi*, ed. M. Galdi. Turin, 1923.

Kern, E. G., *The Influence of Heinsius and Vossius upon French Dramatic Theory*. Baltimore, 1949.

Kitto, H. D. F., *Form and Meaning in Drama*. London, 1969.

Kitto, H. D. F., *Greek Tragedy*. London, 1973.

Knight, R. C., '*Andromaque* et l'ironie de Corneille'. *Actes du Ier congrès international racinien*. Uzès, 1962. Pp. 21–7.

Knight, R. C., 'Corneille's *Suréna*: A Palinode?' *FMLS*, IV (1968), pp. 175–85.

Knight, R. C., '*Cosroès* and *Nicomède*'. *The French Mind. Studies in Honour of Gustave Rudler*, ed. W. G. Moore, R. Sutherland and E. Starkie. Oxford, 1952.

Knight, R. C., 'The Evolution of Racine's *Poétique*'. *MLR*, XXV (1940), pp. 19–39.

Knight, R. C., 'Hippolyte and Hippolytos'. *MLR*, XXIX (1944), pp. 225–35.

Knight, R. C., '*Horace*, première tragédie classique'. *Mélanges d'histoire littéraire (XVIe–XVIIe) siècle) offerts à Raymond Lebègue.* 1959, pp. 195–200.

Knight, R. C., 'A Minimal Definition of Seventeenth-Century Tragedy'. *FS*, X (1956), pp. 297–308.

Knight, R. C., 'Myth and Mythology in Seventeenth-Century French Literature'. *Esprit Créateur*, XVI (1976), pp. 95–104.

Knight, R. C., *Racine et la Grèce*. 1950.

Knight, R. C., 'The Rejected Source in Racine'. *Modern Miscellany*, presented to Eugène Vinaver, ed. T. E. Lawrenson, F. E. Sutcliffe and G. F. A. Gadoffre. Manchester and New York, 1969. Pp. 154–66.

Knight, R. C., 'Le sens de *Pertharite*'. *Mélanges de littérature française offerts à R. Pintard.* Strasbourg, 1975. Pp. 175–84.

Knight, R. C., 'Sophocle et Euripide ont-ils "formé" Racine?' *FS*, V (1951), pp. 126–39.

Kuentz, P., 'Esquisse d'un inventaire des ouvrages de langue française. traitant de la rhétorique entre 1610 et 1715'. *XVIIe siècle*, 80–1 (1968), pp. 133–42.

La Bruyère, J. de, *Œuvres*, ed. G. Servois. 5 vols. 1922.

La Calprenède, G. de Coste de, *Edouard*. 1640.

La Calprenède, G. de Coste de, *La Mort de Mithridate*. 1637.

La Calprenède, G. de Coste de, *La Mort des enfants d'Herodes, ou la Suite de la Mariane*. 1639.

La Mesnardière, J.-H. Pilet de, *La Poétique*. 1639.

Lamy, B., *La Rhétorique ou l'art de parler*. Amsterdam, 1699.

Lancaster, H. C., *A History of French Dramatic Literature in the Seventeenth Century*. 5 parts in 9 vols. Baltimore, 1929–42.

Lanson, G., *Corneille*. 1948.

Lanson, G., *Esquisse d'une histoire de la tragédie française* (1927). 1954.

Lapp, J. C., *Aspects of Racinian Tragedy*. Toronto, 1964.

La Taille, Jean de, *De l'art de la tragédie*, ed. F. West. Manchester, 1939.

Laudun d'Aigaliers, P. de, *Art poëtique françois*. See Lawton, *Handbook*, pp. 91–101.

Lawton, H. W., *Handbook of French Renaissance Dramatic Theory.* Manchester, 1949.

Le Moyne, P., *La Gallerie des femmes fortes*. 1647.

Le Vert, *Aricidie ou le mariage de Tite*. 1646.

Levi, A. H. T., *French Moralists. The Theory of the Passions, 1585 to 1649*. Oxford, 1964.

Livy, *Ab urbe condita*, ed. B. O. Foster et al. 14 vols. London, 1919-59. (LCL)

Lock, W., 'The Use of *Peripeteia* in Aristotle's *Poetics*'. *Classical Review*, IX (1895), pp. 251–3.

Longepierre, H.-B. de R. de, *Médée* (1694), ed. T. Tobari. 1967.
Longepierre, H.-B. de R. de, *Parallèle de M. Corneille et de M. Racine*. See Baillet, vol. ix.
Lucan, *Pharsalia*, ed. J. D. Duff. London, 1928. (LCL)
Lucan, *Pharsalia*. *Dramatic Episodes of the Civil Wars*, trans. R. Graves. Penguin Classics, Harmondsworth, 1956.

MacDowell, D. M., '*Hybris* in Athens'. *Greece and Rome*, XXIII (1976), pp. 14–31.
McFarlane, I. D., 'Notes on the Rhetoric of *Horace*'. *The French Language: Studies presented to L. C. Harmer*. London, 1970, pp. 182–209.
Magnon, J., *Tite*, ed. Herman Bell. Johns Hopkins, 1936.
Mairet, J., *Le Marc-Antoine, ou la Cléopâtre*. 1637.
Mairet, J., *La Sophonisbe*, ed. Ch. Dédéyan. 1945.
Mambrun, P., *Dissertatio Peripatetica de epico carmine*. 1652.
Mantero, R. (ed.), *Corneille critique*. 1964.
Margitić, M. R., *Essai sur la mythologie du 'Cid'*. New York, 1976.
Margitić, M. R., 'La Signification de l'homme cornélien: une proposition à partir de la mythologie du *Cid*'. *Romanic Review*, LXX (1979), pp. 45–55.
Marmontel, J.-F., *Éléments de littérature*. Vols. xii–xv of *Œuvres*. 1819.
Marmontel, J.-F., *Poétique française*. 3 vols. 1763.
Maurens, J., *La Tragédie sans tragique. Le Néo-stoïcisme dans l'œuvre de Pierre Corneille*. 1966.
Mauron, Ch., *L'Inconscient dans l'œuvre et la vie de Racine*. 1969.
May, G., *Tragédie cornélienne, tragédie racinienne. Étude sur les sources de l'intérêt dramatique*, Urbana, 1948.
Mermet, Cl., *La Tragédie de Sophonisbe, reine de Numidie*. Lyons, 1584.
Méron, E., 'De l'*Hippolyte* d'Euripide à la *Phèdre* de Racine: deux concepts du tragique'. *XVIIe siècle*, 100 (1973), pp. 35–54.
Moles, J., 'Notes on Aristotle, *Poetics* 13 and 14'. *Classical Quarterly*, XXIX (1979), pp. 77–94.
Monchrestien, A. de, *Les Tragédies*. Rouen, 1601.
Montaigne, M. de, *Essais*. 7 vols. [1866.]
Montreux, N. de, *Sophonisbe* (1601), ed. D. Stone, Jr. Geneva, 1976.
Moore, W. G., 'Corneille's *Horace* and the interpretation of French classical drama'. *MLR*, XXXIV (1939), pp. 382–95.
Morel, J., 'A propos du plaidoyer d'*Horace*. Réflexions sur le sens de la vocation historique dans le théâtre de Corneille'. *Romanic Review*, LI (1962), pp. 27–32.
Morel, J., 'Rhétorique et tragédie au XVIIe siècle', *XVIIe siècle*, 80–1 (1968), pp. 89–105.
Mornet, D., *Histoire de la littérature française. classique* (1660–1700). 1947.
Mornet, D., *Jean Racine*. 1944.
Mourgues, O. de, *Racine or, The Triumph of Relevance*. Cambridge, 1967.

Nadal, O., *Le Sentiment de l'amour dans l'œuvre de Pierre Corneille*. 1948.

Nelson, R. J., *Corneille: his Heroes and their Worlds*. Philadelphia, 1963.
Newton, W., *Le Thème de Phèdre et d'Hippolyte dans la littérature française*. 1939.
Nurse, P. H., 'A propos de l'heroïsme cornélien'. *RSH*, (1962), pp. 169–73.
Nurse, P. H., 'Quelques réflexions sur la notion du tragique dans l'œuvre de Pierre Corneille'. *Mélanges de littérature française offerts à M. R. Pintard*. Strasbourg, 1975. Pp. 163–74.

Ogier, F., Preface to Jean de Schelandre's *Tyr et Sidon* (1628). See Mantero, pp. 49–58.

Pascal, B., *Œuvres complètes*, ed. L. Lafuma. 1963.
Pfohl, R., *Racine's 'Iphigénie'. Literary Rehearsal and Tragic Recognition*. Geneva, 1974.
Plutarch, *Lives*, ed. B. Perrin. 11 vols. London, 1914–26. (LCL)
Plutarch, *Les Vies des hommes illustres*, trans. J. Amyot, ed. G. Walter. 2 vols. 1937.
Pocock, G., *Corneille and Racine. Problems of Tragic Form*. Cambridge, 1973.
Pommier, J., *Aspects de Racine*. 1954.
Poussin, N., *Lettres*, ed. P. Colombey. 1929.
Poussin, N., *Lettres et propos sur l'art*, ed. A. Blunt. 1964.
Prade, J. Royer de, *Arsace*. 1666.

Quinault, Ph., *Théâtre*. 5 vols. 1715.

Racine, L., *Œuvres*. 6 vols. 1808.
Rapin, R., *Les Réflexions sur la poétique de ce temps*, ed. E. T. Dubois. Geneva and Paris, 1970.
Reinhardt, K., *Sophocles* (1933). Trans. H. Harvey and D. Harvey. Oxford, 1979.
Riddle, L. M., *The Genesis and Sources of Corneille's Tragedies from 'Médée' to 'Pertharite'*. Baltimore, 1926.
Robortello, F., *In librum de arte poetica explicationes*. Florence, 1548.
Rosimond, Cl.-R., de *Le Qui pro quo, ou le valet étourdi*. 1673.
Rotrou, J., *Œuvres*, ed. Viollet-le-Duc. 5 vols. 1820.
Rotrou, J., *Cosroès*, ed. J. Scherer. 1950.
Rotrou, J., *Hercule mourant*, ed. D. A. Watts. Exeter, 1971.
Rudler, G., 'Une source d'*Andromaque? Hercule mourant* de Rotrou'. *MLR* XII (1917), pp. 286 ff., 438 ff.

Sainte-Beuve, Ch.-A. de, *Portraits littéraires*. 2 vols. 1843.
Saint-Évremond, Ch. le Marguetel de, *Œuvres en prose*, ed. R. Ternois. 4 vols. 1962–9.
Saint-Gelais, M. de, *Œuvres complètes*. 3 vols. 1873.
[Saint-Ussans, P. de Saint-Glas, abbé de], *Réponse à la Critique de lu Bérénice' de Racine*. See Granet, vol. ii.

Sarasin, J.-F., *Œuvres*. 1663.
Sarasin, J.-F., *Œuvres*, ed. P. Festugière. 2 vols. 1926.
Scaliger, J. C., *Poetices Libri Septem*. Heidelberg, 1617.
Schaper, E., *Prelude to Aesthetics*. London, 1968.
Scherer, J., *La Dramaturgie classique en France*. 1950.
Scudéry, G. de, *L'Amour tirannique*. 1639.
Scudéry, G. de, *L'Amour tirannique*, ed. V. G. Farinholt. Chicago, 1938.
Scudéry, G. de, *L'Apologie du théâtre*. See Mantero, pp. 79–94.
Scudéry, G. de, *Didon*. 1637.
Scudéry, G. de, *Les Femmes illustres*. 1655.
Scudéry, G. de, *La Mort de César*. 1636.
Scudéry, G. de, *La Mort de César*, ed. H. L. Cook. New York, 1930.
Scudéry, G. de, *Observations sur le 'Cid'*. See Gasté, pp. 71–111.
Sellin, P. R., 'Le Pathétique retrouvé: Racine's catharsis reconsidered'. *Modern Philology*, LXX (1973), pp. 199–215.
Segrais, J. Regnauld de, *Les Nouvelles françoises*. (1657) 1722.
Seneca. *Tragedies*, ed. F. J. Miller, 2 vols. London, 1917. (LCL)
Sophocles. *Works*, ed. F. Storr, 2 vols. London, 1912–13. (LCL)
Stegmann, A., *L'Héroïsme cornélien. Genèse et signification*. 2 vols. 1968.
Stewart, W. McC., 'Charles Le Brun et Jean Racine — Contacts et points de rencontre'. *Atti del Quinto Congresso Internazionale di Lingue e Letterature Moderne (1951)*. Florence, 1953. Pp. 213–29.
Stewart, W. McC., 'Racine's Response to the Stagecraft of Attic Tragedy as Seen in his Annotations'. *Essays presented to H. D. F. Kitto*, ed. M. J. Anderson. London, 1965. Pp. 175–90.
Stewart, W. McC., 'Racine et Sophocle et la norme tragique'. *Actes du IVe congrès international d'esthétique*. Athens, 1960. Pp. 119–27.
Stone, J. A., *Sophocles and Racine. A Comparative Study in Dramatic Technique*. Geneva, 1964.
Suetonius, *Works (The Twelve Caesars)*, ed. J. C. Rolfe. 2 vols. London, 1914 (LCL).
Sweetser, M.-O., *La Dramaturgie de Corneille*. Geneva and Paris, 1977.

Tacitus, *Histories* and *Annals*, ed. C. H. Moore and J. Jackson. 4 vols. London, 1925–37. (LCL)
Tristan l'Hermite, *Théâtre complet*, ed. C. K. Abraham, J. W. Schweitzer and J. van Baelen. University, Alabama, 1975.

Varga, A. Kibédi, 'La Perspective tragique. Eléments pour une analyse formelle de la tragédie classique'. *Revue d'histoire littéraire de la France*, LXX (1970), pp. 918–30.
Varga, A. Kibédi, *Rhétorique et littérature. Étude de structures classiques*. 1970.
Vauquelin de la Fresnaye, J., *Art poétique*, ed. G. Pellissier. 1885.
Vauquelin de la Fresnaye, J., Preface to *'Satyres françoises'*. See Weinberg, *Critical Prefaces*, pp. 271–6.
Velleius Paterculus, *Compendium of Roman History*, ed. F. W. Shipley. London, 1924. (LCL)

Vettori, P., *In primum librum de arte poetica commentarii*. Florence, 1573.
Villars, N.-P.-H. de Montfaucon, abbé de, *La Critique de Bérénice*. See Granet, vol. ii.
Villiers, P. abbé de, *Entretiens sur les tragédies de ce temps*. See Granet, vol. i.
Vinaver, E., *L'Action poétique dans le théâtre de Racine*. Oxford, 1960.
Vinaver, E., 'L'Éclosion du tragique dans le théâtre de Racine'. *Bulletin de l'Académie Royale de langue et de littérature françaises*, XLIV (1966), pp. 111–25.
Vinaver, E., *Racine et la poésie tragique*. 1963.
Virgil, *Works*, ed. H. R. Fairclough. 2 vols. London, 1918–38. (LCL)
Voltaire, *Commentaires sur Corneille*, ed. D. Williams. 3 vols. Banbury, 1974–5.
Vossius, G. J., *De artis poeticae natura ac constitutione liber*. Amsterdam, 1647.
Vossius, G. J., *Poeticarum Institutionum libri tres*. Amsterdam. 1647.

Webster, T. B. L., *Greek Art and Literature*. Oxford, 1939.
Webster, T. B. L., *Greek Tragedy*. Oxford, 1971.
Webster, T. B. L., *An Introduction to Sophocles*. London, 1969.
Weinberg, B., *The Art of Jean Racine*. Chicago, 1963.
Weinberg, B., *Critical Prefaces of the French Renaissance*. Evanston, 1950.

Yarrow, P. J., *Corneille*. London, 1963.

Zuber, R., *Les 'Belles Infidèles' et la formation du goût classique. Perrot d'Ablancourt et Guez de Balzac*. 1968.

Index of Proper Names

This Index is in two parts. The first lists in chronological order the plays of Corneille and Racine mentioned in the text and notes. The second comprises all other proper names with the exception of names of places and of dramatic characters: the latter can be located in the text by referring to the names of dramatists and their plays. Titles of seventeenth-century French plays are included under the names of their authors.

A. The Plays of Corneille and Racine

1. Corneille

Mélite 167, 252
Clitandre 149, 252
La Veuve 161, 167
Médée, 15
L'Illusion comique 252
Le Cid xi, 3–4, 28, 34, 72, 73, 74, 75, 88, 95, 101, 103, 104, 110, 128, 144, 151, 162, 165, 166, 167, 168, 176, 180, 212, 224, 229
Horace xvii, xviii, 2, 4, 6, 14, 15, 20, 26, 34, 60, 62, 83, 84, 95, 100–1, 102–3, 104, 106, 107, 109, 110, 114–15, 121, 130, 135, 137, 195, 212, 222, 227, 228, 232, 233, 235, 238–9, 240, 242–3, 244, 246, 247, 249, 250, 258
Cinna, 2, 6, 12, 20, 55, 83, 84, 100, 102, 104, 106, 107, 110, 121, 144, 154, 166, 167, 177, 180, 182, 183, 190 ff., 195, 198 ff., 202, 208–9, 247, 249, 250, 254, 258
Polyeucte 12, 88, 115, 150, 177, 182, 187, 221, 227, 228, 233, 243, 249, 250
Pompée 12, Chap. II *passim*, 84, 88, 103, 176, 187, 211, 212, 223–4, 225, 227, 228, 232, 233, 236, 249, 254, 260
Le Menteur 88, 252
La Suite du Menteur 82, 177
Théodore 74, 177, 256
Rodogune 11, 14, 26, 35, 48, 64, 66, 88, 121, 151, 166, 172–3, 177, 212, 221, 222, 225, 227, 228, 229, 232, 239, 240, 243, 249, 251, 252, 256, 260
Heraclius, 66, 71, 75, 81, 102, 112, 149, 154, 166, 167, 229, 235, 251, 252

Andromède 81
Don Sanche d'Aragon 252
Nicomède 9–11, 16, 21, 23, 25, 67, 79–80, 84, 85, 88, 105, 119, 157, 161–2, 165, 166, 177, 180, 212, 227, 228, 229, 232, 233, 237, 240, 247, 250, 252, 257
Pertharite 9, 10, 17, 110, 117, 137–8, 151
Œdipe 114, 134, 136, 141, 144, 146, 154, 231, 250
La Toison d'or 81
Sertorius 29, 55, 62, 88, 89, 103, 132, 134, 136, 141 ff., 211, 212, 228, 233, 236, 249, 250, 256, 259
Sophonisbe xix, 5, 7, 11, 12, 19, 28, 34, 44, 62, 66, 67, 88, 103, 134, 141, 211, 212, 233, 251, 259
Othon 101–2, 256
Agésilas 89, 252, 256
Attila 89, 119, 212, 256
Tite et Bérénice xix, 7, 12–13, 19, 56 ff., 245
Suréna 26, 173–4, 180, 187, 211, 212, 224, 225, 228, 233, 236, 240, 250

2. Racine

Les Amours d'Ovide 19
La Thébaïde 103, 128–9, 134 ff., 142, 147, 148, 153, 154, 155, 236, 256
Alexandre le Grand 22, 28, 67, 79, 82, 103, 140–1, 142, 144, 147, 148, 153, 236
Andromaque 2, 3, 5, 6, 7–9, 12, 13, 14, 17, 21, 22, 23–4, 30, 34, 35, 46, 47, 62,

64, 67, 76–7, 78, 79–80, 82, 85, 91,
106, 107 ff., 110, 127, 137–8, 140,
142, 147–8, 151, 153, 154, 156, 157,
158, 161, 174, 176, 181, 182, 183,
195 ff., 205, 208, 210, 211, 215, 222,
224, 226, 227, 232–3, 233–4, 236,
241, 244, 246, 247, 249, 251
Les Plaideurs 195
Britannicus 5, 26, 35, 47, 48, 62, 64, 66,
67, 76–7, 78, 82, 85, 86, 89, 90, 91,
107, 122, 126, 127, 143, 144, 148,
155, 156, 157, 158, 169–70, 172,
174, 176, 181, 182, 201 ff., 211, 226,
233, 236, 241, 250
Bérénice xix, 2, 6, 7, 12–13, 14, 17, 19,
20, 22, 26, Chap. II *passim,* 78–9,
82, 86, 89–90, 91, 106, 107, 119–20,
122, 125–6, 127, 141, 142, 143, 144,
145, 146, 148, 154, 156, 157, 158,
166, 170–1, 172, 173, 174, 182, 183,
205 ff., 210, 216, 217, 222, 226, 227,
233, 234, 236, 241, 244 ff., 247, 249,
250, 252, 254
Bajazet 22, 77, 78, 85, 86, 91, 117–18,
125, 157, 173, 181, 215, 222, 233,
241, 247, 249, 250, 252
Mithridate xix, 7, 10–11, 12, 26, 28, 29,
62, 64, 67, 78, 86, 91, 107, 110, 117,
118–19, 124, 148, 151 ff., 161, 172–
3, 174, 178, 181, 182, 211, 233, 241,
247, 250, 258
Iphigénie en Aulide 14, 24, 62, 67, 76, 78,
79, 80, 81, 82, 85, 90, 107, 110,
111 ff., 118, 119, 124, 125, 157, 167,
213 ff., 212, 216, 224, 233, 241–2,
250, 252
Phèdre 11, 20, 25, 26, 28, 34, 54, 62, 67,
74, 78, 79, 80–1, 85, 88, 91, 106,
107, 110, 117, 124, 125, 127, 157,
170, 171, 173, 174, 177, 181, 182,
201, 209 ff., 211, 212, 224, 233, 242,
246 ff., 249–50, 252
Iphigénie en Tauride 19
Esther 195
Athalie 78, 91, 113–14, 123, 126–7,
167, 171–2, 177, 182, 225–6, 233,
235, 242, 246, 249

B. *General Index of Proper Names*

Achilles 88, 224
Aeneas 44 ff., 60, 74, 102
Aeschylus 240

Agathon 65
Ambrose, Saint 5
Amyot 51
Anchises 60
Appian 14, 36, 51, 253
Aristotle xvi, xvii, xviii, xx, xxi, 2, 7,
13, 14, 15, 16, 17, 26, 28, 29, 34, 48,
65, 68, 72, 73, 76, 77, 79, 80, 82, 84,
86, 87, 88, 89, 90, 94, 96, 97, 98, 99,
104, 105, 136, Chap. VI *passim,* 180,
181, 215, 216–17, 219–21, 223, 225,
226–7, 229, 230, 235, 237, 252, 255
Arnaud, Ch. 256
Aubignac, Abbé d' xvii, 4–5, 14, 17,
30, 44, 61, 62, 65, 66, 72, 73, 74, 75,
82, 88, 96, 97, 99, 104, 105, 106,
111, 121, 129, 130, 131, 132, 134–5,
136, 137, 139, 141 ff., 160, 164, 168,
217, 231, 232, 234, 237, 251, 258,
259, 260
Ault, H. C. 259
Aurelius Victor 55, 254
Baïf, L. de 218, 260
Balzac, Guez de 63, 193, 254, 259
Barber, W. H. 259
Barcillon, J. xiv
Baudelaire xvi
Bell, H. 254
Bellori 27
Bènichou, P. xiv, 22, 252, 254, 256
Benserade:
　　Cléopâtre 41, 49
Bentley, E. xviii, 179, 255, 256
Bidar, 11
　　Hippolyte 74
Blunt, A. 252
Boileau xvi, 95, 255
Boisrobert:
　　La Vraie Didon 44 ff., 65, 74, 253
Bonnard, A. 133
Bouchetel 218, 260
Boyer, 69
　　Clotilde 154
　　Oropaste 164
Bray, R. xvii, 72, 255
Brossette, 258
Brunetière xiii
Butler, P. H xv

Caesar 36, 41, 253
Cassiano dal Pozzo 18
Castelvetro 217, 260
Caussin 25, 195, 258, 259

Chambray 19, 252
Champaigne, J.-B. de 20
Champigny, R. 1
Chantelou 18, 19, 20–1, 27
Chapelain 4, 72, 73, 74, 75, 82, 106, 251, 255
Chaucer 30
Chaulmer:
  La Mort de Pompée 41, 50
Coëffeteau 254
Cook, H. L. 253
Corneille, Th., 69, 95, 224
  Bérénice 65
  Camma 110, 121, 138–9
  Commode 110, 138, 154
  Le Comte d'Essex 195
  Darius 65
  Pyrrhus 65
  Timocrate 65, 121, 195
Couton, G. xv

Davidson, H. M. 258
Dawson, S. W. xxi
Descartes, 25 ff. 29, 30, 222, 248, 252
Desfontaines:
  La Perside 110
Diderot 256
Dido 44 ff., 74, 142
Dio Cassius 36, 42, 54, 55, 56, 224, 253, 254, 255, 260
Dionysius of Halicarnassus 183, 255
Discours à Cliton 33
Donatus 105, 106, 256
Donneau de Visé 149, 257
Doubrovsky, S. xv
Drusilla 47
Dryden 213, 260
Du Bellay 217, 260
Du Bosc 258
Dupleix 42, 49, 57, 254, 260
Dupont, J.-B. 22
Du Ryer:
  Alcionée 160
Du Vair 25

Edwards, M. xv
Euripides xviii, 6, 8, 9, 11, 14, 24, 30, 79, 122, 128, 136, 172, 216–17, 218, 240, 252

Félibien 19 ff., 27, 252
Florus 42
France, P. 17, 252, 258, 259

Fumaroli, M. 251, 252, 258, 259
Furetière 214, 215

Gallagher, M. 260
Ganymedes 39, 42
Garnier, 11, 218
  Cornélie 50
Gasté 255
Gerard, A. S. 254
Germain, G. 130, 256, 260
Ghirardelli 164
Gilbert, 69
  Arie et Petus 154
  Cresphonte 154
  Hippolyte 74
Girdlestone, C. 256
Goldmann, L. xiv
Gossip, C. J. 259
Gouhier, H. 68–9, 255, 260
Goulston 260
Gowing, L. 252
Granet, 254, 256, 257, 259
Graves, R. 36, 253
Grenaille, 11, 16–17
  L'Innocent malheureux 14, 16
Guérin de Bosucal:
  Cléomène 41

Haley, M. P. 256, 258
Hegel xv
Heinsius 97, 160, 218, 219, 257, 260
Herland, L. 253
Herodian 54, 254
Hirtius 253
Homer 24, 64, 79, 128, 224
Horace 82
House, H. 163, 258
Howarth, W. D. 255

Ixion 88

Jasinski, R. xiv
Jodelle 218
Jones, J. xxi, 256, 260
Josephus 14, 47, 53, 253, 254
Justin 9, 14, 67

Kern, E. 255
Kitto, H.D.F. 248, 256, 260
Knight, R. C. xxi, 22, 97, 129, 251, 253, 254, 255, 256, 257, 258, 259, 260
Krantz, E. 260

Kuentz, P. 258, 259

La Bruyère xiii, 29, 95, 106, 256
La Calprenède:
 Édouard, 16
 Mithridate xix, 7, 11, 12, 62
 La Mort des enfants d'Hérode 11, 16
La Mesnardière 17, 72, 95, 105, 106,
 160, 161, 177, 227, 252, 256, 259,
 260
Lamy 259
Lancaster, H. C. 255, 257
Lanson, G. xiii, 96, 223, 255, 260
Lapp, J. C. 255
La Rochefoucauld 29
La Taille, Jean de 17–18, 218, 252,
 260
Laudun d'Aigaliers 218, 260
Lawton, H. W. 256, 260
Le Brun 20
Le Moyne 11, 64, 65, 251, 254
Le Vasseur 19, 20–1, 136
Le Vert:
 Aricidie 12, 47, 251
Levi, A. H. T. 252
Lock, W. 258
Longepierre, xiii, 95, 106, 259
 Médée, 255
Louis XIII xiv
Louis XIV xiv, 195
Lucan 36, 37, 38, 42, 49, 50, 61, 65,
 253, 254

Maccabees 14
MacDowell, W. M. 240, 260
McFarlane, I. D. 258
Magnon:
 Tite 57 ff., 254
Mairet, 255
 Marc-Antoine 3, 41, 49
 Sophonisbe xix, 5, 7, 12, 19, 41, 95
Mambrun 84, 255, 260
Mantero 259, 260
Margitić, M. R. xiv
Marmontel 121, 130, 233, 256, 260
Martino, P. 134
Maurens, J. xiv
Mauron, Ch. xiv
May, G. xiii, xix, xx, 97, 98, 99, 106,
 111, 255, 256, 260
Medea 88
Melin de Saint-Gelais 12
Mercury 46

Méré 29
Mermet 12
Méron, E. 252
Mesnard, P. 53, 60, 254
Milton 252
Moles, J. 260
Montaigne 50, 254
Montchrestien 12
Montemayor 22
Montmort, Mme de 21
Montreux 12
Moore, W. G. xxi
Morel, J. 195, 211, 258, 259
Mornet, D. xiii, 113, 255
Mourgues, O. de 255

Nadal, O. xiv, 260
Nelson, R. J. xv
Nero 55
Nurse, P. H. 251, 255, 259

Octavius 49
Ogier 216, 259
Orestes 64

Pascal xvi, 18, 29, 93, 213, 252
Pazzi 253
Perrault 20
Pertinax 54
Pfohl, R. 260
Picard, R. 19, 252, 256
Plato xvi
Plautus 141
Plutarch 36, 51, 253, 254, 255
Pocock, G. xiii, xv
Pommier, J. 251
Pothinus 41, 42
Poussin 18 ff., 27 ff., 30, 252
Pure, Abbé de 73, 251

Quinault 69, 95, 223, 224
 Astrate 65, 66
 Cyrus 65, 154
 Le Feint Alcibiade 164

Racine, L. 19, 46, 53, 56, 253
Rapin viii, 105, 256
Reinhardt, K. 260
Richelet 215
Riddle, L. M. 253
Robortello 87
Rosimond:
 Les Qui pro quo 161

Rotrou, 139, 223, 255
  Antigone 136, 137
  Cosroès 10, 21, 23
  Hercule mourant 10, 95, 110, 137
  Iphigénie 112
Rudler, G. 137, 251, 257

Sainte-Beuve 96, 255
Saint-Évremond xiii, xvi, xxi, 18, 95,
  96, 140, 252, 257
[Saint-Ussans] 159
Sandrart 252
Sarasin 72, 160, 216, 219, 256, 259,
  260
Sarcey 255
Scaliger 93, 105, 219, 260
Schaper, E. xxi, 79, 255, 260
Schelandre 216
Scherer, J. 136, 158, 160 ff., 170, 256,
  257, 258
Scudéry, G. de, 4, 57 ff., 64, 65, 221,
  257, 260
  L'Amour tyrannique 106
  Didon 45
  La Mort de César 40, 41
Segrais 22
Seelin, P. R. 257, 260
Senault 25
Seneca 11, 136, 253
Severus 54
Sextus Pompeius Magnus 41
Shakespeare 94
Sophocles xviii, 6, 7, 14, 99–100, 111,
  114, 122 ff., 130, 131, 136, 162, 163,
  164–5, 166–7, 168, 170–1, 174,
  183, 242, 245, 252, 258
Stefonio 11, 164
Stegmann, A. xv, 12, 97, 251, 253,
  254, 259
Stella, J. 27
Stewart, W. McC. 252, 256
Stone, J. A. 256
Suetonius 22, 44, 49, 51, 53, 55, 56,
  57–8, 67, 254
Sweetser, M.-O. 255

Tacitus 47, 53, 254
Tasso 27
Terence 141
Theodotus 40, 42, 253
Trissino 12
Tristan l'Hermite, 11
  Mariane 41, 50, 160, 176

Varga, A. K. 258
Vauquelin de la Fresnaye 220, 260
Velleius Paterculus 37
Vespasian 46, 47, 54, 55, 61
Vettori 15, 29, 166, 237
Villiers, Abbé de 65, 254
Vinaver, E. xix, 15, 29, 129, 162, 165,
  166, 238, 255, 257, 258, 260
Virgil 3, 9, 22, 44 ff., 60, 61, 64, 65,
  67, 74, 79
Vologeses I 47
Voltaire 9, 50, 106, 137
Vossius 84, 255, 257, 260

Walton, C. L. 253
Watts, D. A. 257, 259
Webster, T. B. L. 130, 238, 240, 256,
  260
Weinberg, B. xv, 258, 260

Xiphilinus 56

Zuber, R. 254